Chemistry for Engineers

Chemistry for Engineers

Teh Fu Yen

University of Southern California, USA

Imperial College Press

ICP

Published by

Imperial College Press
57 Shelton Street
Covent Garden
London WC2H 9HE

Distributed by

World Scientific Publishing Co. Pte. Ltd.
5 Toh Tuck Link, Singapore 596224
USA office: 27 Warren Street, Suite 401-402, Hackensack, NJ 07601
UK office: 57 Shelton Street, Covent Garden, London WC2H 9HE

British Library Cataloguing-in-Publication Data
A catalogue record for this book is available from the British Library.

CHEMISTRY FOR ENGINEERS

Copyright © 2008 by Imperial College Press

ISBN-13 978-1-86094-774-2
ISBN-10 1-86094-774-3
ISBN-13 978-1-86094-775-9 (pbk)
ISBN-10 1-86094-775-1 (pbk)

Printed in Singapore.

This book is dedicated to my wife, Shiao-Ping Siao Yen, Ph.D., formerly (1959-1969) at the Mellon Institute (Fellow, Basic Polymer Science Research), currently (1969-present) at the Jet Propulsion Laboratory, California Institute of Technology (Senior Technical Staff), for her support, understanding, and sacrifice.

FOREWORD

The phrase "Think cosmologically, act globally" conveys what I feel regarding the need for engineers to study science. We all live in this vast universe which is ever expanding its limits and challenging human dimensions. But our actions are limited to his global sphere we inhabit, occupy and can reach out for. Thus codes and regulations are an important instrument for guiding the actions of engineers. But what lies beyond and underneath these regulations is science. Thus to truly understand the workings of these processes, our thinking should be oriented towards the laws of the cosmos, the study of science. Science is involved with knowing the nature and origins of matter. At the microscopic level it is called **physics** and at larger and more complex systems, it translates to **biology**. **Chemistry** is the connecting link in between, which defines the integrating reactions and interactions in this chain.

Presently, I cannot locate any book that addresses the needs of engineers when they want to understand the chemistry, or the science which yields the processes they deal with. This was my motivation to write such a book and making available such a resource. I believe that engineers should be trained in basic science because engineering is an applied science, and it should return to science at higher levels of advanced engineering studies. This book aims at bridging the concepts and theory of chemistry with examples from fields of practical application, thus reinforcing the connection between science and engineering.

As a trained chemist and spending years of experience with research institutes, industries and universities in engineering, I identified the various fields of chemistry that will interest engineers. My doctoral training is in synthetic organic chemistry; other graduate experience is in chemical engineering, biochemistry and mathematics. My post doctorate training is in polymer chemistry and geochemistry. During the last four decades of teaching, my research effort has been in environmental engineering intermixed with petroleum and fuel chemistry.

What chemistry to teach engineers? This answer requires some difficult thought. A straight legitimate way to consider the current curriculum for an engineering program (such as BS in general) requires both general chemistry and calculus based general physics, with at least a two semester sequence of study in either area. As per ABET ruling, a number of engineering disciplines do not require even general chemistry. This is sad especially for the senior level of engineers or post graduate engineers. It is indeed disastrous that a well known military leader cannot construct an emergency runway without the supply of cement.

Perhaps the more adequate definition of engineering is from an ancient Chinese scholar, Jia Yi. He claimed nature such as Heaven and Earth is only a testing tool (reactor) for engineers, and that the pursuit of evolution and understanding of life processes is engineering. This reflects that engineering is logically the reasoning to critically resolve the fundamental problems of great importance.

In this book, specialized topics of chemistry are revisited. The first five chapters, Physical Chemistry, Inorganic Chemistry, Organic Chemistry, Analytical Chemistry, and Surface Chemistry are a review of general chemistry covered in first year undergraduate engineering courses.

For materials, metallurgical and mechanical engineering students, material chemistry, polymer chemistry, and inorganic chemistry become important. For bioengineering and agricultural engineering, biochemistry becomes important, and similarly, for geological engineering, geochemistry becomes important. For manufacturing and petroleum engineering, fuel chemistry is essential. For civil engineering, cement chemistry and asphalt chemistry are important. For environmental engineering and chemical engineering, all twelve chapters are required. According to the real needs of the students, the instructors can arbitrarily delete certain topics.

Since this book in intended to be used worldwide without borders, free selection of the material depends on regional need.

                                                                    Teh Fu Yen

PREFACE

Having decided to go on this project of making a chemistry book for engineers, the main problem faced was deciding what to write. There was no similar treatise which I could select or look at. The first requirement what I thought for a long while is a good review of the necessary chemistry which links high school chemistry and college chemistry and also a short introduction of the chemistry undergraduate major requirement. These are a total of five different subjects – physical chemistry, inorganic chemistry, organic chemistry, analytical chemistry and surface chemistry. Could the basic concepts of these five subjects become sufficient for review of the essential part of chemistry for the need or basic tools for an accomplished engineer?

The next effort is in determining to what extent any subject should be covered. For example, for organic chemistry, there are thousands of important compounds or nomenclature, thousands of chemical synthesis, thousands of named reactions and thousands of important mechanisms. The question is whether any engineer needs to learn these or familiarize with them. So one has to select, shorten and abstract the necessary basic principles for engineers, with limited space to cover the entire subject. One can always be blamed for being arbitrary and subjective. The author's forty years of experience in teaching engineering students and contacts with engineering societies has solved some of the difficulties which were encountered by others.

I would like to thank the policy of the School of Engineering at the University of Southern California to have this chance of imparting global education without borders. I would also like to acknowledge many colleagues and students without whom this book would not be possible. In particular, I express my gratitude to Ngo Yeung Chan, Zixuan Chen, Neelakshi Hudda, Wanda Tan, and Nishant Vijayakumar.

Credits of chapters 1, 3, 4 and 5 go to the publishers Pierce Education Corp and Prentice Hall, who had published these four chapters in Environmental

Chemistry and granted the present publication the necessary rights. Other credits are listed below.

Credit to Oxford University Press and to C.M.Dobson, J.A.Jerrard and A.J.Pratt, *Foundation to Chemical Biology*, 2001, Fig. 1.2 for Fig. 6-1 ; Figs. 1.7 and 1.8 for Fig. 6-2; Fig. 2.2 for Fig. 6-4; Fig. 2.9 for Fig. 6-8; Fig. 3.2 for Fig. 6-9; Fig. 3.7 for Fig. 6-10; Fig. 3.8 for Fig. 6-11; Fig. 3.11 for Fig. 6-13; Fig. 3.12 for Fig. 6-15; Fig. 3.13 for Fig. 6-16. Credit to Springer-Verlag Inc. and to B.D.Hames and N.M.Hooper, *Biochemistry*, 4th ed., 2000, p. 48, Fig. 4 for Fig. 6-14; p. 137, Fig. 1 and Fig. 6-24; p. 138, Fig. 2 for Fig. 6-25. Credit to Elsevier Science Publishers and D.Koruga, S.Hameroff, J.Withers, R.Loutfy and M.Sundareshan, *Fullerene*, 1993, F: 5. 1-2 (p. 142) for Fig. 6-26 and Fig. 6-27; F: 5. 1-3 (p. 143) for Fig. 6-28; F: 5.24 (p. 150) for Fig. 6-29; F: 5.3-4 (p. 158) for Fig. 6-30. Credit to John Wiley and Sons and to D.Voet and J.G.Voet, *Biochemistry*, 1990, Fig. 32-2 for Fig. 6-31; Fig. 28-34 for Fig. 6-32; Fig. 28-35 for Fig. 6-33; Fig. 32-38 for Fig. 6-33A; Fig. 32-11 for Fig. 6-34; Fig. 32-12 for Fig.6-35. Credit to Cambridge University Press and to K.E.Peters, C.C.Walter and J.M.Moldowan, *The Biomarker Guide*, 2nd ed., II *Biomarkers and Isotopes in Petroleum Systems and Earth History*, 2005; Fig. 12.4 for Fig. 7-9; Fig. 18.85 for Fig. 7-11; Fig. 18.118 for Fig. 7-13; Fig. 19.1 for Fig. 7-14; Fig. 19.16 for Fig. 7-15; Fig. 18.131 for Fig. 7-16; Fig. 18.138 for Fig. 7-17; Fig. 13.63 for Fig. 7-32. Credit to Columbia University Press and to G.Ottonello, *Principles of Geochemistry*, 1997, Fig. 11.14 for Fig. 7-23; Fig. 11.16 for Fig. 7-24; Fig. 11.27 for Fig. 7-26; Fig. 11.29 for Fig. 7-27. Credit to John Wiley and Sons and to G.Faure, *Isotope Geology*, 2nd ed., 1986, Fig. 6.2 for Fig. 7-25. Credit to Cornell University Press and P.-G. de Gennes, *Scaling Concepts in Polymer Physics*, 1979, Fig. 4-1 used for Fig. 9-7. Credit to Nature Publishing Corp. and to J. Klein, *Nature*, **271**, 143, 1978, Fig. 2 used for Fig. 9-8. Credit to John Wiley and Sons and to S.C.Rosen, *Fundamental Principles of Polymer Materials*, 2nd ed., 1993, Fig. 14.1 used as Fig. 9-14; Fig. 14.2 as Fig. 9-15; Fig. 14.4 as Fig. 9-16; Fig. 18.1 as Fig. 9-20; Fig. 18.15 as Fig. 9-23; Fig. 18.20 as Fig. 9-24 and Fig. 18.21 as Fig. 9-25.

CONTENTS

CHAPTER 1

PHYSICAL CHEMISTRY

*P*hysical chemistry is concerned with the study of the physical properties and structure of matter using the laws of chemical interaction. Generally, the purpose of physical chemistry is threefold:

- to collect the appropriate data required to define the properties of matter
- to establish the energy relations in physical and chemical transformations
- to predict the extent and rate of the transformation taking place and identify its controlling factors

For our concern as engineers, the principles of physical chemistry could lead to an understanding of such concepts as the identification of compositions in aqueous solutions, the effects of additives on water purification, the extent and prevention of corrosion in piping, and so on. There are two common approaches to understanding physical chemistry. The first is the **synthetic approach**, which begins with the study of the structure and behavior of matter from subatomic particles, electrons, and nuclei, to atoms and molecules, and then proceeds to their states of aggregation and subsequent chemical reactions. The other is the **analytical approach**, which begins with the investigation of large objects, such as biosystems and bodies of water, and works its way back to atoms and particles. The analytical approach will be used in this book. In this chapter, we will discuss the subjects of gas-liquid-solid phase behavior, thermodynamics, and kinetics. At the end we will discuss some of the commonly used units and conventions.

1.1 GAS-LIQUID-SOLID

All matter exists in one of three states of aggregation: gaseous, liquid, or solid. The particular state of a substance is determined by the pressure and temperature under which it exists. Because the gaseous phase is the most random form of the three states, we can easily understand liquid and solid state behavior by fully comprehending the gaseous state. Thus, our attention will be focused here on the gaseous phase.

There are several basic laws or generalizations that are important to the study of gases. The first one is **Boyle's Law**: The volume of any definite quantity of gas at constant temperature varies inversely with the pressure on the gas. Expressed mathematically

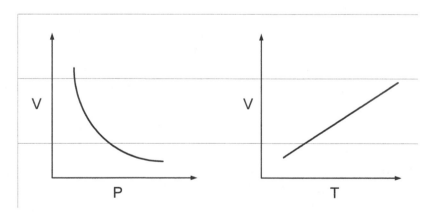

Figure 1-1. Illustration of Boyle's Law and Charles's Law.

$$V = \frac{K_1}{P} \quad (T = \text{constant}) \qquad [1\text{-}1]$$

If V is plotted versus P at a constant temperature, it will exhibit a hyperbolic characteristic curve, as shown in Figure 1-1.

Charles's or **Gay-Lussac's Law** states that the volume of a definite quantity of gas at constant pressure is directly proportional to the absolute temperature (Fig. 1-1), or

$$V = K_2 T \quad (P = \text{constant}) \tag{1-2}$$

To obtain a simultaneous variation of the volume with pressure and temperature, we proceed as follows:

$$V = f(P, T)$$

$$dV = \left(\frac{\partial V}{\partial P}\right)_T dP + \left(\frac{\partial V}{\partial T}\right)_P dT \tag{1-3}$$

From Equation [1-1] by differentiation, we get

$$\left(\frac{\partial V}{\partial P}\right)_T = -\frac{K_1}{P^2} = -\frac{PV}{P^2} = -\frac{V}{P} \tag{1-4}$$

Similarly, from Equation [1-2] we get

$$\left(\frac{\partial V}{\partial T}\right)_P = K_2 = \frac{V}{T} \tag{1-5}$$

Combining Equations. [1-4] and [1-5] into Equation [1-3]

$$dV = -V/P \, dP + V/T \, dT$$

or

$$dV/V + dP/P = dT/T$$

$$\ln V + \ln P = \ln T + \ln C$$

$$PV = \text{constant} \times T$$

$$PV = RT \quad \text{or} \quad PV = nRT \qquad [1\text{-}6]$$

Equation [1-6] is the **ideal gas equation**, where R is a universal constant for all ideal gases. In general, $R=Nk$ where N is Avogadro's number and k is Boltzmann's constant in terms of individual molecules. The value of R can be found from the fact that 1 mole of any ideal gas at **standard conditions** — that is, at 0°C and 1 atmosphere of pressure — occupies a volume of 22.414 liters. Therefore,

$$R = \frac{PV}{nT} = \frac{(1\,\text{atm})(22.414\,\text{L})}{(1\,\text{mole})(273.16\,\text{K})}$$

$$= 0.082 \text{ liter-atm/K/mole}$$

R can be expressed in any set of units representing work or energy. As an exercise, the following can be derived:

$$R = 8.314 \text{ joule/degreeK/mole}$$

$$= 1.987 \text{ cal/degreeK/mole}$$

As stated above, k is Boltzmann's constant in terms of individual molecules. In this instance, k can be easily computed out as

$$k = 1.38 \times 10^{-23} \text{ J/K}$$

if the Avogadro's number

$$N = 6.02 \times 10^{23}/\text{mole}$$

As shown in Equation [1-3], V is considered a function of T and P. The partial derivatives in the equation have definite physical meanings and are measurable quantities. There are three commonly tabulated properties:

- compressibility coefficient

$$\kappa = -\frac{1}{V}\left(\frac{\partial V}{\partial P}\right)_T \qquad\qquad \text{[1-6a]}$$

- expansion coefficient

$$\alpha = \frac{1}{V}\left(\frac{\partial V}{\partial T}\right)_P \qquad\qquad \text{[1-6b]}$$

- pressure coefficient

$$\beta = \frac{1}{P}\left(\frac{\partial P}{\partial T}\right)_V \qquad\qquad \text{[1-6c]}$$

If a gas obeys the ideal gas law, it can be easily found that

$$\alpha = \kappa\beta P \qquad\qquad \text{[1-6d]}$$

[Example 1-1] Thermometers are frequently broken in the laboratory by over-heating. If a thermometer is exactly filled with mercury at 50°C, what pressure will be developed within the thermometer if it is heated to 52°C? (For mercury, the expansion coefficient is 1.8×10^{-4} per degree and the compressibility coefficient is 3.9×10^{-6} per atm.)

$$V = f(P,T)$$

$$dV = \left(\frac{\partial V}{\partial T}\right)_P dT + \left(\frac{\partial V}{\partial P}\right)_T dP$$

$$V = \text{constant}, \quad dV = 0$$

$$\left(\frac{\partial P}{\partial T}\right)_V = \frac{-(\partial V/\partial T)_P}{(\partial V/\partial P)_T} = \frac{\alpha}{\kappa}$$

For Hg, $\alpha = 1.8 \times 10^{-4}$ degree^{-1}, $\kappa = 3.9 \times 10^{-6}$ degree^{-1}. Thus,

$$\left(\frac{\partial P}{\partial T}\right)_V = 46 \text{ atm/degree}$$

1.2 THERMODYNAMICS

Thermodynamics is the study of the energy accompanying physical and chemical processes and the transformation of energy from one form to another. The two most important words in thermodynamics are **heat** and **work**, which are related forms of energy. Heat energy can do work, and work energy can generate heat. In dealing with thermodynamic problems, the term **"system"** is frequently employed. A system is defined as any parts of the world selected for study. In turn, the portion of the universe excluded from the system is called the surroundings or environment. A **boundary** (real or imaginary) separates the system from the surroundings. An **open system** can exchange both matter and energy with its surroundings, while an **isolated system** cannot. **A closed system** can exchange energy, but it cannot change matter. As an example, Figure 1-2 gives a general representation of a natural water system treated as an open system.

1.2.1 Temperature, Heat, and Work

The **zeroth law of thermodynamics** can be stated as: systems in thermal equilibrium have the same temperature. If two systems are in thermal equilibrium with a third system, then they all are in equilibrium with each other, and they all have the same temperature. Heat is a form of energy that passes from one body to another solely as a result of temperature difference. If the temperature of a system is kept constant, $dT = 0$, it is said to be under **isothermal conditions**. An **adiabatic process** is one for which there is no heat transfer between it and its surroundings, $dq = 0$. The basic unit of heat is the **calorie**, defined as the heat required to raise the temperature of one gram of water by one Celsius degree. In engineering practices, it is common to measure heat in **Btu**, which is the heat required to raise one pound of water by one Fahrenheit degree. The unit Btu equals 252 calories.

It is possible to have a system of methods to transfer energy to a system as work. **Mechanical**, or **pressure-volume work** is the most familiar. For a closed system, it is equivalent to the pressure times the change in volume:

$$dw = PdV \qquad [1\text{-}7]$$

Since heat and work are both forms of energy, they can be equated.

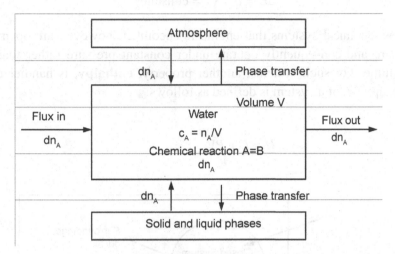

Figure 1-2. General representation of natural water systems is treated as an open system. The system receives fluxes of matter from the surroundings and undergoes chemical changes symbolized by the reaction A=B. The time invariant condition is represented by $dC_A/dt=0$. (Stumm and Morgan)

1.2.2   The First Law of Thermodynamics

The **first law of thermodynamics**, or the **law of conservation of energy**, states that energy can be neither created nor destroyed. The mathematical statement of the first law could be written as

$$\Delta E = q - w \qquad [1\text{-}8]$$

where

ΔE = change in internal energy of the system

q = heat flowing into the system

w = work done by the system

By convention, q has a positive value if heat is absorbed by the system, and a negative vatlue otherwise. If the system does work on the surroundings, w has a positive value, as shown in Figure 1-3.

In chemical systems, expansion work is usually the work performed. For a special case, if the volume of the system remains constant, then no expansion work can be done (i.e., $w = 0$). For this case

$$\Delta E = q_v \quad (V = \text{constant}) \tag{1-9}$$

Most chemical systems that engineers encounter, however, are open to the atmosphere and consequently operate under constant pressure rather than constant volume. For such systems, another property, **enthalpy**, is handier to use. The enthalpy, H, of a system is defined as follows:

$$H = E + PV \tag{1-10}$$

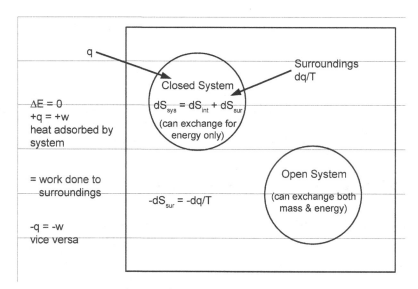

Figure 1-3. Illustration of a closed system, surroundings, and an isolated system or "universe" of a system plus surroundings. Heat transferred to the system, q, is positive, and that lost from the surroundings is $-q$. The entropy change of the system, dS_{sys}, is the sum of an internal change dS_{int} and a flow from the surroundings dS_{sur}.

Assuming that a chemical reaction takes place at a constant pressure and temperature, the change in the internal energy of the system would be

$$\Delta E = E_2 - E_1 = q_p - w = q_p - PdV = q_p - P(V_2 - V_1)$$

where q_p is the heat absorbed at constant pressure. By rearranging the terms, then

$$(E_2 + PV_2) - (E_1 + PV_1) = q_p = H_2 - H_1$$

or

$$\Delta H = q_p \quad (\text{Constant } T, P) \qquad [1\text{-}11]$$

It is a universal practice to designate ΔH as **heat of reaction**. Chemical changes that are accompanied by the absorption of heat, which makes ΔH positive, are called **endothermic reactions**. **Exothermic reactions** are those with a negative ΔH value. The heat of reaction for a chemical reaction can be easily calculated from standard enthalpies, usually available in handbooks, of the species involved. Calculation of the heat of combustion of methane gas is illustrated in the following example:

$$CH4(g) + 2O2(g) \quad \rightarrow \quad CO2(g) + 2H2O(l)$$
$$\Delta H^0_{298} \quad -17.89 \quad\quad 0 \quad\quad\quad\quad -94.05 \quad 2(-68.32)$$

Hence, H^0_{298} for combustion = $(-94.05) + 2(-68.32) - (-17.89) - 0$

$$= -212.80 \text{ Kcal/mole of methane}$$

1.2.3 The Second Law of Thermodynamics and Entropy Pollution

The first law, stating that energy must be conserved in all ordinary processes, imposes no direction on energy transformations. However, certain restrictions do exist. For example, heat always flows from a higher temperature level to a lower one and never in the reverse direction. Also, the efficiency of transformation from one form of work to another, such as from mechanical to electrical in an electrical generator, can be made to approach 100%. The transformation of heat (thermal energy) into work is especially inefficient, the highest efficiency being 45% for a modern turbine, as shown in Table 1-1. This inefficiency indicates that various forms of energy have different qualities. In this sense, work can be termed as energy of a higher quality than heat.

The concept of **entropy** was developed to serve as a general criterion of spontaneity for physical and chemical changes, or for direction of energy transformation. For the given state of a system, entropy, S, can be defined quantitatively as

$$S = k \ln \Omega \qquad\qquad [1\text{-}12]$$

where

k = Boltzmann's constant

Ω = the thermodynamic probability, defined as the number of ways that the particles of the system can be arranged among the energy levels accessible to them.

In most textbooks, entropy is defined by the following differential equation:

$$dS = dq_{rev} / T \qquad\qquad [1\text{-}13]$$

where the quantity q_{rev} is the amount of heat that the system absorbs if a chemical change is brought about in a reversible manner. If a spontaneous change occurs in a system, it will always be found that the total entropy change, including everything involved, is a positive value. A calculation of entropy change can help us determine whether a chemical or physical transformation could occur. If $S > 0$, the change will occur spontaneously; if $S < 0$, the change usually occurs in the reverse direction; and if $S = 0$, the system is at equilibrium. In general, the **second law of thermodynamics** states that all processes in nature tend to occur only with an increase in entropy and that the direction of change is always such as to lead to the entropy increase.

The first law tells us that the energy of the universe is constant, while the second law indicates that energy has different qualities and the entropy of the universe tends toward a maximum. Table 1-2 lists the quality of various forms of energy in terms of entropy — the higher the entropy per unit energy, the lower the quality of the energy. Those forms, having small amounts of entropy per unit energy, tend to transform into others with higher values, thereby producing greater quantities of entropy.

Table 1-1. Efficiency of Energy Converters

Type	Efficiency	Type of Conversion From	Type of Conversion To
Electric generator - Electric motor	99	Mechanical	Electrical
Low	62	Electrical	Mechanical
High	92	Electrical	Mechanical
Dry cell battery	90	Chemical	Electrical
Large steam boiler	88	Chemical	Thermal
Home furnace - Gas	85	Chemical	Thermal
Home furnace - Oil	65	Chemical	Thermal
Storage battery (lead acid)	72	Electrical	Chemical
		Chemical	Electrical
Fuel cell	60	Chemical	Electrical
Man on bicycle	50	Chemical	Mechanical
Liquid fuel rocket (H_2)	47	Chemical	Thermal
Turbine - Steam	45	Thermal	Mechanical
Turbine - Gas (aircraft or industrial)	35	Chemical	Thermal
Electric Power plant - Fossil fueled	40	Chemical	Thermal
Nuclear fueled	32	Nuclear	Thermal
		Thermal	Mechanical
		Mechanical	Electrical
Internal combustion engine - Diesel	37	Chemical	Thermal
Otto cycle (automobile)	25	Thermal	Mechanical
Wankel (rotary)	18		
Lasers	30-10	Electrical	Radiant
Lamps - High intensity	32	Electrical	Radiant
Lamps - Fluorescent	20	Electrical	Radiant
Lamps - Incandescent	4	Electrical	Radiant
Unaided walking man	12	Chemical	Mechanical
Solar cell	10-15	Radiant	Electrical
Steam locomotive	8	Chemical	Thermal
		Thermal	Mechanical
Thermocouple	6	Thermal	Electrical

Source: Much of the data of this table was obtained from C.M. Summers, Sci. Amer., 224(3), 155 (Sept. 1971).

Table 1-2. Quality of Various Forms of Energy

Form of Energy	Entropy Per Unit Energy
Gravitational	0
Nuclear	10^{-4}
Thermal	
Stars (106 K)	10^{-3}
Earth (102 K)	10^{2}
Chemical	1-10
Radiant	
Sunlight (visible)	1-10
Cosmic microwave	10^{4}

Source: Data adapted from F.J. Dyson, Sci.Amer., 224(3), 50-59 (Sept. 1971)

Therefore, any energy crisis is not the result of a lack of energy, because energy cannot be created or destroyed. Instead, it has resulted from the production of entropy associated with the conversion of high-grade energy sources into lower-grade ones. The production of entropy has been referred to as **entropy pollution,** because it is a kind of measure of the extent to which the universe has been irreversibly degraded. Table 1-3 gives the major energy resources available at the surface of the earth; each resource will be discussed further in later chapters.

1.2.4   Free Energy

To use entropy change as a criterion of spontaneous change requires taking both the system and all its surroundings into consideration. Usually, it is more convenient to limit our attention to the system only. This concentration can be accomplished in the following way:

$$\Delta S_{universe} = \Delta S_{system} + \Delta S_{surrounding}$$

$$= \Delta S_{system} + \frac{q_{surrounding}}{T} \qquad [1\text{-}14]$$

$$\geq 0$$

Table 1-3. Estimated Energy Resources*

Type	Total World Supply [a]	Economically Available (at No More Than Double Current Costs)	
		World	U.S.
Depletable supplies			$(10^{21}$ J$)$
Fossil (chemical)			
Tar sands	2.1	-----	-----
Natural gas	3.8	2.2 - 3.8	0.6 - 1.1
Petroleum	6.0	3.2 - 6.0	0.6 - 1.1
Oil shale	13.3	-----	-----
Coal and lignite	185.0	21.1 - 31.5	5.0 - 7.2
Total Fossil	210.2	26.5 - 41.3	6.2 - 9.4
Nuclear			
Ordinary fission	$2 \times 10^{4\,b}$	14.0	1.4
Breeder fission	$6 \times 10^{6\,b}$	4000.0	400.0
Fusion (D-T)	215	-----	-----
Fusion (D-D)	1×10^{10}	-----	-----
Continuous supplies			$(10^{21}$ J/ year$)$
Solar	899	-----	$50.0^{\,c}$
Tidal	0.094	-----	0.009
Geothermal	0.010	-----	0.001

a Total supply including amount consumed to date.

b Estimated from total quantity of uranium and thorium within 1 mile of land surface assume 1% to be able for mining (Int. At. Energy Ag. Bull., 14(4), 11, 1972).

c Total supply is listed because no cost figures are available.

* Source: Data adapted from C. Starr, Sci. Amer., 224(3), 43, (Sept. 1971).

Because,

$$\Delta H = q_{system} = -q_{surroundings}$$

$$\Delta S_{universe} = \Delta S_{system} - \frac{\Delta H_{system}}{T} \geq 0 \qquad\qquad [1\text{-}15]$$

Multiplying both sides by $-T$

$$-T\Delta S_{universe} = \Delta H_{system} - T\Delta S_{system} = \Delta G_{system} \leq 0 \qquad [1\text{-}16]$$

ΔG_{system} is defined by Equation [1-16] and is known as the change in **Gibbs free energy** of the system. So, if $\Delta G < 0$, the reaction will go spontaneously; if $\Delta G = 0$, the system is in equilibrium. ΔG is also defined as

$$G = H - TS \qquad [1\text{-}17]$$

At constant T and P

$$\begin{aligned} \Delta G &= \Delta H - T\Delta S \\ &= \Delta E + P\Delta V - T\Delta S \\ &= q - w + P\Delta V - T\Delta S \end{aligned} \qquad [1\text{-}18]$$

If the system change is brought about reversibly, then q becomes q_{rev} and w becomes w_{max}, and the maximum quality of work that can be obtained is

$$\Delta G = q_{rev} - w_{max} + P\Delta V - q_{rev}$$

and

$$-\Delta G = w_{max} - P\Delta V$$

$P\Delta V$ gives the portion of the work that must be wasted; therefore, the $-\Delta G$ indicates the useful work available for the system change.

$$-\Delta G = w_{useful} \quad (\text{constant } T, P) \qquad [1\text{-}19]$$

In principle, any spontaneous process can be made to do useful work as shown in Equation [1-19]. To find the relationship between the free energy and **equilibrium constant**, we proceed as follows:

$$dG = -SdT + VdP \qquad [1\text{-}20]$$

At constant temperature

$$dG = VdP = RT(dP/P) = RT \, d(\ln P) \qquad [1\text{-}21]$$

To give a reference point for the free energy as G^o at 25°C and 1 atmosphere,

$$G - G^o = RT \ln(P/P^o) = RT \ln \alpha \qquad [1\text{-}22]$$

where $\alpha \propto P \propto$ concentration
where α is the activity. Consider the following reaction:

$$aA + bB \rightarrow cC + dD$$

$$\Delta G = \Sigma G_{products} - \Sigma G_{reactants} \qquad [1\text{-}23]$$

also $Gi = Gi° + RT \ln \alpha i$

The free energy of this reaction is given by the following equation by substituting Equation [1-22] into Equation [1-23]:

$$\Delta G = \Delta G° + RT \ln \left(\frac{\{C\}^c \{D\}^d}{\{A\}^a \{B\}^b} \right)$$

As the reaction proceeds to a state of equilibrium, ΔG will be zero, and thus

$$\Delta G° = - RT \ln \left(\frac{\{C\}^c \{D\}^d}{\{A\}^a \{B\}^b} \right)_{equilibrium} \qquad [1\text{-}24]$$

$$= -RT \ln K$$

where K is the **equilibrium constant**. Values for the standard free energies of various substances can be found from engineering or chemistry handbooks. By calculating the standard free energy change for the reaction, K is easily determined from the preceding equation.

1.2.5 Thermodynamic Properties

Thermodynamic systems may consist of one or more parts called **phases**, in which they are physically and chemically homogeneous. Thermodynamic systems are characterized by a small number of properties. The properties may be divided into two types: extensive and intensive. **Extensive properties** are additive; that is, they depend on the amount of substances present. Examples of such properties are total mass, volume, and energy. On the other hand, **intensive properties** are those whose values are independent of the total amount, but which depend on the concentration of the substance(s) in a system. Examples of intensive properties are pressure, temperature, molar volume, and chemical potential. Figure 1-4 illustrates a simplified model for the thermodynamic description of natural water systems. P and T are intensive variables, and the mole numbers, n_i, in each phase are extensive variables that together determine the volume, mass, composition, and other properties of the system.

Various forms of thermodynamic work are available. Basically, they can all be expressed as an intensive property times an extensive variation, as shown in Table 1-4.

Many thermodynamic properties and their relationships have been derived, making it a difficult task to memorize them all. To solve the problem, a handy scheme has been devised for memorizing important thermodynamic relations. A detailed description of the scheme based on Yen is given in Figure 1-5.

There are many thermodynamic functions that can be derived from the code sentence. "**T**he **G**ibbs **P**otential **H**as **S**hown **E**ndless **V**aluable **A**pplications" for the functions of T, G, P, H, S, E, V, and A. This is actually based on the principle of symmetry of thermodynamic potentials; e.g., Jacobian. In order to obtain dE as a function of appropriate independent variables, look at the two diagonal lines toward the base E, taking the signs

$$dE = -PdV + TdS$$

Similarly

$$dG = VdP - SdT$$

$$dH = VdP + TdS$$

$$dA = -PdV - SdT \qquad [1-24a]$$

Again for triangular relationship, for example, ΔSVT

$$\left(\frac{\partial E}{\partial S}\right)_V = T$$

the variable V is situated at the right angle while the independent variable and answer are at the other apexes.

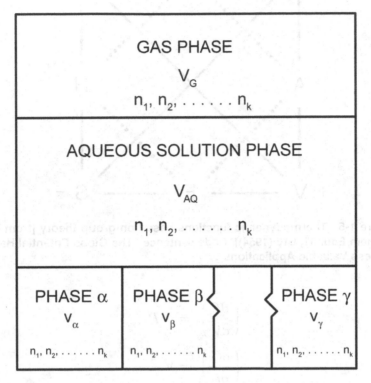

Figure 1-4. Simplified model for the thermodynamic description of natural water systems.

Table 1-4. Expressions for the Thermodynamic Work dW Done on a System

Type	Intensive Property	Extensive Variation	Expression
Expansion	Pressure, P	Volume, dV	$-PdV$
Electrical	Potential, E	Charge, de	$-Ede$
Gravitational	Force, mg	Height, dh	$mg\ dh$
Chemical	Chemical potential, μ	Moles, dn	μdn
Surface	Interfacial tension, γ	Area, dA	γdA

Source: Stumm and Morgan

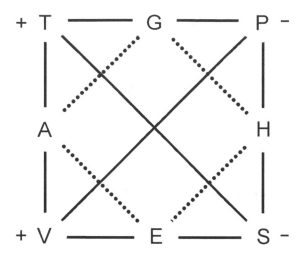

Figure 1-5. Thermodynamic functions based on group theory [from T.F. Yen, J. Chem Edu. 31, 610 (1954)]. Code sentence "The Gibbs Potential Has Shown Endless Valuable Applications".

$$\left(\frac{\partial E}{\partial V}\right)_S = -P$$

$$\left(\frac{\partial G}{\partial P}\right)_T = V$$

$$\left(\frac{\partial G}{\partial T}\right)_P = -S$$

$$\left(\frac{\partial H}{\partial S}\right)_P = T \qquad\qquad [1\text{-}24\text{b}]$$

$$\left(\frac{\partial H}{\partial P}\right)_S = V$$

$$\left(\frac{\partial A}{\partial V}\right)_T = -P$$

$$\left(\frac{\partial A}{\partial T}\right)_V = -S$$

The Maxwell relationship can be obtained by considering the right triangle with the same base; for example, ΔTVS, ΔPSV

$$\left(\frac{\partial T}{\partial V}\right)_S = -\left(\frac{\partial P}{\partial S}\right)_V$$

$$-\left(\frac{\partial P}{\partial T}\right)_V = -\left(\frac{\partial S}{\partial V}\right)_T$$

$$-\left(\frac{\partial S}{\partial P}\right)_T = \left(\frac{\partial V}{\partial T}\right)_P$$

[1-24c]

$$\left(\frac{\partial V}{\partial S}\right)_P = \left(\frac{\partial T}{\partial P}\right)_S$$

from the corners of the diagram

$$\left(\frac{\partial V}{\partial S}\right)_E = \frac{T}{P}$$

$$\left(\frac{\partial T}{\partial V}\right)_A = \frac{-P}{S}$$

[1-24d]

$$-\left(\frac{\partial P}{\partial T}\right)_G = \frac{-S}{V}$$

$$-\left(\frac{\partial S}{\partial P}\right)_H = \frac{V}{T}$$

from the trapezoid $EATS$, $AGPV$, and so on.

$$E - A = TS$$

$$A - G = -PV$$

[1-24e]

$$G - H = -ST$$

$$H - E = VP$$

and chemical potentials such as

$$\mu = \left(\frac{\partial E}{\partial n}\right)_{S,V} = \left(\frac{\partial A}{\partial n}\right)_{V,T} = \left(\frac{\partial G}{\partial n}\right)_{T,P} = \left(\frac{\partial H}{\partial n}\right)_{P,S} \qquad [1\text{-}25]$$

The symmetry is based on group theory.

A definite quantity of heat is required to raise the temperature of a given mass of any material by one Celsius degree. This quantity is called the heat capacity of the system, C. Thus,

$$C = \frac{dq}{dT} \qquad [1\text{-}26]$$

If a unit mass is taken as the basis for heat capacity, the equation becomes $dq = m\,C\,dT$, where m represents mass and C is now called **specific heat capacity**. The specific heat capacity of water is approximately $1(\text{cal})/(\text{g})(^{\circ}\text{C})$. When the volume is held constant, by combining the preceding equation and $\Delta E = q - w$ — that is, Equation [1-8] and [1-26] — C_v at a constant volume can be obtained:

$$C_V = \left(\frac{\partial E}{\partial T}\right)_V \qquad [1\text{-}27]$$

Similarly, by some simple manipulation, we can get C_p at constant pressure.

$$C_P = \left(\frac{\partial H}{\partial T}\right)_P \qquad [1\text{-}28]$$

The general equation to describe the C_p and C_v is

$$C_P - C_V = \left[\left(\frac{\partial E}{\partial V}\right)_T + P\right]\left(\frac{\partial V}{\partial T}\right)_P \qquad [1\text{-}29]$$

The actual value of the difference for an ideal gas could easily be found to equal R. Table 1-5 lists the molar heat capacities for various gases.

Table 1-5. Molar Heat Capacities of Gases*

Gas	C_P	C_V	$C_P/ C_V = \gamma$
Argon, A	4.97	2.98	1.67
Helium, He	4.97	2.98	1.67
Mercury, Hg	4.97	2.98	1.67
Hydrogen, H_2	6.90	4.91	1.41
Oxygen, O_2	7.05	5.05	1.40
Nitrogen, N_2	6.94	4.95	1.40
Chlorine, Cl_2	3.25	6.14	1.34
Nitric oxide, NO	7.11	5.11	1.39
Carbon monoxide, CO	6.97	4.97	1.40
Hydrogen chloride, HCl	7.05	5.01	1.41
Carbon dioxide, CO_2	8.96	6.92	1.29
Nitrous oxide, N_2O	9.33	7.29	1.28
Sulfur dioxide, SO_2	9.40	7.30	1.29
Ammonia, NH_3	8.63	6.57	1.31
Methane, CH_3	8.60	6.59	1.31
Ethane, C_2H_4	12.71	10.65	1.19
Dimethyl ether, C_2H_4O	15.89	13.73	1.16

*in cal deg^{-4} $mole^{-1}$ at 25°C

1.2.6 Applications to Solid Systems

The following is a discussion leading to the understanding of some properties of metals, ceramic materials, and rubbers. From Figure 1-5, we can easily obtain the **Helmholtz free energy** equation, as shown here

$$A = E - TS$$

or

$$dA = dE - TdS - SdT \qquad [1\text{-}24f]$$

Also from Figure 1-5

$$E = H - VP$$

$$dE = dH - VdP - PdV \qquad [1\text{-}24\text{g}]$$

and Figure 1-5

$$dH = VdP + TdS$$

Thus, combining this relation with Equation [1-24f]

$$dE = TdS - PdV$$

Then, substituting into Equation [1-24e], it follows that

$$dA = -PdV - SdT \qquad [1\text{-}24\text{h}]$$

For an elastomer, **chain extension** will take place isothermally with a change of ΔL from the length of polymeric chains, and $dW = -PdV$ from Equation [1-24h], as shown here

$$\therefore dA_T = dW = FdL \qquad [1\text{-}24\text{i}]$$

where F is the force required to extend or to compress the chain segment. Thus one can rewrite the Helmholtz free energy equation as

$$F = \left(\frac{\partial A}{\partial L}\right)_{T,V} = \left(\frac{\partial E}{\partial L}\right)_{T,V} - T\left(\frac{\partial S}{\partial L}\right)_{T,V} \qquad [1\text{-}30]$$

because deformation takes place with no volume change. For the preceding equation, the first term becomes important in the case of metals and ceramics

$$F = \left(\frac{\partial E}{\partial L}\right)_{T,V} \qquad [1\text{-}30\text{a}]$$

However, internal energy causes heatup and fracture with extension for rubber. The opposite conditions apply to elasticity as entropy decreases with extension and a large change of chain segment conformations

$$F = -T\left(\frac{\partial S}{\partial L}\right)_{T,V}$$ [1-30b]

which indicates that the second term becomes important for rubber.

Under deformation the chain segment can be analyzed with an end-to-end distance of the polymer chain. The end-to-end distance can be related to a probability function $P(r)$, and the entropy of this chain segment is as follows (here distance r is proportional to L):

$$S = S_0 + k \, lnP(r)$$

$$P(r) = \frac{\exp\left[-\left(\frac{r}{\rho}\right)^2\right]}{\left(\rho\pi^{\frac{1}{2}}\right)^3}$$

which has a Gaussian distribution, and where ρ is density. Hence,

$$S = S_o - k\left[3\ln(\pi^{\frac{1}{2}}\rho) + \left(\frac{r}{\rho}\right)^2\right]$$

Thus,

$$(\frac{\partial S}{\partial r}) = \frac{-2kr}{\rho^2}$$

or,

$$F = \frac{2kTr}{\rho^2} = kr$$ [1-31]

after substituting with

$$F = -T\left(\frac{\partial S}{\partial L}\right)_{T,V} \qquad\qquad [1\text{-}30b]$$

which means the force to extend or compress the chain segment is to distance r and is directly proportional to temperature. The chain acts like a spring.

1.3 KINETICS

Thermodynamics is able to tell us what can happen, and to what extent, but it is unable to tell us how a change will actually occur. **Chemical kinetics** searches for the factors that influence the rate of reaction and brings a time factor into consideration. The rate of reaction depends on the nature of the reacting substances, the temperature, and the concentration of the reactants. The **rate of a chemical reaction** is the rate at which the concentrations of reacting species vary with time; that is, $-dC/dt$, where C is the concentration of the reactant. The sum of all the exponents to which the concentrations in the rate equation are raised is the **order** of the chemical reaction. Thus, a rate equation is expressed as

$$-\frac{dC}{dt} = kC_1^{n_1}\, C_2^{n_2}\, C_3^{n_3}\,.... \qquad\qquad [1\text{-}32]$$

where

k = rate constant

n = order of the reaction $(n_1 + n_2 + ...)$

1.3.1 First-Order Reactions

A **first-order reaction** is one in which the rate of reaction is proportional to the concentration of the reactant. For example, the following reaction is a first-order reaction:

$$N_2O_5 \rightarrow 2\,NO_2 + 1/2\,O_2$$

Therefore

$$-\frac{dC}{dt} = kC \qquad \text{[1-33]}$$

If the initial concentration, at $t = 0$, is C_o, the concentration (C) at some later time (t) can be found by integrating the preceding equation, which gives

$$-\int_{C_0}^{C} \frac{dC}{C} = k \int_{0}^{t} dt$$

and

$$-\ln \frac{C}{C_o} = \ln \frac{C_o}{C} = kt$$

or

$$C = C_o \exp(-kt) \qquad \text{[1-34]}$$

The **half-life** of the reaction can be determined by inserting the requirements that at $t = t_{1/2}$ and the concentration $C = \frac{1}{2} C_o$ into Equation [1-34], that gives

$$t_{1/2} = \frac{\ln 2}{k} = \frac{0.693}{k} \qquad \text{[1-35]}$$

1.3.2 Second-Order Reactions

For a **second-order reaction**; for example, A + B $\rightarrow$ products, the rate equation can be expressed as follows

$$-\frac{dC_A}{dt} = -\frac{dC_B}{dt} = kC_A C_B$$

or

$$\frac{dX}{dt} = k(a-x)(b-x)$$

where

x = amount of the reactants consumed

a = initial concentration of A

b = initial concentration of B

If $a \neq b$, the following equation can be obtained through simple mathematical manipulation:

$$k = \frac{2.303}{t(a-b)} \log \frac{b(a-x)}{a(b-x)}$$ [1-36]

The order of reaction can be evaluated by a graphical method (see Appendix A).

1.3.3 Consecutive Reactions

Chemical reactions such as

$$A \xrightarrow{k_1} B \xrightarrow{k_2} C$$

which proceed from reactants to products through one or more intermediate stages are called **consecutive reactions**. The rate equations are as follows:

$$-\frac{dC_A}{dt} = k_1 C_A$$

$$\frac{dC_B}{dt} = k_1 C_A - k_2 C_B$$

$$\frac{dC_C}{dt} = k_2 C_B$$

If at $t = 0$ we have $C_A = C_{A0}$, $C_B = C_C = 0$, then a solution for the concentration of each component at time t is as follows (it is a good exercise to solve the following equations):

$$C_A = C_{A0} e^{-k_1 t}$$ [1-37]

$$C_B = \frac{k_1 C_{A0}}{k_2 - k_1} \left\{ e^{-k_1 t} - e^{-k_2 t} \right\}$$ [1-38]

$$C_c = C_{A0}\left\{1 + \frac{k_1 e^{-k_2 t}}{k_2 - k_1} - \frac{k_2 e^{-k_1 t}}{k_2 - k_1}\right\} \qquad \text{[1-39]}$$

Consecutive reactions are of great importance in engineering. Bacterial nitrification of ammonia can be described by a consecutive reaction. Ammonia is oxidized by *Nitrosomonas* bacteria to nitrite, which is then oxidized by *Nitrobacter* bacteria to nitrate as indicated here:

$$NH_3 \xrightarrow{\textit{Nitrosomonas}} NO_2^- \xrightarrow{\textit{Nitrobactor}} NO_3^-$$

The changes in nitrogen forms are shown in Figure 1-6, where the concentrations of nitrite and nitrate were set equal to zero when $t = 0$, and k_1 was assumed to be equal to $2k_2$.

There are some other types of complex reactions, such as **parallel types**, in which two reacting species compete with each other to react with a third reacting species.

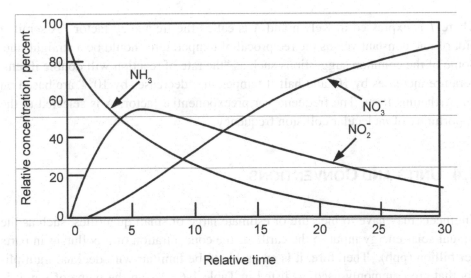

Figure 1-6. Nitrogen changes during nitrification, assuming consecutive first-order reactions.

1.3.4   Temperature Dependence of Reaction Rates

Chemical and biological reaction rates generally increase with increasing temperature. Most of the reactions follow the **Arrhenius-type of temperature dependence**

$$\frac{d \ln k}{dT} = \frac{E_a}{RT^2}$$ [1-40]

where E_a is a constant termed the activation energy. By integrating the preceding equation, we get

$$\ln k = \frac{-E_a}{RT} + \ln A$$ [1-41]

or

$$k = A \exp\left(-\frac{E_a}{RT}\right)$$

where T is expressed in Kelvin and A is called the frequency factor. A semilog plot of rate constant versus the reciprocal of temperature should be a straight line. Some of the common suggestions such as "the rate of reaction will double if temperature increases by $10°$ and half if temperature decreases by $10°$ " are based on the Arrhenius form. The frequency or preexponential factor, A is selected to the opportunity of molecular collision frequency.

1.4 UNITS AND CONVENTIONS

Engineers may have to measure or estimate large or small quantities, such as the annual solar energy input to the earth, or the concentration of a pollutant in parts per billion (ppb). Therefore, it is advisable to be familiar with decimal multipliers that are commonly used, as listed in Table 1-6. When the name of a unit is preceded by one of these prefixes, the size of the basic unit is modified by the decimal multiplier. For example, 1 Tg stands for 10^{12} grams and 1 Pg is for 10^{15} grams. In the field of energy, there are special notations for large energy values, where 1 **Quad** $= 10^{15}$ Btu (British thermal unit) and 1 **Quin** $= 10^{18}$ Btu. Accordingly, 1 PBtu = 1 Quad, or 1 EBtu = 1 Quin and 1 Quin = k Quad.

One important conversion factor for air pollution study is that for the conversion of ppm to $\mu g/g^3$ or mg/cm^3, or vice versa. One **ppm** means 1 volume in 1 million volumes. For instance, to convert 1 ppm of SO_2 at 1 atmosphere and 0°C, we would proceed as follows:

$$1 \text{ ppm} = (1/10^6)(1 \text{ mole}/22.4 \text{ liter})(64 \text{ gram/mole})(10^6 \text{ } \mu g/g)(10^3 \text{ liter/m3})$$

$$= 2857 \text{ } \mu g/m^3$$

or

$$= 2612 \text{ } \mu g/m^3 \text{ at } 25 \text{ °C}$$

Conversion factors for several air pollutants are given in Table 1-7.

Conversely, 1000 $\mu g/m^3$ of CO can be calculated as

$$(1000\mu g/m^3)(1g/10^6\mu g)(1 \text{mole}/28g)(22.4 \text{L/mole})(1m^3/10^6cm^3)(1cm^3/10^3L)(10^6 \text{L/L})$$

$$= 0.8 \text{ ppm}$$

A table of common unit conversions is included in Table 1-8. Throughout the book both English and metric system are used separately. In many cases, special units have been designated for the convenient usages. Many names of elements and constants are based on the International Union of Pure and Applied Chemistry (IUPAC) recommendation. Some useful constants are listed in Table 1-9.

Table 1-6. Decimal Multipliers that Serve as SI Unit Prefixes

Prefix	Origin	Symbol	Multiplying Factor
yotta	Greek or Latin *octo*, "eight"	Y	10^{24}
zetta	Latin *septem*, "seven"	Z	10^{21}
exa	Greek *hex*, "six"	E	10^{18}
peta	Greek *pente*, "five"	P	10^{15}
tera	Greek *teras*, "monster"	T	10^{12}
giga	Greek *gigas*, "giant"	G	10^{9}
mega	Greek *megas*, "large"	M	10^{6}
kilo	Greek *chilioi*, "thousand"	k	10^{3}
hecto	Greek *hekaton*, "hundred"	h	10^{2}
deka	Greek *deka*, "ten"	da	10^{1}
deci	Latin *decimus*, "tenth"	d	10^{-1}
centi	Latin *centum*, "hundred"	c	10^{-2}
milli	Latin *mille*, "thousand"	m	10^{-3}
micro	Latin *micro-* (Greek *mikros*), "small"	μ	10^{-6}
nano	Latin *nanus* (Greek *nanos*), "dwarf"	n	10^{-9}
pico	Spanish *pico*, "a bit," Italian *piccolo*, "small"	p	10^{-12}
femto	Danish-Norwegian *femten*, "fifteen"	f	10^{-15}
atto	Danish-Norwegian *atten*, "eighteen"	a	10^{-18}
zepto	Latin *septem*, "seven"	z	10^{-21}
yocto	Greek or Latin *octo*, "eight"	y	10^{-24}

Table 1-7. Conversion Factors for Air Pollutants

	Temperature (°C)	Pressure (mm)	1 ppm equivalence in $\mu g/m^3$
Carbon monoxide (CO)	0	760	1,250
	25	760	1,145
Nitric oxide (NO)	25	760	1,230
Nitrogen dioxide (NO$_2$)	25	760	1,880
Ozone (O$_3$)	0	760	2,141
	25	760	1,962
PAN {CH$_3$(CO)O$_2$NO$_2$}	0	760	5,938
	25	760	4,945
Sulfur dioxide (SO$_2$)	0	760	2,860
	25	760	2,620

Source: H.C. Perkins, Air Pollution, McGraw Hill, 1974, p.385.

Table 1-8. Common Unit Conversions

Gas Constant	Volume	Density
0.082 liter-atm/mole K	1 ft^3 = 28.316 liter	1g/cm^3 = 1000 kg/m^3
62.36 liter-mm Hg/mole K	= 7.481 gal	= 62.428 lb/ft^3
8.314 Joule/g-mole K	1 in.^{3} = 16.39 cc	= 8.345 lb/gal
1.314 atm-ft^3/lb-mole K	= 5.787 x 10^{-4} ft^3	= 0.03613 lb/in.^{3}
1.987 cal/g-mole K	1 gal = 3.785 liter	
1.987 Btu/lb-mole °R	= 8.34 lb H$_2$O	
0.73 atm-ft^3/lb-mole °R	1 m^3 = 35.32 ft^3	
10.73 psi-ft^3/lb-mole °R	= 264.2 gal	
1545 ft-lb$_f$/lb-mole °R		
Length	**Viscosity**	**Conversion Factor**
1 mile = 1609 m = 5280 ft	1 poise	1 cal/g-mole = 1.8Btu/lb-mole
1 ft = 30.48 cm = 12 in.	= 6.7197×10^{-2} lb$_m$/ft-sec	1 amu = 1.66063 × 10^{-24}g
1 in. = 2.54 cm	= 2.0886×10^{-3} lb$_f$-sec/ft^2	1 eV = 1.6022 × 10^{-12}erg
1 m = 3.2808 ft	= 2.4191×10^2 lb$_m$/ft-hr	1 radian = 57.3°
= 39.37 in.	= 1 g/cm-sec	1 cm/sec = 1.9685 ft/min
1 nm = 10^{-9}m = 10 A		1 rpm = 0.10472 radian/sec

Table 1-8. Continued

Pressure	Constant	Mass
1 atm = 101325 N/m^2	$h = 6.6262 \times 10^{-27}$erg-sec	1 kg = 2.2046 lb
= 14.696 psi	$k = 1.38062 \times 10^{-16}$erg/K	1 lb = 453.59 g
= 760 mmHg	$N_0 = 6.022169 \times 10^{23}$	1 ton = 2000 lb
= 29.921in.Hg	$C = 2.997925 \times 10^{10}$cm/sec	= 907.2 kg
(32 °F)	F = 96487 coul/eq	1 B ton = 2240 lb
= 33.91 ftH$_2$O	$e = 1.60219 \times 10^{-19}$coul	= 1016 kg
(39.1 °F)	g = 980.665 cm/sec^2	1 tonne = 2205 lb
= 2116.2 lb$_f$/ft^2	=32.174 ft/sec^2	= 1000 kg
= 1.0133 bar		1 slug = 32.2 lb
= 1033.3 g$_f$/cm^2		= 14.6 kg

Area	Power	Force
1 m^2 = 10.76 ft^2 = 1550 in.^{2}	1 HP = 550 ft-lb$_f$/sec	1N = 1 kg-m/sec^2
1 ft^2 = 929.0 cm^2	= 745.48 watt	= 10^5 dyne
	1 Btu/hr = 0.293 watt	= 0.22481 lb$_f$
		= 7.233 lb$_m$-ft/sec^2

Transfer Coefficient	Energy & Work	Stress
1 Btu/hr-ft^2 °F	1 cal = 4.184 Joule	1 MPa = 145 psi
= 5.6784 Joule/sec-m^2 K	1 Btu = 1055.1 Joule	1 MPa = 0.102 kg/mm^2
= 4.8825 Kcal/hr-m^2 K	= 252.16 cal	1 Pa = 10 dynes/cm^2
= 0.45362 Kcal/hr-ft^2 K	1 HP-hr = 2684500 Joule	1 kg/mm^2 = 1422 psi
= 1.3564x10^{-4}cal/sec-cm^2 K	= 641620 cal	1 psi = 6.90 × 10^{-3} MPa
1 lb/hr-ft^2	= 2544.5 Btu	1 kg/mm^2 = 9.806 MPa
= 1.3562 × 10^{3}kg/sec-m^2	1 KW-hr = 3.6 × 10^6 Joule	1 dyne/cm^2 = 0.10 Pa
= 4.8823 kg/hr-m^2	= 860565 cal	1 psi = 7.03 × 10^{-4} kg/mm^2
= 0.45358 kg/hr-ft^2	= 3412.75 Btu	1 psi in.$^{1/2}$ = 1.099 × 10^{-3} MPa m$^{1/2}$
1 cal/g °C = 1 Btu/lb$_m$ °F	1 l-atm = 24.218 cal	1 MPa m$^{1/2}$ = 910 psi in.$^{1/2}$
= 1 Pcu/lb$_m$ °C	1 ft-lb$_f$ = 0.3241 cal	
1 Btu/hr-ft °F	1 Pcu = 453.59 cal	
= 1.731 W/m K	1 kg-m = 2.3438 cal	
= 1.4882 kcal/hr-m K		

Table 1-9. Some Useful Constants

Atomic mass	$m_u \approx 1.6605402 \times 10^{-27}$
Avogadro's number	$N \approx 6.0221367 \times 10^{23} \, mol^{-1}$
Boltzmann's constant	$k \approx 1.380658 \times 10^{-23} \, J \cdot K^{-1}$
Elementary charge	$e \approx 1.60217733 \times 10^{-19} \, C$
Faraday's constant	$F \approx 9.6485309 \times 10^4 \, C \cdot mol^{-1}$
Gas (molar) constant	$R = k \cdot N \approx 8.314510 \, J \cdot mol^{-1} \cdot K^{-1}$
	$\approx 0.08205783 \, L \cdot atm \cdot mol^{-1} \cdot K^{-1}$
Gravitational acceleration	$g = 9.80665 \, m \cdot s^{-2}$
Molar volume of an ideal gas at 1 atm and 25°C	$\overline{V}_{ideal \; gas} \approx 24.465 \, L \cdot mol^{-1}$
Permittivity of vacuum	$\varepsilon_0 = 8.854187 \times 10^{-12} \, C \cdot V^{-1} \cdot m^{-1}$
Planck's constant	$h \approx 6.6260755 \times 10^{-34} \, J \cdot s$
Zero of the Celsius scale	$0°C = 273.15 \, K$

Source: IUPAC, 1988.

REFERENCES

1-1. J. W. Moore and E. A. Moore, *Environmental Chemistry*, Academic Press, New York, 1976.

1-2. S. H. Maron and C. F. Prutton, *Principles of Physical Chemistry*, 4th ed., Macmillan, New York, 1965.

1-3. J. M. Smith and H. C. Van Ness, *Introduction to Chemical Engineering Thermodynamics*, 3rd ed., McGraw-Hill, New York, 1975.

1-4. J. S. Winn, *Physical Chemistry*, Harper Collins College Press, New York, 1995.

1-5. R. A. Albert and R.J. Silbey, *Physical Chemistry*, 2nd ed., Wiley, New York, 1997.

1-6. E. Grunwald, *Thermodynamics of Molecular Species*, Wiley, New York, 1997.

1-7. F. D. Rossini, *Chemical Thermodynamics*, Wiley, New York, 1950.

1-8. C. E. Dykstra, *Physical Chemistry, A Modern Introduction*, Prentice-Hall, Upper Saddle River, New Jersey, 1997.

1-9. J. B. Hudson, *Thermodynamics of Materials: A Classical and Statistical Synthesis*, Wiley, New York, 1996.

1-10. C. B. Skimmer, *Introduction to Chemical Kinetics*, Academic Press, New York, 1974.

1-11. S. W. Benson, *Foundations of Chemical Kinetics*, McGraw-Hill, New York, 1960.

1-12. K. G. Denbigh, *Thermodynamics of the Steady State*, Methuen, London, 1951.

1-13. A. M. Klotz, *Chemical Thermodynamics, Basic Theory and Methods*, Prentice-Hall, Englewood Cliffs, New Jersey, 1957.

1-14. H. J. M. Bowen, *Environmental Chemistry of the Elements*, Academic Press, London, 1979.

PROBLEM SET

1. Generally for gases, the compressibility coefficient, k, the expansion coefficient, a, and the pressure coefficient, b, can be expressed by

$$\kappa = -\frac{1}{V}\frac{\partial V}{\partial P} \qquad \alpha = \frac{1}{V}\left(\frac{\partial V}{\partial T}\right)_P \qquad \beta = \frac{1}{P}\left(\frac{\partial P}{\partial T}\right)_V$$

a) What are their values if those gases are ideal gases and

b) Prove

$$a = kbP$$

c) Show that

$$\left(\frac{\partial P}{\partial T}\right)_V \left(\frac{\partial T}{\partial V}\right)_P \left(\frac{\partial V}{\partial P}\right)_T = -1$$

2. Calculate the maximum work done by the isothermal expansion of one mole of ideal gas at 0°C from 2.24 L to 22.4 L.

3. For a Carnot cycle as shown in the figure attached.

$$Q_2 = \int_a^b dQ = \int_a^b PdV = \int_a^b \frac{RT_2 dV}{V} = RT_2 \ln\left(\frac{V_b}{V_a}\right)$$

$$Q_1 = -\int_a^b dQ = RT_1 \ln\left(\frac{V_c}{V_d}\right)$$

Prove the efficiency is

$$n = 1 - \frac{Q_1}{Q_2} = 1 - \frac{T_1}{T_2}$$

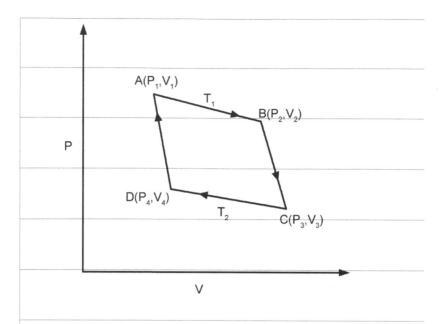

The Carnot Cycle

A to B is an isothermal expansion at temperature T_1, volume change is V_1 to V_2, and heat absorbed is Q_1.

B to C is adiabatic expansion, temperature is T_1 to T_2, volume change is V_2 to V_3, and heat absorbed is 0.

C to D is isothermal compression, temperature is T_2, volume change is V_3 to V_4, and heat absorbed is Q_2.

D to A is adiabatic expansion, temperature is T_2 to T_1, volume change is V_4 to V_1, and heat absorbed is 0.

4.   For a consecutive reaction

$$A \xrightarrow{\;k_1\;} B \xrightarrow{\;k_2\;} C$$

If $A(0) = A_0$ and $B(0) = 0 = C(0)$

and

$$\frac{dA}{dt} = -k_1 A$$

$$\frac{dB}{dt} = k_1 A - k_2 B$$

$$\frac{dc}{dt} = k_2 B$$

calculate A, B, and C.

5. For ideal gas, derive

$$C_P - C_V = R$$

and

$$\gamma = \frac{C_P}{C_V} = 1.67$$

CHAPTER 2

INORGANIC CHEMISTRY

*I*norganic chemistry generally deals with compounds not containing carbon and usually derived from minerals. It is concerned with most non-carbon elements of the periodic table (see Appendix B). The elements in the periodic table are arranged horizontally in periods and vertically in groups according to their chemical similarities. Starting from the simplest atom, hydrogen, which contains one proton and one electron, we can use the quantum chemistry approach to discuss the probable locations of electrons with quantum numbers. From the atomic orbitals and the building-up principle through hybridization, molecular bonding and shapes can be developed. This will embody rings and clusters as well as ligand fields involving the d-electrons. Finally, through the crystalline solid state, the metals and alloys will be reviewed for their practical value.

2.1 STRUCTURE OF HYDROGEN ATOM AND PERIODICITY

In the periodic table, elements are arranged by increasing atomic number from left to right and from top to bottom. The horizontal rows are called **periods**; the vertical columns of related elements are called **groups**. Across the table there is a general trend from metallic to non-metallic; down a group there is an increase in atomic size and in electropositive behavior. The members of a group have similar behavior because of similarities in their electronic configurations.

The simplest atom is hydrogen, which consists of one electron, one proton and no neutrons in the nuclei. A simple mathematic solution of hydrogen can be illustrated here (Figure 2-1).

Columbic attraction of Z and e and the resulting potential energy is

$$V = -\frac{e^2}{r} \qquad [2\text{-}1]$$

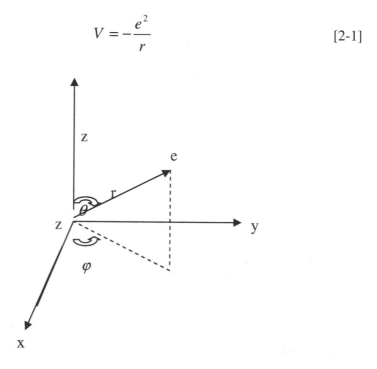

Figure 2-1. The polar coordinates of the hydrogen atom. Notice one electron from the center of the nucleus.

From wave mechanics

$$\frac{\partial^2 f(x)}{\partial x^2} = -\frac{4\pi^2}{\lambda^2} f(x) \quad \text{or} \quad \nabla^2 \psi + \left(\frac{2\pi}{\lambda}\right)^2 \psi = 0 \qquad [2\text{-}1A]$$

Since $\lambda = \dfrac{h}{\mu v}$ of the DeBroglie's wavelength

$$\therefore \nabla^2 \psi + \frac{4\pi^2 \mu^2 v^2}{h^2} \psi = 0$$

where μ is reduced mass or

$$\mu = \frac{M_e M_n}{\left(M_e + M_n\right)}$$

If v is velocity, then the total energy can be expressed as:

$$E = \frac{1}{2}\mu v^2 + V$$

where

$$\frac{1}{2}\mu v^2$$

is kinetic energy and V is potential energy.

Solving for v^2 and substitute into Equation [2-1A]

$$\nabla^2 \psi + \frac{8\pi^2 \mu}{h^2}\left(E - V\right)\psi = 0 \qquad [2\text{-}2]$$

Equation [2-2] is known as the Schrödinger equation. Change Eq. [2-2] into polar coordinates of

$$\begin{cases} x = r\sin\theta\cos\phi \\ y = r\sin\theta\sin\phi \\ z = r\cos\theta \\ \text{and} \\ r = \left(x^2 + y^2 + z^2\right)^{\frac{1}{2}} \end{cases}$$

Polar form of Eq. [2-2] is

$$\frac{1}{r^2}\frac{\partial}{\partial r}\left(r^2\frac{\partial\psi}{\partial r}\right)+\frac{1}{r^2\sin\theta}\frac{\partial}{\partial\theta}\left(\sin\theta\frac{\partial\psi}{\partial\theta}\right)+\frac{1}{r^2\sin^2\theta}\frac{\partial^2\psi}{\partial\phi^2}+\frac{8\pi^2\mu}{h^2}\left(E+\frac{e^2}{r}\right)\psi=0 \quad \text{[2-3]}$$

assuming separation of variable as

$$\psi(r,\theta,\phi)=R(r)\Theta(\theta)\Phi(\phi) \qquad\qquad \text{[2-4]}$$

and Eq. [2-3] is multiplied by

$$\frac{r^2\sin^2\theta}{R\Theta\phi}$$

then

$$\frac{\sin^2\theta}{R\partial r}\frac{\partial}{\partial r}\left(r^2\frac{\partial R}{\partial r}\right)+\frac{\sin\theta}{\Theta}\frac{\partial}{\partial\theta}\left(\sin\theta\frac{\partial\Theta}{\partial\theta}\right)+\frac{1}{\Phi}\frac{\partial^2\Phi}{\partial\phi^2}+r^2\sin^2\theta\frac{8\pi^2\mu}{h^2}\left(E+\frac{e^2}{r}\right)=0 \quad \text{[2-5]}$$

The third term in Eq. [2-5] is function of ϕ only, Eq. [2-5] can be zero only when the third term is constant or

$$\frac{1}{\Phi}\frac{\partial^2\Phi}{\partial\phi^2}+m^2=0 \qquad\qquad \text{[2-6]}$$

where m is a constant.
Solution of Eq. [2-6] is

$$\Phi=Ae^{im\phi} \qquad\qquad \text{[2-7]}$$

where $m=0,\pm1,\pm2,\ldots$ which is **magnetic quantum number** and $A=(2\pi)^{-\frac{1}{2}}$

Substituting Eq. [2-6] into Eq. [2-5] and dividing by $\sin^2\theta$

$$\frac{1}{R}\frac{\partial}{\partial r}\left(r^2\frac{\partial R}{\partial r}\right)+\frac{1}{\Theta\sin\theta}\frac{\partial}{\partial\theta}\left(\sin\theta\frac{\partial\Theta}{\partial\theta}\right)-\frac{m^2}{\sin^2\theta}+\frac{r^2 8\pi^2\mu}{h^2}\left(E+\frac{e^2}{r}\right)=0$$

Collecting $\begin{cases} \text{r terms and letting it equal } \beta \\ \theta \text{ terms and letting it equal } -\beta \end{cases}$

$$\frac{1}{R}\frac{\partial}{\partial r}\left(r^2\frac{\partial R}{\partial r}\right)+\frac{r^2 8\pi^2\mu}{h^2}\left(E+\frac{e^2}{r}\right)=\beta \qquad \text{[2-8]}$$

$$\frac{1}{\Theta\sin\theta}\frac{\partial}{\partial\theta}\left(\sin\theta\frac{\partial\Theta}{\partial\theta}\right)-\frac{m^2}{\sin^2\theta}=-\beta \qquad \text{[2-9]}$$

Now an associated Legandre equation (see Appendix C) is

$$\left(1-x^2\right)D^2 y-2xDy+\left\{n(n+1)-\frac{m^2}{1-x^2}\right\}y=0$$

let $x=\cos\theta$, $y(x)=\Theta(\theta)$ be substituted in Eq. [2-9] and rearranged, then Eq. [2-9] can be written as

$$\frac{1}{\sin\theta}\frac{\partial}{\partial\theta}\left(\sin\ \theta\frac{\partial\Theta}{\partial\theta}\right)+\left(\beta-\frac{m^2}{\sin^2\theta}\right)\Theta=0$$

after replacing

$$\frac{\partial\Theta}{\partial\theta}=\frac{\partial y}{\partial x}\frac{\partial x}{\partial\theta}=-\sin\theta\frac{\partial y}{\partial x}$$

the first term becomes

$$\frac{1}{\sin\theta}\frac{\partial}{\partial\theta}\left(-\sin^2\theta\frac{\partial y}{\partial x}\right) \quad \text{or} \quad \frac{\partial}{\partial x}\left(\sin^2\theta\frac{\partial y}{\partial x}\right)$$

since

$$\frac{\partial}{\partial\theta}=\frac{\partial}{\partial x}\frac{\partial x}{\partial\theta}=-\sin\theta\frac{\partial}{\partial x}$$

Now the first term can be written as $\left(1-x^2\right)D^2 y-2xDy$, which can be solved. The solution is

$$\Theta = P_l^m(\cos\theta) \qquad\qquad\qquad\qquad\qquad [2\text{-}10]$$

$$\beta = l(l+1)$$

where $l = |m|,\ |m|+1,\ |m|+2,\ ...\ m \le l$

l is the **azimuthal** or **angular momentum quantum number**.

To solve Eq. [2-8], set

$$\frac{\alpha^2}{4} = -\frac{8\pi^2\mu}{h^2}E;\quad b = \frac{8\pi^2\mu e^2}{h^2\alpha};\quad \rho = r\alpha$$

and [2-8] becomes

$$\rho^2\frac{\partial^2 R}{\partial\rho^2} + 2\rho\frac{\partial R}{\partial\rho} + \left[b\rho - \frac{1}{4}\rho^2 - l(l+1)\right]R = 0 \qquad [2\text{-}11]$$

This equation can be solved based on the associated Laguerre polynomials (see Appendix C). The solution for the associated Laguerre polynomial equation:

$$x^2D^2u + 2xDu + \left\{\left[k - \frac{1}{2}(p-1)\right]x - \frac{x^2}{4} - \frac{1}{4}(p^2-1)\right\}u = 0$$

is

$$u = G_{k_\delta}^p(x) = e^{-\frac{x}{2}}x^{\frac{p-1}{2}}L_k^p(x)$$

Hence, Eq. [2-11] can be written as

$$R(\rho) = G_k^p(\rho) = e^{-\frac{1}{2}\rho}\rho^{\frac{1}{2}(p-1)}L_k^p(\rho)$$

$$o \le p \le k$$

provided that

$$\frac{p^2-1}{4} = l(l+1) \quad \text{or} \quad p = 2l+1$$

$$k - \frac{1}{2}(p-1) = b \quad \text{or} \quad k - l = \frac{8\pi^2 \mu e^2}{h^2 \alpha}$$

but

$$\frac{\alpha^2}{4} = -\frac{8\pi^2 \mu}{h^2} E = \frac{16\pi^4 \mu^2 e^4}{h^4 (k-l)^2}$$

also

$$E_n = -\frac{2\pi^2 \mu e^4}{n^2 h^2} \qquad \text{[2-12]}$$

$$n = k - l$$

The solution is

$$R_n^l(\rho) = e^{-\frac{1}{2}p} \rho^l L_{n+l}^{2l+1}(\rho) \qquad \text{[2-13]}$$

where

$$\rho = \alpha r = \frac{8\pi^2 \mu e^2 r}{nh^2}$$

Here n is 1, 2, 3 … and is called the **principal quantum number**.

The Schrödinger equation can be solved from Eq. [2-4] by the individual equations [2-7], [2-10] and [2-13].

$$\psi_{n,l}^m(r, \theta, \phi) = R(r)\Theta(\theta)\Phi(\phi)$$

$$= AR_n^l(r)P_l^m(\cos\theta)e^{im\phi}$$

$$= Ae^{-\frac{1}{2}r} r^l L_{n+l}^{2l+1}(r)Pe^m(\cos\theta)e^{im\phi} \qquad \text{[2-14]}$$

where

$$n = 1, 2, 3, \ldots$$

$$l = n - 1, n - 2, ..., 0 \qquad\qquad 0 \le l < n$$
$$m = -l, -l + 1, ..., l - 1, l \qquad -l \le m \le l$$

are all integers.

This is the **eigen function** of an electron or the probability of the electron in relation to the nucleus of the hydrogen atom. The three quantum numbers are n, l, and m.

Some simple wave solutions of the lower values of n, l and m of the hydrogen-like wave functions of K shell are listed below:

$$\psi_{1S} = \pi^{-\frac{1}{2}} \left(\frac{Z}{a_o} \right)^{\frac{3}{2}} e^{-\frac{Zr}{a_o}} \qquad\qquad [2\text{-}15]$$

for $n = 1$, $l = 0$, and $m = 0$

Z is atomic number and a_o is defined as

$$a_o = \frac{h^2}{4\pi^2 \mu e^2}$$

Usually a_o is the Bohr radius, which is equivalent to 0.529Å and the energy turns out to the –13.6 eV for Eq. [2-12]. A graph depicting Eq. [2-15] is illustrated in Fig. 2-2(a). The probability distribution of the electron is called orbital.

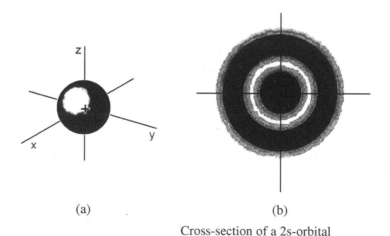

(a)   (b)

Cross-section of a 2s-orbital

Figure 2-2. The wave function for the lowest energy level (ground state) of the hydrogen atom.

Solutions for the L-shell are listed here

$$\psi_{2S} = \left[\frac{(2\pi)^{-\frac{1}{2}}}{4}\right]\left(\frac{Z}{a_o}\right)^{\frac{3}{2}}\left(2 - \frac{Zr}{a_o}\right)e^{-\frac{Zr}{2a_o}}$$ [2-16]

$n = 2$, $l = 0$, $m = 0$

$$\psi_{2P_z} = \left[\frac{(2\pi)^{-\frac{1}{2}}}{4}\right]\left(\frac{Z}{a_o}\right)^{\frac{3}{2}}\left(\frac{Zr}{a_o}\right)e^{-\frac{Zr}{2a_o}}\cos\theta$$ [2-17]

$n = 2$, $l = 0$, $m = 0$

$$\psi_{2P_x} = \left[\frac{(2\pi)^{-\frac{1}{2}}}{4}\right]\left(\frac{Z}{a_o}\right)^{\frac{3}{2}}\left(\frac{Zr}{a_o}\right)e^{-\frac{Zr}{2a_o}}\sin\theta\cos\phi$$ [2-18]

$$\psi_{2P_y} = \left[\frac{(2\pi)^{-\frac{1}{2}}}{4}\right]\left(\frac{Z}{a_o}\right)^{\frac{3}{2}}\left(\frac{Zr}{a_o}\right)e^{-\frac{Zr}{2a_o}}\sin\theta\sin\phi$$ [2-19]

$n = 2$, $l = 1$, $m = \pm 1$

The $2s$ electron orbital is similarly displaced in Fig. 2-2(b). As for the p-orbitals, the p_z, p_x and p_y are directionally dependent. They are geometrically illustrated in Fig. 2-3.

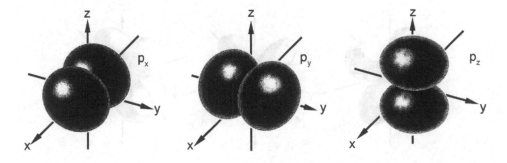

Figure 2-3. The angular dependence of the boundary surfaces of the hydrogen p-orbitals.

The angular dependence of d-orbitals is more complex. The following only suggests that the radial behavior of each Eigen function is changed. (see Appendix D)

$$d_{xy} = \frac{1}{2}(15)^{\frac{1}{2}} \sin^2 \theta \sin 2\phi \qquad [2\text{-}20]$$

$$d_{x^2-y^2} = \frac{1}{2}(15)^{\frac{1}{2}} \sin^2 \theta \cos 2\phi \qquad [2\text{-}21]$$

$$d_{xz} = (15)^{\frac{1}{2}} \sin \theta \cos \phi \cos \psi \qquad [2\text{-}22]$$

$$d_{yz} = (15)^{\frac{1}{2}} \sin \theta \cos \theta \sin \psi \qquad [2\text{-}23]$$

$$d_{z^2} = \frac{1}{2}(5)^{\frac{1}{2}}(3\cos^2 \theta - 1) \qquad [2\text{-}24]$$

The five d-orbitals are illustrated in Fig. 2-4.

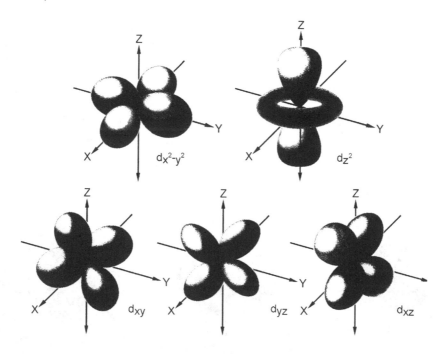

Figure 2-4. The angular dependence of the boundary surfaces of the hydrogen *d*-orbitals.

A fourth quantum number, an electron spin quantum number, s, is required to specify each electronic wave function. The s number can be $+ \frac{1}{2}$, or $- \frac{1}{2}$. No two electronic wave functions can have the same former quantum number, so each orbital can accommodate a maximum of two electrons. The pattern of allowed combinations of all quantum numbers, n, l, m, and s, is shown in Table 2-1.

Table 2-1. The allowed combinations of quantum numbers for atoms, and the associated nomenclature of the corresponding orbitals. Each orbital can contain a maximum of two electrons, one each with electron spin quantum number $+ \frac{1}{2}$ and $- \frac{1}{2}$.

Shell	Sub-shell			Maximum no.	Maximum no.
n value	l value	designation	m_l values	of electrons per sub-shell	of electrons per shell
$n = 4$	$l = 3$	4f	-3, -2, -1, 0, 1, 2, 3	14	32
	$l = 2$	4d	-2, -1, 0, 1, 2	10	
	$l = 1$	4p	-1, 0, 1	6	
	$l = 0$	4s	0	2	
$n = 3$	$l = 2$	3d	-2, -1, 0, 1, 2	10	18
	$l = 1$	3p	-1, 0, 1	6	
	$l = 0$	3s	0	2	
$n = 2$	$l = 1$	2p	-1, 0, 1	6	8
	$l = 0$	2s	0	2	
$n = 1$	$l = 0$	1s	0	2	2

The **Pauli exclusion principle** is the outcome of the electron configuration using the basis of quantum numbers. It states that no more that two electrons can occupy a given atomic (or molecular) orbital, and when two electrons occupy one orbital, spins must be paired. In this manner, the electron configurations of other atoms are built up. This is called the **Aufbau principle**, as demonstrated in Table 2-2.

Another rule used for building up by filling up individual atomic orbitals is called **Hund's rule**. This rule states that the ground state electron configuration of an atom maximizes the number of unpaired spins. The filling of shells can be demonstrated by Fig. 2-5, which reflects the order of energy at the appropriate point. This fashion of addition can create blocks in the periodic table (Fig. 2-6).

Table 2-2. Electron configuration of ground-state atoms up to K ($Z = 19$).

H	$(1s)^1$
He	$(1s)^2 = [\text{He}]$
Li	$[\text{He}]\,(2s)^1$
Be	$[\text{He}]\,(2s)^2$
B	$[\text{He}]\,(2s)^2\,(2p)^1$
C	$[\text{He}]\,(2s)^2\,(2p)^2$
N	$[\text{He}]\,(2s)^2\,(2p)^3$
O	$[\text{He}]\,(2s)^2\,(2p)^4$
F	$[\text{He}]\,(2s)^2\,(2p)^5$
Ne	$[\text{He}]\,(2s)^2\,(2p)^6 = [\text{Ne}]$
Na	$[\text{Ne}]\,(3s)^1$
Mg	$[\text{Ne}]\,(3s)^2$
Al	$[\text{Ne}]\,(3s)^2\,(3p)^1$
Si	$[\text{Ne}]\,(3s)^2\,(3p)^2$
P	$[\text{Ne}]\,(3s)^2\,(3p)^3$
S	$[\text{Ne}]\,(3s)^2\,(3p)^4$
Cl	$[\text{Ne}]\,(3s)^2\,(3p)^5$
Ar	$[\text{Ne}]\,(3s)^2\,(3p)^6 = [\text{Ar}]$
K	$[\text{Ar}]\,(4s)^1$

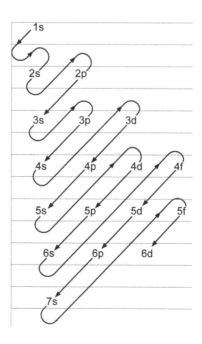

Figure 2-5. A diagram showing the order of filling of orbitals in the periodic table.

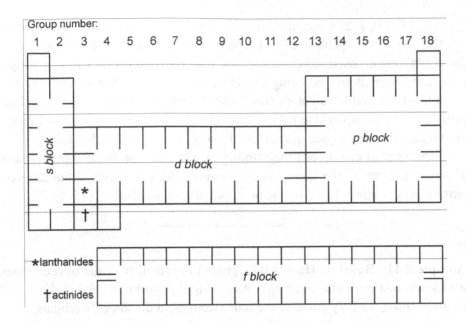

Figure 2-6. Structure of the periodic table, showing the *s*, *p*, *d*, and *f* blocks.

2.1.1 Atomic Term Symbols

The electronic configuration of any atom, either ground or excited, may be represented by an **atomic term symbol**. The atomic term symbol is written in the form

$$^{2S+1}L_J$$

which means:

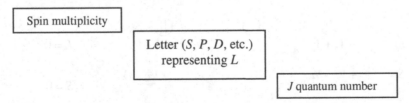

Here S is the total spin quantum number, and L is the total magnetic quantum number.

$$S = \Sigma s$$
$$L = \Sigma m$$

When $L = 0, 1, 2, 3$, etc., the corresponding letter used in the term symbol is S, P, D, F, etc. The quantity $2S + 1$ is known as the multiplicity. Thus, for an atom containing no unpaired electrons, $S = 0$ and the multiplicity is 1 (a **singlet state**), and for an atom containing one electron, $S = \frac{1}{2}$ and the multiplicity is 2 (a doublet state), and for an atom containing 2 unpaired electrons, $S = 1$ and the multiplicity is 3 (a triplet state). The $(2S + 1)$ degeneracy can be read as a multiplet: singlet, doublet, triplet, quartet, etc. Finally, $J = L + S, L + S - 1$, etc.

Term symbol can predict the stability of a given atom for the ground state and different forms of the excited state. Example 2-1 will illustrate the different forms of oxygen atoms based on the principle of term symbol.

[Example 2-1] Based on Hund's rule, predict the stability of the oxygen atom. For O atom, one can write $1s^2 2s^2 2p^4$. Assuming the S-orbitals (both $1s^2$ and $2s^2$) have been filled, we only consider the four electrons in the three $2p$ orbitals.

m	$+1, +1$	$0, 0$	-1	$L = 1$	
S	$+\frac{1}{2}, -\frac{1}{2}$	$+\frac{1}{2}$	$+\frac{1}{2}$	$S = 1$	$^3P_2$

m	$+1, +1$	$0, 0$		$L = 2$	
S	$+\frac{1}{2}, -\frac{1}{2}$	$+\frac{1}{2}, -\frac{1}{2}$		$S = 0$	$^1D_2$

m	$+1, +1$		$-1, -1$	$L = 0$	
S	$+\frac{1}{2}, -\frac{1}{2}$		$+\frac{1}{2}, -\frac{1}{2}$	$S = 0$	$^1S_0$

There the common three forms of the oxygen atom, $^3P_2$, $^1D_2$, and $^1S_0$, can have a stability as

$$^3P_2 > {}^1D_2 > {}^1S_0$$

⟵

stability

In reality photo disassociation of oxygen molecules by UV can produce these unstable species

$$O_2 \xrightarrow{\;200nm\;} 2O\left(^3P\right)$$

$$O_2 \xrightarrow{\;<176nm\;} O\left(^3P\right) + O\left(^1D\right)$$

$$O_2 \xrightarrow{\;<92nm\;} 2O\left(^1S\right)$$

The shorter the wavelength of the UV, the stronger is the dissociation energy.

Generally, the subscript of the term symbol can be omitted. Similarly, one can evaluate the term symbol of a nitrogen atom or O^+ ion. This can be used as an exercise.

The ground state term symbols of the elements have been presented in Fig. 2-7, which gives periodic variations. Notice both the spin multi-properties alternate from odd to even across any row as the number of electrons varies from even to odd.

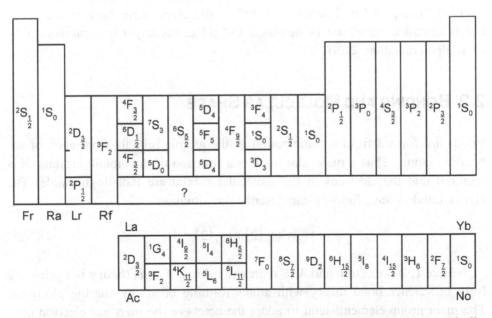

Figure 2-7. The ground state term symbols of the elements show periodic variations. The term symbols for the elements 105-109 (denoted by ?) are unknown. Also, those for the final few actinides are pressured to follow their lanthanide counterpart.

In traditional **x-ray crystallography**, the sources are no more than a target that is bombarded in vacuum with electrons having kinetic energies in the hundreds of kiloelectron-volt range. When the electrons strike the target, their energy is released in the metal as they undergo collisions. The electron strikes an atom and transfers to it enough energy to remove one of the atom's core electrons. For a Cu target, the process is

$$e^- + Cu(1s^2 2s^2 2p^6 ...) \rightarrow Cu^+(1s^1 2s^2 2p^6 ...) + 2e^-$$

where one of the electrons produced is the ejected core electron and the other is simply the incident electron. The Cu^+ left is highly energized and is in the excited stage. One way this ion can spontaneously lose its excitation energy is called photon emission process.

$$Cu^+(1s^1 2s^2 2p^6 ...) \rightarrow Cu(1s^2 2s^2 2p^5 ...) + hv$$

hv here is photon, which is in the x-ray region. Usually the Cu target is provided with a Ni filter so that only a monochromatic source of α doublet at 1.544 and 1.541 Å. The β of Cu K series, at 1.392 Å and others have been filtered. The bonds lengths in carbon are at the ranges in Cu Kα. For larger size, targets can be Cr K alpha radiation, 2.291 Å.

2.2 BONDING AND MOLECULAR SHAPE

Bonds are formed from a combination of the atomic orbitals from each of the bonding atoms. Thus a molecular bond is a function of the atomic orbitals. It is required that the electrons in the molecular orbital are indistinguishable. The Heiter-London wave function can describe this situation

$$\psi = \psi_A(1)\psi_B(2) + \psi_A(2)\psi_B(1) \qquad \text{[2-25]}$$

Here 1, 2 is electron and *A*, *B* represent orbital. **Lewis theory** is a primitive form of valence bond theory with atoms forming bonds by sharing electrons. The main group elements tend to adopt the octets or the inert gas electron configuration, although some elements, such as B or Be are stable with incomplete octets. Lewis theory recognizes the sharing of electrons to form the covalent bonds. Many large atoms also display higher valency where it is energetically favorable for more than eight valence electrons (see Fig. 2-8).

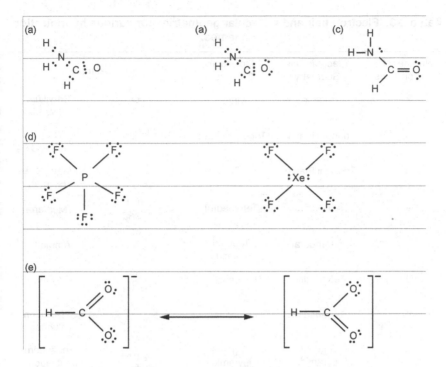

Figure 2-8. (a)-(c) Development of a Lewis bonding scheme for $HCONH_2$. Examples of (d) hyper-valency, and (e) resonance hybridization.

Another complementary approach to the problem of bonding is the **valence shell electron pair repulsion** (VSEPR) **theory**. The theory treats each atom or molecule in isolation and describes the geometry of the bonds and bonding electron pairs around it. See Table 2-3 for the details of the molecular geometry. Therefore geometry is modified by the variations in repulsion strengths between electron pairs.

Another process combining pure atomic orbitals is called **hybridization**, which circumvents the rigid geometry of pure orbitals that is usually required. The common hybrids (orbitals) are listed in Table 2-4.

2.2.1 Rings and Clusters: Huckel Approach and Wade's Rules

Rings are often formed by nonmetallic elements with directional covalent bonding. Homo-element bonding, such as S_8 or C_6H_6 (benzene), are common. Borazine, $B_3N_3H_6$, is iso-electric with benzene. **Clusters** are polyhedral arrangements or atoms formed in the nonmetals, such as P_4, boranes, Pb_5^{2-}, etc. as listed in Fig. 2-9.

Table 2-3. Electron pair and molecular geometries for various hybridization schemes.

Number of Electron Pairs	Electron Pair Geometry	Molecular Geometry	Example	
2	Linear	Linear	H — Be — H	Beryllium dihydride
3	Trigonal Planar	Trigonal Planar		Boron trifluoride
3	Trigonal Planar	Bent		Methylene (singlet)
4	Tetrahedral	Tetrahedral		Methane
4	Tetrahedral	Trigonal Pyramidal		Ammonia
4	Tetrahedral	Bent		Water
4	Tetrahedral	Linear		Hydrogen fluoride
5	Trigonal Bipyramidal	Trigonal Bipyramidal		Phosphorus pentafluoride
5	Trigonal Bipyramidal	See-saw		Sulfur tetrafluoride
5	Trigonal Bipyramidal	T-shaped		Chlorine trifluoride
5	Trigonal Bipyramidal	Linear		Xenon difluoride
6	Octahedral	Octahedral		Sulfur hexafluoride
6	Octahedral	Square-based Pyramidal		Chlorine pentafluoride
6	Octahedral	Square Planar		Xenon tetrafluoride
7	Pentagonal Bipyramidal	Pentagonal Bipyramidal		Iodine heptafluoride
7	Pentagonal Bipyramidal	Distorted Octahedral		Xenon hexafluoride
8	Square Anti-prism	Square Anti-prism		Xenon octafluoride
9	Tri-capped Trigonal Prism	Tri-capped Trigonal Prism		Rhenium nonahydrive

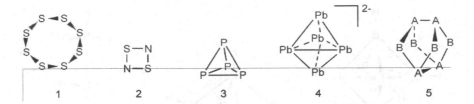

Figure 2-9. Examples of compounds with rings and clusters.

Obviously for S_8 or P_4, single electron σ bonds can be expanded for the S atom, with each associated with two nonbonding electron pairs; and for P atom, each is associated with one pair. Benzene has a framework of σ bonds formed by sp^2 hybrid. There are six delocalized MO for six π - electrons in the ground state. The Hückel $4n + 2$ rule helps to predict the stability where n is a whole number of 2, 6, 10... but not 4, 8, ... One consequence of this prediction is $(C_5H_5)^-$, which is a 6π -electron anion and an important ligand.

Fig. 2-10 illustrates three types of boranes. The **closo boranes** are closed polyhedral structures with triangular faces, the trigonal bipyrimid (5 vertices), octahedron (6 vertices) and isosahedron (12 vertices). These polyhedra belong to the deltahedra class. A good example is $(B_6H_6)^{2-}$ in Fig. 2-10(a). The other class is called **nido** (nest-like) **borane**, generally $(B_nH_n)^{2-}$, which are formed with one vertice missing for the n vertex deltahedron. A good example is B_5H_4 or B_nH_{n+4} of Fig. 2-10(b). The last class is called **arachno** (web-like) **boranes**, which can be imagined as detahedron with two vertices missing. The example in Fig. 2-10(c) is B_4H_{10} or B_nH_{n+6}.

Table 2-4. Hybrid AO geometries.

Hybrid	Geometry	Molecular example
sp	linear	BeH_2
sp^2	trigonal planar	BF_3
sp^3	tetrahedral	CH_4
dsp^2	square planar	$PdBr_4^{2-}$
dsp^3	trigonal bipyramidal	PF_5
d^2sp^3	octahedral	SF_6

The Wade rule can be illustrated by the boranes in Fig. 2-10.

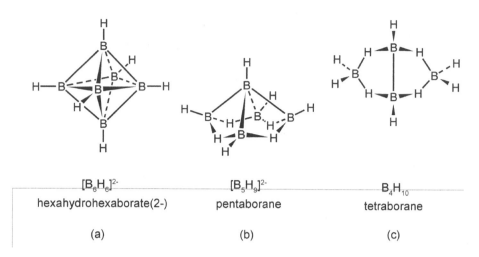

$[B_6H_6]^{2-}$	$[B_5H_9]^{2-}$	B_4H_{10}
hexahydrohexaborate(2-)	pentaborane	tetraborane
(a)	(b)	(c)

Figure 2-10. The three types of boranes.

The Wade rules are made to predict the number of **skeletal bonding electrons** (SBE). For the closo boranes, the rules for SPE is $2n + 2$, for the nido boranes, $2n + 4$ and for the arachno boranes, $2n + 6$; n is the number of boron atoms. The rules may be applied to naked clusters formed by p-block metals, such as $Pb_5{}^{2-}$ or $Sn_9{}^{4-}$. Verifications of the borane classes are listed in Table 2-5.

Table 2-5. Computation of SBE*.

	(a) Valence electron	(b) Electron for 2-center bonds	(c) SBE
Closo	$3n + n + 2$	$2n$	$2n + 2$
Nido	$3n + n + 4$	$2n$	$2n + 4$
Arachno	$3n + n + 6$	$2n$	$2n + 6$

* (a) – (b) = (c)

Transition metal complexes such as π-acceptor ligands, e.g., CO, can accept electron density from filled metal d-orbitals. These polynuclear complex compounds are known as binary carbonyls usually having metal-metal bonds. Some binary carbonyls and ions formed by 3d transition metals are listed in Table 2-6.

Most stable carbonyls obey the **18-electron rule**. Usually the estimation is the summation of the group number of the transition metal and the number of carbonyl group times two. For example, for $Fe(CO)_5$, Fig. 2-11(a), $8 + (2 \times 5) = 18$. Sometimes this is referred to as the **effective atomic number**

(EAN) **rule**, i.e., the noble gas configuration of Kr, EAN = 36. For other series, such as 4d or 5d series, Xe core, EAN = 54; and Ra core, EAN = 86. Let us illustrate the computation for $Fe(CO)_5$ here again: atomic number of Fe is 26 and the addition of 5 CO contribution is $5 \times 2 = 10$; hence EAN = 36.

Table 2-6. Binary carbonyls and ions formed by 3d series elements.

Group No.	5	6	7	8	9	10
	$[V(CO)_6]^-$	$Cr(CO)_6$	$Mn_2(CO)_{10}$	$Fe(CO)_5$	$Co_2(CO)_8$	$Ni(CO)_4$
			$[Mn(CO)_6]^+$	$[Fe(CO)_4]^{2-}$	$[Co(CO)_4]^-$	
				$Fe_2(CO)_9$		
				$Fe_3(CO)_{12}$		

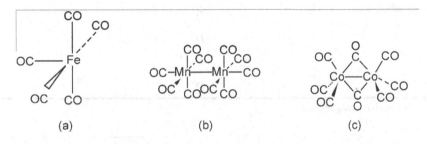

(a) (b) (c)

Figure 2-11. Some complexes with π-acceptor ligands.

The 18-electron rule can be a useful guide to stable **organometallics**, especially when π-acceptor ligands are present. For organic ligands, the hapticity and their electron number should be accounted for. **Hapticity** is the number of carbon atoms bonded directly to the metal, often using the symbol η^2, meaning one carbon, or η^3, meaning three carbon atoms (see Table 2-7.)

A number of organometallics are shown in Fig. 2-12. Many ferrocene "sandwich" compounds and "piano-stool" structures obey the 18-electron rule, with some exceptions.

Metallocenes of $M(\eta^5 - C_5H_5)_2$ structures, compound (b) in Fig. 2-12, are known for the 3d series elements, V, Ni with 15-20 valence electrons. Ferocene (M = Fe) with 18 electrons is by far the most stable; cobaltocene (M = Co) with 19 electrons is a very strong reducing agent that easily forms the 18-electron ion $[Co(\eta^5 - C_5H_5)_2]^+$. The compound (e) in Fig. 2-12 has an electron count of 22 from Fe($\eta^5 - C_5H_5$)($\eta^1 - C_5H_5$)(CO)$_2$. With the loss of CO ligand it is possible to gain stability.

Table 2-7. Some organic ligands, classified according to hapticity and electron number.

Ligand name	Ligand formula	Hapticity	Electron number
Methyl, akyl	CH_3, RCH_2	η^1	1
Alkylidene	R_2C	η^1	2
Alkylidyne	RC	η^1	3
Ethylene (ethane)	C_2H_4	η^2	2
Allyl (propenyl)	CH_2CHCH_2	η^1	1[a]
		η^3	3
Cyclopentadienyl	C_5H_5	η^1	1[a]
		η^3	3[a]
		η^5	5
Benzene	C_6H_6	η^6	6

[a] Uncommon bonding arrangements.

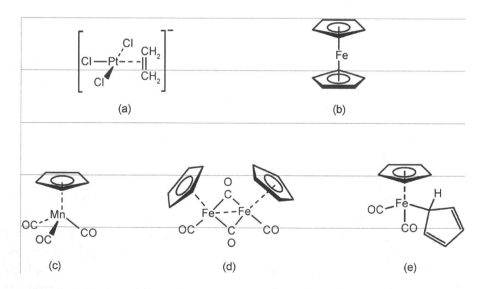

Figure 2-12. Ferocene sandwich and piano-stand compounds that follow the 18-electron rule.

2.2.2 Ligand Field Theory

In a free atom or ion the five d-orbitals with different values of magnetic quantum number, m, have the same energy. For samples of the field d-orbitals, refer to Figure 2-4. However, in any compound those orbitals interact with the surrounding ligands and a **ligand field splitting** is produced. The most common coordination is an octahedron with surrounding ligands. Two of the d-orbitals known as the e_g set, namely d_z and $d_{x^2-y^2}$, are formed at a higher energy level than the other three, d_{xy}, d_{yz}, and d_{xz}, known as the t_{2g} set.

The splitting between t_{2g} and e_g orbitals (Fig. 2-13) gives rise to colors and the transition energy is denoted by Δ_0, which can be measured experimentally. The ligand field splitting comes from orbital overlapping of donor and acceptor interactions. Ligands with containing π-acceptors having empty antibonding π-orbitals, such as CO, will lower the energy of t_{2g}, and Δ_0 certainly will increase. For strong σ donor, ligands will also increase the Δ_0. Thus, the order increase Δ_0 for a number of ligands are as follows:

$$I^- < Br^- < Cl^- < F^- < OH^- < H_2O < NH_3 < PPh_3 < CN < CO$$

This is called the spectro chemical series.

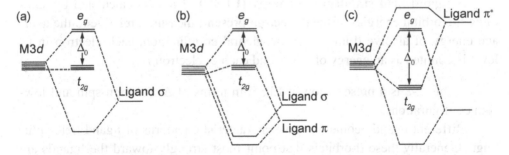

Figure 2-13. Partial MO diagram showing the splitting of ligand field energy with (a) σ -donor only, (b) π -donor, and (c) π -acceptor ligands.

For transition metals, two or three d-electrons will occupy the t_{2g} orbitals with paralleled spins, but with four or more there will be different possibilities. If an extra repulsion coming from spin pairing is large enough, the ground state will be of the high-spin type, keeping electrons in separate orbitals as far from each other as possible. On the other hand, if Δ_0 is larger than the spin pairing energy, the favored configuration will be the low-spin type, placing as many electrons as possible in the t_{2g} even if they are paired (as shown in Fig. 2-14).

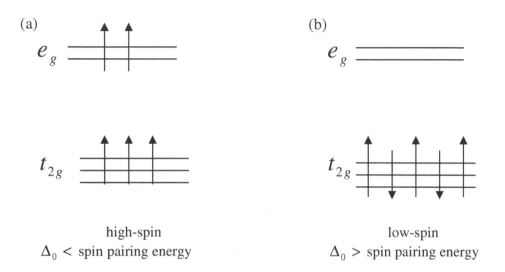

Figure 2-14. Electron configurations for d^5 in (a) high-spin, and (b) low-spin octahedral complex.

A **ligand field stabilization energy (LFSE)** of ions is calculated by summing the orbital energies of the d-electron present, measured relative to the average energy of all five d-levels. In octahedral coordination, each electron in t_{2g} level is counted as an energy of $-\frac{2}{5}\Delta_0$ and each e_g electron is $+\frac{3}{5}\Delta_0$.

Table 2-8 is a presentation of LFSE in terms of Δ_0 for high-spin and low-spin configurations.

Different ligand geometries give characteristic patterns of ligand field splitting. Generally these d-orbitals that point most strongly toward the ligands are raised in energy relative to others. For example, a square-planar configuration, e.g., d^8 configuration, the d-orbital pointing towards the ligand x-y plane is higher than the others. Figure 2-15 presents different ligand field splitting patterns.

As shown in Fig. 2-15 for the tetrahedral coordination, the splitting is in the opposite direction of an octahedral coordination. The distortion from the octahedral geometry usually is termed the **Jahn-Teller distortion**, such as the $d^4(Cu^{2+})$ and the high spin $d^4(Cr^{2+})$ leading to tetrahedrally distorted octahedron.

Table 2-8. Electron configurations for d^n high- and low-spin octahedral complexes, with corresponding ligand field stabilization energies.

n	High-spin Configuration	LFSE	Low-spin Configuration	LFSE
0	–	0	–	–
1	$(t_{2g})^1$	$-\dfrac{2}{5}\Delta_0$	–	–
2	$(t_{2g})^2$	$-\dfrac{4}{5}\Delta_0$	–	–
3	$(t_{2g})^3$	$-\dfrac{6}{5}\Delta_0$	–	–
4	$(t_{2g})^3(e_g)^1$ [a]	$-\dfrac{3}{5}\Delta_0$	$(t_{2g})^4$	$-\dfrac{8}{5}\Delta_0$
5	$(t_{2g})^3(e_g)^2$	0	$(t_{2g})^5$	$-\dfrac{10}{5}\Delta_0$
6	$(t_{2g})^4(e_g)^2$	$-\dfrac{2}{5}\Delta_0$	$(t_{2g})^6$	$-\dfrac{12}{5}\Delta_0$
7	$(t_{2g})^5(e_g)^2$	$-\dfrac{4}{5}\Delta_0$	$(t_{2g})^6(e_g)^1$ [a]	$-\dfrac{9}{5}\Delta_0$
8	$(t_{2g})^6(e_g)^2$	$-\dfrac{6}{5}\Delta_0$	–	–
9	$(t_{2g})^6(e_g)^3$ [a]	$-\dfrac{3}{5}\Delta_0$	–	–
10	$(t_{2g})^6(e_g)^4$	0	–	–

[a] Configurations susceptible to Jahn-Teller distortion.

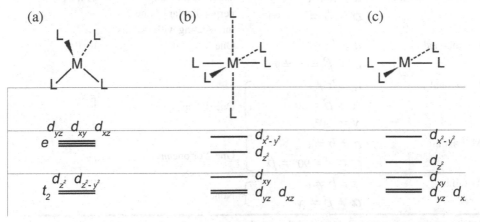

Figure 2-15. Ligand field splitting patterns for (a) tetrahedral, (b) tetragonally distorted octahedral, and (c) square-planar complexes.

2.3 THE SOLID STATE

Both solid state chemistry and physics deal with the types and classification of structures and bondings, especially various properties of solids such as mechanical, electrical, magnetic, optical and thermal properties. Solid or condensed matter can be categorized into two classes, the **crystalline** solid or **non-crystalline amorphous** or **glassy** solid. Although both solids contain short-range local chemical bonding arrangements, the amorphorous compounds do not have a **unit cell**. Crystalline compounds can be studied by x-ray diffraction, which is dependent on the fact that regular atomic spacing in the unit cell of crystals is close to the wavelength of the x-rays. Degree of crystallinity can be determined by evaluation of the long-range order. However, there is no degree that can be measured for amorphorous compounds. The **mesomorphic** material usually possesses either one-dimensional order (such as asphaltene) or two-dimensional order (liquid crystals). Crystals are usually classified into seven crystal systems

Table 2-9. Crystal systems and Bravais lattices.

System	Unit cell specification	Essential symmetry[a]	Bravais lattice[b]
Cubic	$a = b = c$ $\alpha = \beta = y = 90°$	Four 3s; sc, bcc, tcc	P, I, F
Tetragonal	$a = b \neq c$ $\alpha = \beta = y = 90°$	One 4 or $\bar{4}$	P, I
Orthorhorhombic	$a \neq b \neq c$ $\alpha = \beta = y = 90°$	Three 2s mutually perpendicular or one 2 intersecting with two ms	P, I, C, F
Rhombohedral	$a = b = c$ $\alpha = \beta = y \neq 90°$	One 3	R (P)
Hexagonal	$a = b \neq c$ $\alpha = \beta = 90°$ $y = 120°$	One 6, hcp	P
Monoclinic	$a \neq b \neq c$ $\alpha = y = 90° \neq \beta$	One 2 or one m	P C
Triclinic	$a \neq b \neq c$ $\alpha \neq \beta \neq y$	None	P

[a] 3, 4 etc. are the rotation axes; $\bar{3}$, $\bar{4}$ etc. are the inversion aces; m is mirror plane.
[b] P = primitive lattice containing lattice points at the corners of the unit cell; F = face-centered lattice; I = body-centered lattice.

(Table 2-9). In terms of lattice, there are 14 distinct **Bravais lattice** types that fill the three-dimensional space (Fig. 2-16). Note the unit cell scale lengths are a, b, and c with angles of α, β, and γ.

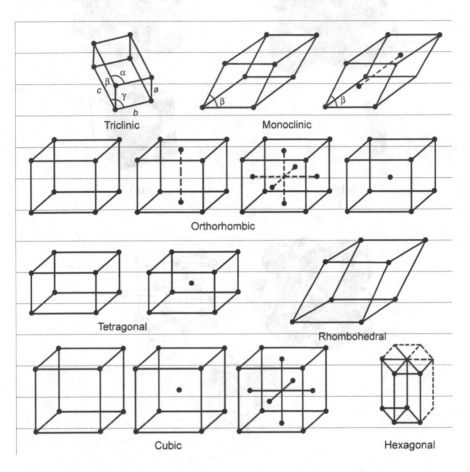

Triclinic   Monoclinic

Orthorhombic

Tetragonal   Rhombohedral

Cubic   Hexagonal

Figure 2-16. The 14 Bravais lattices, which can be classified into the seven crystal types in Table 2-9. Note the three types of cubic systems: sc, simple cubic (or primitive cubic) system, lattice points at corners of a cube having points at (0, 0, 0), (a, a, a) (a, 0, 0), etc; bcc, body-centered cubic lattice has these points plus one at ($\frac{a}{2}, \frac{a}{2}, \frac{a}{2}$) at center of the cube; fcc, face-centered cubic lattice contains all points of sc plus these at ($\frac{a}{2}, \frac{a}{2}$, 0), ($\frac{a}{2}$, 0, $\frac{a}{2}$), ($\frac{a}{2}, \frac{a}{2}$, a) etc. at the center of the cube faces.

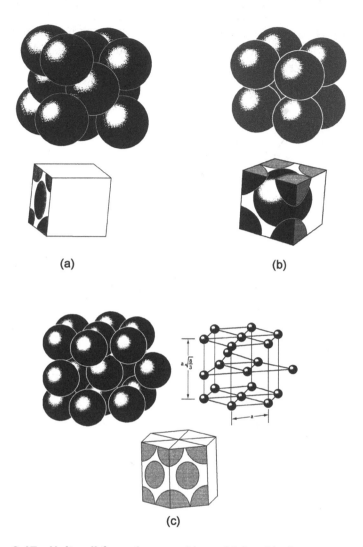

(a)   (b)

(c)

Figure 2-17. Unit cell for sphere packing of (a) cubic face centered unit cell, (b) cubic body centered unit cell, and (c) hexagonal closet packed (hcp) unit cell

The unit cell, in reality, is a sliced portion of the atoms that contribute to the packing of the lattice positions. For example, the unit cell for sphere packing of fcc lattice is shown in Fig. 2-17(a). Evidently, the atoms in the unit cell are $(\frac{1}{8}$ atoms per corner)(8 corners) + $(\frac{1}{2}$ atom per face)(6 faces) = 4 atoms.

From a unit cell one can easily evaluate the packing fractions, which defines the atom sphere volume in the unit cell to the total volume of the unit cell. The structural distinction of packing between near-neighbor (bonded) and the next-near-neighbor (non-bonded) becomes useful parameters to indicate the metallic tendency for heavy elements. For example, the ratio indicated in Table 2-10 of Sb and Te become nearly equal, which suggests that the influence of directional bonding persists in the metallic state.

Table 2-10. The ratio of next-near-neighbor to near-neighbor distances in some solid p-block elements.

P	1.787	S	1.81	Cl	1.65
As	1.33	Se	1.49	Br	1.46
Sb	1.16	Te	1.21	I	1.33

[Example 2.2] Compare the packing fraction of some common types of close packing, assuming the spheres are touching.

For fcc,

$$\text{face diagonal} = \sqrt{2}a = 4r \text{, or } a^3 = 16\sqrt{2}r^3$$

If r is the radius of the sphere,

$$Pf_{(fcc)} = \frac{4\left(\pi r \frac{3}{3}\right)}{a^3} = \frac{4\left(4\pi r \frac{3}{3}\right)}{1b\sqrt{2}r^3} = \frac{\pi}{3\sqrt{2}} = 0.7405$$

Similarly for

$$Pf_{(sc)} = \frac{\pi}{6} = 0.5236$$

$$Pf_{(bcc)} = \frac{\sqrt{3}\pi}{8} = 0.6802$$

$$Pf_{(hcp)} = 0.7405$$

In some unusually ionic crystals, such as B_2O_3, the smaller ions can occupy a trigonal planar interstitial site. The B^{3+} ions are co-planar with three touching O^{2-} ions as shown in Fig. 2-18. Assuming the radius of the interstitial ion, B^{3+} is r', then from simple geometry of an isosceles triangle,

$$\frac{\sin 120°}{2r} = \frac{\sin 30°}{r+r'}$$

or

$$\frac{\sqrt{3}}{2r} = \frac{1}{(r+r')}$$

thus

$$\frac{r'}{r} = \frac{(2-\sqrt{3})}{\sqrt{3}} = 0.1547$$

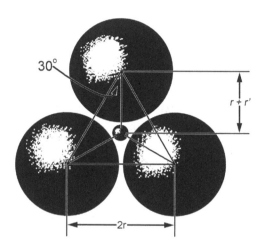

Figure 2-18. A hypothetical B_2O_3 structure. The small ion is in the interstitial site among the large ions.

Different packing of lattice for the elements in the periodic table can be seen in Fig. 2-19.

Bonding in crystals can be roughly broken down into five types. Table 2-11 is useful in understanding structure-property relations in solids. In real situations, solids may exhibit features of more than one type of bonding. We will explore the metallic state of metals further.

2.3.1 Metals

Bonding in solids involves molecular orbitals forming energy bonds rather than discrete energy levels. Two complementary theories form the basis of bond theory: the tightly bound electron model and the free electron model. The former model is an extension of the **linear combination** of the **atomic orbitals** (LCAO) of its constituent elements, which can be illustrated in Fig. 2-20, of a linear chain of H atoms. As the number of atoms increase, the lowest energy molecular orbital is formed when all the H(1s) atomic orbitals are in there. The highest energy orbital is formed when all the atomic orbitals are out of place.

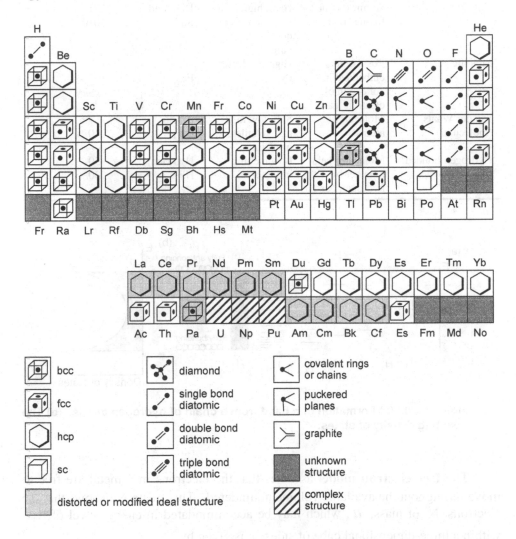

Figure 2-19. Crystal structure or bonding motifs in periodic table presentation.

Table 2-11. Types of solids.

Type	Units present	Characteristics	Examples	Approximate cohesive energy, kJ mol^{-1}
Ionic	Positive and negative ions	Brittle, insulating, and fairly highly melting	NaCl LiF	795 1010
Covalent	Atoms (bonded to one another)	Hard, high melting, and nonconducting (when pure)	Diamond SiC	715 1010
Metallic	Positive ions embedded in a collection of electron 'gas'	High conductivity	Na Fe	110 395
van der Waals (molecular)	Molecules or atoms	Soft, low melting, volatile and insulating	Argon CH$_4$	7.6 10
Hydrogen-bonded	Molecules held together by hydrogen bonds	Low melting insulators	H$_2$O (ice) HF	50 30

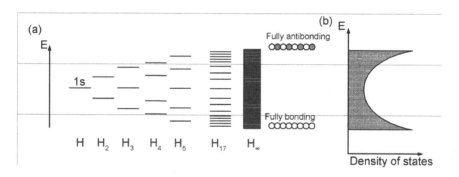

Figure 2-20. (a) Formation of a band from a chain of hydrogen atoms. (b) The resulting density of states.

The **free electron model** assumes that the electrons in a metal are free to move throughout the available volume unhindered. This treatment shows that the electrons, N, of mass, μ, which may be accommodated in energy level of E$_{max}$ within a three-dimensional cube of sides, a, is given by

$$N = \frac{8\pi a^3}{3}\left(\frac{2\mu E_{max}}{h^2}\right)^{\frac{3}{2}}$$

[2-26]

or, in another way

$$E_{max} = \frac{h^2}{2\mu}\left(\frac{3\rho}{8\pi}\right)^{\frac{2}{3}}$$ [2-27]

The density of states diagram is shown in Fig. 2-21.

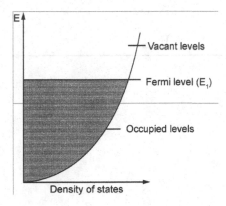

Figure 2-21. Density of states diagram resulting from the free electron model.

It is interesting to compare the E_{max} calculated from the free electron theory with experimental data from the measurements by **x-ray photoelectron spectroscopy** (XPS) as in Table 2-12.

Table 2-12. Value of E_{max} for metals calculated from free electron model and measured by XPS.

E_{max} (eV)	Na	Mg	Al
Calculated	3.2	7.2	12.8
Experimental	2.8	7.6	11.8

For ionic solids, **Coulombic effects** by electrostatic forces are important. Such energy required for the binding process is known as the **lattice enthalpy** (or crystal energy). The long range Coulomb interaction between charges at different distances in the crystal structure is

$$U_c = \frac{NAz_+z_-e^2}{4\pi\varepsilon_o r_o}$$ [2-28]

where N is Avagadro's constant, A is **Madelung constant**, e is charge on elec-
tron, z_+, z_- are numbers of charges on a pair of ions of r_0 apart, ε_0 is the vacuum
permittivity. Madelung constant depends on the structure and the coordination
number and is derived from the long-range summation of ionic interactions. This
lattice energy can be compared experimentally from a Born-Haber cycle based on
Hess' law in thermochemistry. The **Born-Haber cycle** is illustrated as follows in
Fig. 2-22.

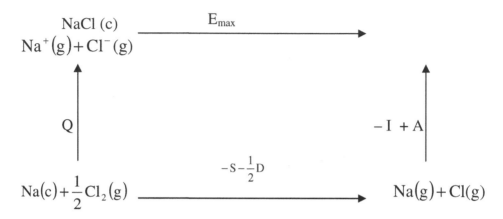

E_{max} = lattice enthalpy
Q = standard heat of formation of crystalline sact
S = heat of sublimation of metallic Na
I = Ionization potential of Na
A = Electron affinity of Cl
D = heat of dissociation of $Cl_2(g)$ into atom

Figure 2-22. Born-Haber cycle for evaluation of the lattice enthalpy of NaCl.

The experimental evaluation of E_{max} for a number of crystal for various salts
are listed in Table 2-13. Actually, the computation of E_{max} is based on the correction and extended form of Born-Landé model

$$U = U_{max}\left(1 - \frac{1}{n}\right)$$ [2-29]

where n is a constant that varies in the range of 7-12 depending on the ions. The
condition is for the repulsion of the short-range overlap, which is proportional to
$\frac{1}{r^n}$.

Table 2-13. The Born-Haber cycle*.

Crystal	$-Q$	I	S	D	A	E_c	E_c**
NaCl	99	117	26	54	88	181	190
Nabr	90	117	26	46	80	176	181
NaI	77	117	26	34	71	166	171
KCl	104	99	21	54	88	163	173
Kbr	97	99	21	46	80	160	166
KI	85	99	21	34	71	151	159
RbCl	105	95	20	54	88	159	166
RbBr	99	95	20	46	80	157	161
RbI	87	95	20	34	71	148	154

* Symbols for energy term follow that of Fig. 2-22 and the unit is in Kcal/mole.
** Calculated, Eq. [2.29].

Referring to the free electron model, the energy levels in a band are progressively filled. At zero Kelvin, the electrons only occupy the lowest energy level available. As the temperature increases, the electrons are thermally excited to a higher level according to Boltzmann distribution. The highest energy level usually is referred to as the Fermi level or Fermi energy E_+. The Fermi-Dirac statistics can be seen in Fig. 2-23 as a distribution function of probability.

$$f(E) = \left(1 + \exp\left[\frac{E - E_f}{k_B T}\right]\right)^{-1}$$
[2-30]

where k_B is Boltzmann constants.

The conductivity of a solid is proportional to the number and mobility of the charge carriers. In a metal or a metallic conductor, one or more of bands are only partially filled, giving the electron vacant orbitals into which they can move freely (Fig. 2-24a). Metallic property may also result when a band is prevented from being filled by the overlap of a second band (Figs. 2-24a and 2-24b). An **insulator** is in the core when the highest occupied band is unequally separated from the lowest unoccupied band (Fig. 2-24c). A **semiconductor** has a small band gap to allow the electron to move (Fig. 2-24d).

Most semiconductors use extrinsic semiconductors, which employ the **doping principle**, rather than the intrinsic semiconductor made of pure material. Doping Si with P replaces some tetrahedrally banded Si atoms in the lattice with P. Each replacement provides one extra valence electron, which requires a small amount of energy to escape into the conduction band of Si. This is termed the **n-**

type semiconductor. On the other hand, replacing an Si atom with Al creates a hole (missing electron) which may move in the valence band, giving a **p-type semiconductor**. Both n- and p-type semiconductors can be illustrated in Fig. 2-25, which finds greater use in the photoelectric energy source devices.

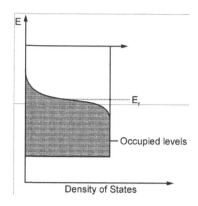

Figure 2-23. Fermi-Dirac distribution of electrons in an idealized band.

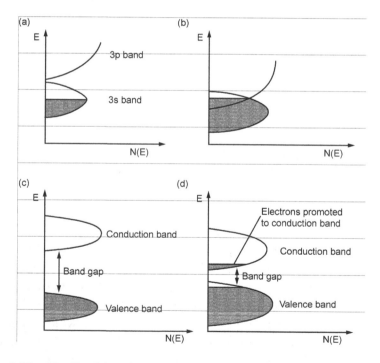

Figure 2-24. Idealized band structures. (a) Metallic conductor with partially filled band resulting from partially filled atomic orbitals; (b) metallic conductor with partially filled band resulting from band overlap; (c) an insulator; (d) a semiconductor.

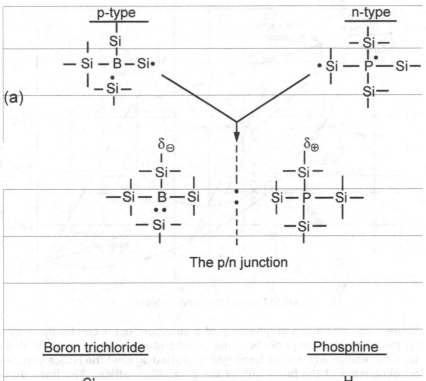

Figure 2-25. (a) Origin of barrier field; (b) field creation similar to chemical complexation.

Band-gap engineering involves the preparation of semi-conductors by correlation of energy gap to the lattice constant as shown in Fig. 2-26.

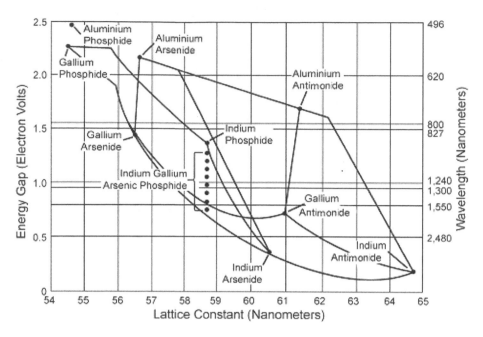

Figure 2-26. Band-gap engineering of a semiconductor can be done by varying the elemental components of the semiconductor alloy in a controlled way.
Here, the energy across the band gap is plotted against the lattice constant, or
the dimension of the basic unit of the crystalline lattice. The lines that con-
nect points on the graph show how the band gap and the lattice constant vary
for mixtures of the binary compounds to which the points correspond. Such
mixtures are ternary compounds. For example, the ternary compound gallium
aluminum arsenide can be made to have any band gap between about 1.4 and
2.2 electron volts by varying the ratio of gallium to aluminum between the
band-gap extremes represented by gallium arsenide and aluminum arsenide.
Materials of differing composition that have roughly the same lattice constant,
such as these two binary compounds, can be deposited in alternating layers
to create a single crystal lattice.

2.3.2 Alloy System

Referring to the Hildebrand solubility parameter values in Table 2-14, approximately close values of the metals in the crystal classes (vertical columns) will
form a continuous series of solid solutions without any change in structure. Examples are Cu-Au and Ag-Au.

For two metals with solubility parameters apart and in two different crystal
types such as Cu and Zu in the brass system, pure Cu crystallizes in a face-
centered cubic structure, which is in the α-phase. This Cu dissolves up to about
38% Zn to become body-entered cubic β-phase. Continuous addition of Zn to

58% a complex cubic structure begins to form, called α-brass, which is hard and brittle. At about 67% Zn, the lexagonal closest packed ε-phase arises. Finally, further addition of Zn will end in η-phase, a distorted hcp arrangement of pure Zn. Actually, this is a general transition when using the alloy system.

Table 2-14. Solubility parameters of metals grouped into crystal classes. Classes are as follows: I, body-centered cubic; II, hexagonal close packed; III, face-centered cubic; IV, diamond structure; and V, sheet structure.

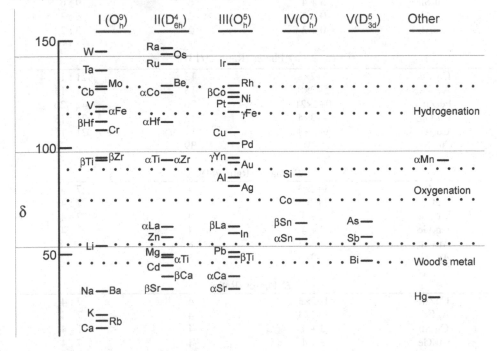

An empirical rule is formulated to account for this regular behavior. This principle is based on the constant ratio of valence electrons to atoms in relation to each phase whish is called the **Hume-Rothery Rule**. The phase transitions for a number of alloy systems are illustrated in Table 2-14. The valence electrons of all the transition metals are assumed to be zero. In Table 2-14 the transition of phases from β to α and from α to ε are listed.

From Table 2-15, it is obvious that Wood's metal, which contains Sn, Bi, Pb and Cd, are compatible with a melting point below 80°C and can be used as a constant temperature bath medium. Most catalysis, either for oxidation or reduction, is based on the principle of the close values of solubility parameters of metals. The intermetallics of the Sb-Sn system for the construction of filters for desulfurization are also based on Fig. 2-26.

Table 2-15. Electron compounds illustrating Hume-Rothery Rule.

Alloy*	Electrons	Atoms	Ratio
β Phases (Ratio 3/2)			
CuZn	1 + 2	2	3 : 2
AgCd	1 + 2	2	3 : 2
CuBe	1 + 2	2	3 : 2
AuZn	1 + 2	2	3 : 2
Cu_8Al	3 + 3	4	6 : 4
Cu_5Sn	5 + 4	6	9 : 6
CoAl	0 + 3	2	3 : 3
FeAl	0 + 3	2	3 : 2
γ Phases (Ratio 21/13)			
Cu_5Zn_8	5+2×8	13	21 : 13
Ag_5Zn_8	5+2×8	13	21 : 13
Fe_5Zn_{21}	0+2×21	26	42 : 26
Cu_9Ga_4	9+3×4	13	21 : 13
Cu_9Al_4	9+3×4	13	21 : 13
$Cu_{31}Sn_8$	31+4×8	39	63 : 39
ε Phases (Ratio 7/4)			
$CuZn_8$	1+2×3	4	7 : 4
$AgCd_8$	1+2×3	4	7 : 4
Cu_8Sn	3 + 4	4	7 : 4
Cu_8Ge	3 + 4	4	7 : 4
Au_5Al_8	5+3×3	8	14 : 8
Ag_5Al_8	5+3×3	8	14 : 8
ε Phases (Ratio 7/4)			
$CuZn_8$	1+2×3	4	7 : 4
$AgCd_8$	1+2×3	4	7 : 4
Cu_8Sn	3 + 4	4	7 : 4
Cu_8Ge	3 + 4	4	7 : 4
Au_5Al_8	5+3×3	8	14 : 8
Ag_5Al_8	5+3×3	8	14 : 8

* The alloy composition is variable within a certain range, but the nomial compositions listed always fall within the range.

REFERENCES

2-1. P. A. Cox, The Elements: Their Origin, Abundance and Distribution, Oxford University Press, Oxford, UK, 1989.

2-2. P. A. Cox, The Electronic Structure and Chemistry of Solids, Oxford University Press, Oxford, UK, 1987.

2-3. P. A. Cox, Inorganic Chemistry, Bios Scientific Publishers Ltd., Oxford, UK, 2000.

2-4. B. Douglas, D. H. McDaniel, and J. J. Alexander, Concepts and Models of Inorganic Chemistry, 2nd ed., Wiley, New York, 1983.

2-5. J. Emsley, The Elements, 2nd ed., Clarendon Press, Oxford, UK, 1991.

2-6. B. N. Figgis and M. A. Hittchman, Ligand Field Theory and Its Applications, Wiley-VCH, New York, 2000.

2-7. M. Gerloch, Magnetism and Ligand-field Analysis, Cambridge University Press, Cambridge, UK, 1983.

2-8. S. F. A. Kettle, Symmetry and Structure: Readable Group Theory for Chemists, 2nd ed., Wiley, Chichester, UK, 1995.

2-9. K. M. Mackay and R. A. Mackay, Introduction to Modern Inorganic Chemistry, 4th ed., Blackie, Glasgow, UK, 1989.

2-10. J. N. Murrel, S. F. A. Kettle and J. M. Tedder, The Chemical Bond, Wiley, Chichester, UK, 1978.

2-11. U. Müller, Inorganic Structural Chemistry, Wiley, Chichester, UK, 1993.

2-12. D. R. Shriver and P. W. Atkins, Inorganic Chemistry, 3rd ed., Oxford University Press, Oxford, UK, 1999.

2-13. R. Thompson, Industrial Inorganic Chemicals: Production and Uses, Royal Society of Chemistry, London, UK, 1995.

2-14. A. G. Whittaker, A. R. Mount and M. R. Heal, Physical Chemistry, 2nd ed., Bios Scientific Publishers Ltd., Oxford, UK, 2000.

PROBLEM SET

1. Give the term symbol of nitrogen atom or the O^+.

2. An aryl – substituted hexacarbonyl (hexapentaene) diiron complex, e.g. I (R = Ph, 4 – Me C_6H_4) undergo facile intramolecular C=C bond cleavage on treatment with $Fe_3(CO)_{12}$ to yield a bis($\mu 3$-η-2-allennylidene) iron complexes II. What is the structure of II.

3. Predict the maximum rate of hydrogen uptake if given atomic percentage of ruthenium is dissolved in platinum as hydrogenation catalyst.

4. For Copper Kα radiation in XRD, what Cu atom energy change is necessary to produce a photon with the same wavelength?

CHAPTER 3

ORGANIC CHEMISTRY

*O*rganic chemistry deals with compounds of **carbon**, the organic materials usually present in liquid, solid, and gaseous wastes that an engineer will encounter in practice. Therefore, it is important that engineers have a fundamental knowledge of organic chemistry. The aspects of organic chemistry with which an engineer is concerned differ considerably from those with which an organic chemist is concerned. Organic chemists are more interested in the synthesis of compounds and the mechanisms of organic reactions. Engineers, on the other hand, focus their attention on the removal, reduction, or the degradation of those compounds. In this chapter, emphasis will be made on the structures and properties of organic compounds, their reactivity and stability, their functional groups, and the methods of studying their structure-activity relationships.

3.1 STRUCTURE AND PROPERTIES

Carbon, being the first element of Group IVA in the periodic table, normally has four covalent bonds that allow it to link with other atoms in a wide variety of ways through covalent bonding. In this manner, a great variety of compounds can be formed, the most common of which are **hydrocarbons**. The bond angle between two covalent bonds in a tetrahedron is 109°28'. This value can be obtained by the following calculation.

Here, point O is located in the center of a cube, as indicated in Figure 3-1. Because

$$\overrightarrow{AO} \cdot \overrightarrow{BO} = |\overrightarrow{AO}||\overrightarrow{BO}|\cos\phi \qquad\qquad [3\text{-}1]$$

$$\overrightarrow{AO} \cdot \overrightarrow{BO} = (1,1,1)(-1,-1,1) = -1-1+1 = -1 \qquad\qquad [3\text{-}2]$$

$$|\overrightarrow{AO}| = |\overrightarrow{BO}| = (1^2 + 1^2 + 1^2)^{1/2} = \sqrt{3}$$

Thus,

$$\phi = \cos^{-1}(-1/3) = 109°28' \qquad\qquad [3\text{-}3]$$

or from

$$|\overrightarrow{AO} \times \overrightarrow{BO}| = |\overrightarrow{AO}||\overrightarrow{BO}|\sin\phi \qquad\qquad [3\text{-}4]$$

$$\overrightarrow{AO} \times \overrightarrow{BO} = \left(\begin{vmatrix} 1 & 1 \\ -1 & 1 \end{vmatrix}, \begin{vmatrix} 1 & 1 \\ 1 & -1 \end{vmatrix}, \begin{vmatrix} 1 & 1 \\ -1 & -1 \end{vmatrix} \right) = (2,-2,0) \qquad\qquad [3\text{-}5]$$

$$\left|\overrightarrow{AO}\right| \times \left|\overrightarrow{BO}\right| = \left(2^2 + (-2)^2 + 0^2 \right)^{\frac{1}{2}} = 2\sqrt{2} \qquad\qquad [3\text{-}6]$$

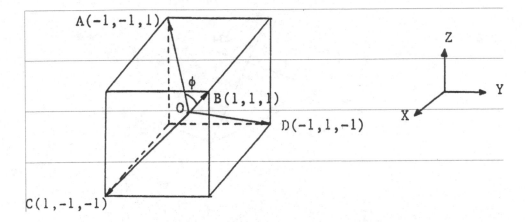

Figure 3-1. The tetrahedral angle at the center of a cube.

$$\phi = \sin^{-1}\left(2\sqrt{\tfrac{2}{3}}\right) = 109°28'$$

[3-7]

The tetrahedral angle by vector analysis is illustrated in Figure 3-1. This tetrahedral angle is important for developing the spatial relationships of chemical compounds. From elementary geometry, we know that any three points in a space determine a plane. As shown in Figure 3-2, for notation, r_{ij} is the bond length, ϕ_{ijk} is the bond angle, and τ_{ijkl} is the torsional angle between planes ijk and jkl. Multiple atoms show the property of having the least distance between them, forming a chain structure that often consists of conformations behaving as a helical structure. For a helix where

a = radius of helix

n = atoms in one turn

l = identity period

θ = projection angle $(2\pi/n)$

There exist the following relationships:

$x = a \cos \theta$

$y = a \sin \theta$

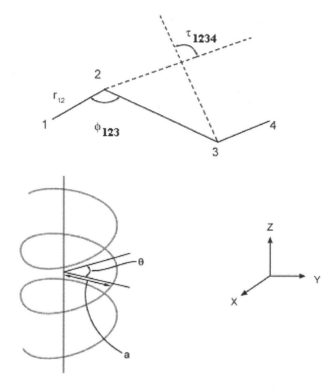

Figure 3-2. The torsional angle and projection angle of a helical structure.

$$z = l/n \qquad\qquad [3\text{-}8]$$

and

$$\vec{A} = (a \cos \theta,\ a \sin \theta,\ l/n) \qquad\qquad [3\text{-}9]$$

$$\vec{B} = (a \cos 2\theta,\ a \sin 2\theta,\ 2\ l/n) \qquad\qquad [3\text{-}10]$$

$$\vec{C} = (a \cos 3\theta,\ a \sin 3\theta,\ 3\ l/n) \qquad\qquad [3\text{-}11]$$

$$\vec{r}_{12} = (\vec{B} - \vec{A}) = (a \cos 2\theta - a \cos \theta, a \sin 2\theta - a \sin \theta, l/n) \qquad [3\text{-}12]$$

$$|\vec{r}_{12}| = \left\{ 2a^2 (1 - \cos \theta) + \frac{l^2}{n^2} \right\}^{\frac{1}{2}} \qquad [3\text{-}13]$$

$$\phi_{123} = \cos^{-1} \frac{a^2 (2 \cos \theta - 2 \cos^2 \theta + 1) \dfrac{l^2}{n^2}}{2a^2 (1 - \cos \theta) + \dfrac{l^2}{n^2}} \qquad [3\text{-}14]$$

$$\tau_{1234} = \cos^{-1} \left(\frac{\vec{N}_2 \cdot \vec{N}_2}{|N_1||N_2|} \right) \qquad [3\text{-}15]$$

Each plane is made from three successive atoms. The torsional angle is between the normals of two successive planes where

$$\vec{N}_1 = \vec{r}_{12} \times \vec{r}_{23} \qquad [3\text{-}16]$$

$$\vec{N}_2 = \vec{r}_{23} \times \vec{r}_{34} \qquad [3\text{-}17]$$

In this manner, for successive atoms in given molecules, the conformation of the molecules can be determined; for example, the torsional angles and the tacticities of the carbon chain can be determined. For the coordinates X_i, Y_i, Z_i transferred to i - 1 (the coordinates X_{i-1}, Y_{i-1}, Z_{i-1}) then

$$X_{i-1} = A X_i + B \qquad [3\text{-}18]$$

$$A = \begin{bmatrix} -\cos \phi & -\sin \phi & 0 \\ \sin \phi \cos \tau & -\cos \phi \cos \tau & \sin \tau \\ \sin \phi \sin \tau & -\cos \phi \sin \tau & \cos \tau \end{bmatrix} \qquad [3\text{-}19]$$

$$B = \begin{bmatrix} r \\ o \\ o \end{bmatrix} \qquad\qquad [3\text{-}20]$$

for n-paraffins, $\tau = 180°$ (trans) or $\tau = \pm 60°$ (gauch); if $\tau = 180°$, then

$$A = \begin{bmatrix} -\cos\phi & -\sin\phi & 0 \\ -\sin\phi & \cos\phi & 0 \\ 0 & 0 & -1 \end{bmatrix} \qquad [3\text{-}21]$$

Matrix A can be reduced to a trans (or syndiotactic) configuration for polymers. A relation for θ, τ, and ϕ can be

$$1 + 2\cos\theta = \cos\tau - \cos\phi\cos\tau - \cos\phi \qquad [3\text{-}22]$$

For example, a polyethylene chain with a 3_1 helix will have a τ angle value of 60°, because

$$4\cos^2\tau - 16\cos\tau + 7 = 0 \qquad\qquad [3\text{-}23]$$

In the following pages, a method will be discussed for characterizing carbon compounds according to their bond types and the relationship of bond types to properties.

For a carbon compound there are four types of carbon atoms that may be characterized by their bonding to other carbon atoms: **primary, secondary, tertiary,** and **quaternary carbon** atoms. The numbers of carbon atoms for each bond type in a certain carbon compound are noted as k_1, k_2, k_3, and k_4 respectively. This carbon compound can be specified as $[k_1, k_2, k_3, k_4]$. For 2,2-dimethyl-3-ethyl heptane, the structure is shown in Figure 3-3, Structure A. By counting the numbers of different bond types, the structure can be represented as [5,4,1,1]. Table 3-1 lists some general formulae. The total number of carbon atoms in a compound, C, may be expressed as

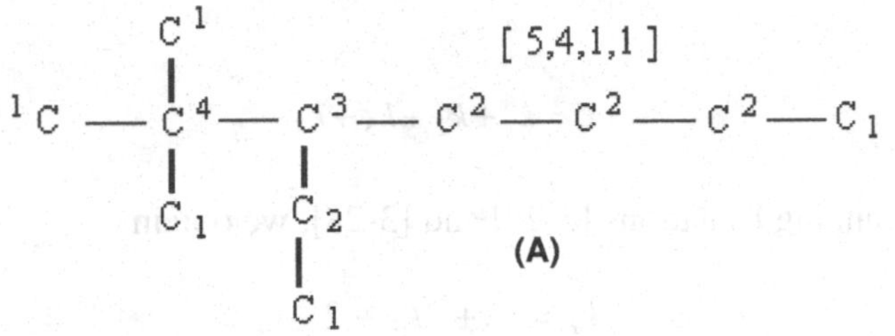

$$
\begin{array}{c}
\overset{1}{C} \\
| \\
\overset{1}{C} - \overset{4}{C} - \overset{3}{C} - \overset{2}{C} - \overset{2}{C} - \overset{2}{C} - \overset{}{C}_1 \\
| \quad\quad | \\
\overset{}{C}_1 \quad \overset{}{C}_2 \\
\quad\quad | \\
\quad\quad \overset{}{C}_1
\end{array}
\qquad [\,5,4,1,1\,]
$$

(A)

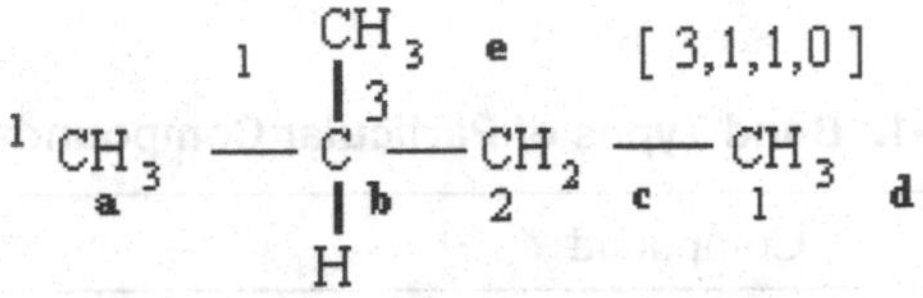

$$
\begin{array}{c}
\quad\quad\overset{1}{C}H_3 \;\; e \qquad [\,3,1,1,0\,] \\
\quad\quad |\,3 \\
\overset{1}{C}H_3 \underset{a}{{}^3} - \overset{}{C} - \underset{2}{C}H_2 \underset{c}{{}} - \underset{1}{C}H_3 \; d \\
\quad\quad |\,b \\
\quad\quad H
\end{array}
$$

(B)

Figure 3-3. Structure of (A), 2,2' -dimethyl-3-ethyl heptane and (B), 2-methyl butane or isopentane.

$$
C = \sum_{i=1}^{4} k_i \qquad\qquad [3\text{-}24]
$$

If N_i is the summation of the i type bond, then

$$
N_i = n_{ij} + \sum_{j=1}^{4} n_{ij} = i k_i \qquad\qquad [3\text{-}25]
$$

The total number of bonds can be obtained

$$
C - 1 = \frac{1}{2}\sum_{i=1}^{4} N_i = \frac{1}{2}\sum_{i=1}^{4} i k_i \qquad\qquad [3\text{-}26]
$$

so

$$
C - 1 = \frac{1}{2}\left(k_1 + 2k_2 + 3k_3 + 4k_4\right) \qquad\qquad [3\text{-}27]
$$

but

$$C = k_1 + k_2 + k_3 + k_4 \qquad [3\text{-}28]$$

By rearranging Equations [3-27] and [3-28], we obtain

$$k_1 = k_3 + 2k_4 + 2 \qquad [3\text{-}29]$$

Table 3-1. Bond Types of Particular Compounds

Bond Type	Compound
$[k_1, 0, 0, 0]$	ethane
$[0, k_2, 0, 0]$	monocyclic naphthene
$[0, 0, k_3, 0]$	cubane
$[0, 0, 0, k_4]$	diamond
For combinations of two $k_i \neq 0$ there are	
$[k_1, k_2, 0, 0]$	n-paraffin
$[k_1, 0, k_3, 0]$	all isobutyl paraffin
$[k_1, 0, 0, k_4]$	all tert-butyl-substituted paraffin (neopentylparaffin)
$[0, k_2, k_3, 0]$	multicyclic naphthene
$[0, k_2, 0, k_4]$	tricyclane
$[0, 0, k_3, k_4]$	polyspirocyclopentadiene
For $(3 - k_i) \neq 0$ there are	
$[k_1, k_2, k_3, 0]$	isoparaffins
$[k_1, k_2, 0, k_4]$	all ethyl-substituted paraffins
$[k_1, 0, k_3, k_4]$	all isopropyl-substituted paraffins
$[0, k_2, k_3, k_4]$	caged-fused naphthene
For $(4 - k_i) \neq 0$	(branched paraffins)
$[k_1, k_2, k_3, k_4]$	

[Example 3-1] The elemental analysis for an isoparaffin is 84.9%C and 15.1%H, and the methyl content based on IR analysis is 42.5%. Find the number of branches (which are related to its solubility) and the methylene content (-CH$_2$-) of this isoparaffin.

The general formula for isoparaffins is C_nH_{2n+2}.

$$H/C = (2n + 2)/n = 2 + 2/n$$

$$n = 2 /[(12\%H / \%C) - 2] = 14.9 \sim 15$$

$$C_{methyl}/C = (\%methyl/15) / (\%C /12) = (42.5/15)/(84.9/12) = 0.4$$

$$k_1 = C_{methyl} = 0.4n = 6$$

$$n = 15$$

for isoparaffin $k_4 = 0$

$$k_1 = k_3 + 2k_4 + 2 = k_3 + 2$$

$$k_3 = k_1 - 2 = 6 - 2 = 4$$

thus four branches
$$n = k_1 + k_2 + k_3 + k_4$$

$$15 = 6 + k_2 + 4 + 0$$

$$k_2 = 5$$

$$C_{methylene}/C = 5/15 = (\%methylene/14)/(84.9/12)$$

$$\% \text{ methylene} = 33.01$$

$$\text{methylene content} = 14k_2/[12n + (2n + 2)] = 33\%$$

In inorganic chemistry, a molecular formula is specific for one compound. This is not true for organic chemistry, for most molecular formulas do not represent any particular compound. Compounds having the same molecular formula are called **isomers**. Table 3-2 shows how the number of isomers increases with the carbon number.

Isomers of the same molecular formula often have different physical properties. One way to represent the difference between isomers is by their **bond type matrices**, as developed by Tatevskii. N_{ij} is the number of carbon bonds between i and j type atoms. The bond type matrix, N, is as follows:

Table 3-2. Carbons and Their Isomer Numbers

No. of Carbon	Name	No. of Isomer
1	methane	1
2	ethane	1
3	propane	1
4	butane	2
5	pentane	3
6	hexane	5
7	heptane	9
8	octane	18
9	nonane	35
10	decane	75
20	eicosane	366,319
30	triacontane	4.11×10^9

$$N = \begin{bmatrix} n_{11}, & n_{12}, & n_{13}, & n_{14} \\ n_{21}, & n_{22}, & n_{23}, & n_{24} \\ n_{31}, & n_{32}, & n_{33}, & n_{34} \\ n_{41}, & n_{42}, & n_{43}, & n_{44} \end{bmatrix} \qquad [3\text{-}30]$$

Actually $n_{11} = 0$ and $n_{ij} = n_{ji}$, so there are nine significant values in the matrix.

$$N = \begin{bmatrix} n_{12}, & n_{13}, & n_{14} \\ n_{22}, & n_{23}, & n_{24} \\ 0 & n_{33} & n_{34} \\ 0 & 0 & n_{44} \end{bmatrix} \qquad [3\text{-}31]$$

For example, for 2,2-dimethyl-3-ethyl heptane, the bond type matrix is

$$\begin{bmatrix} 2 & 0 & 3 \\ 2 & 2 & 0 \\ 0 & 0 & 1 \\ 0 & 0 & 0 \end{bmatrix} \qquad\qquad [3\text{-}32]$$

The properties of carbon compounds are related to their structures by the general formula

$$P = \sum_{ij=1}^{4} n_{ij} P_{ij} \qquad\qquad [3\text{-}33]$$

where

   n_{ij} = the element of bond type matrix

   P_{ij} = the element of property matrix

The data for a property matrix can often be found in some handbooks. Although the isomers can be represented by matrices, the product of Equation [3-33] is not by matrix multiplication, but rather by products of corresponding individual components. This is known as Tatevskii's method.

[Example 3-2]   Given the molar volume matrix, V_{ij}^{20}, as

$$\begin{bmatrix} V_{12} & V_{13} & V_{14} \\ V_{22} & V_{23} & V_{24} \\ V_{32} & V_{33} & V_{34} \\ V_{42} & V_{43} & V_{44} \end{bmatrix} = \begin{bmatrix} 41.472 & 33.979 & 29.695 \\ 16.002 & 6.479 & 1.003 \\ - & -5.356 & -12.624 \\ - & - & -22.596 \end{bmatrix}$$

find the densities of C_7 isomers.

Let us calculate the isomer, 2,2,3-trimethylbutane first.

$$
\begin{array}{cc}
C & C \\
| & | \\
C-C-C-C \\
| \\
C
\end{array}
$$

The bond type matrix of this isomer is

$$
N = \begin{bmatrix} 0 & 0 & 2 & 3 \\ 0 & 0 & 0 & 0 \\ 0 & 0 & 0 & 1 \\ 0 & 0 & 0 & 0 \end{bmatrix}
$$

$$
d_4^{20} = \frac{M}{\displaystyle\sum_{i \le j=1}^{4} n_{ij} v_{ij}^{20}} \tag{3-34}
$$

$$
\sum_{i \le j=1}^{4} n_{ij} v_{ij} = 2(33.979) + 3(29.659) - 12.624 = 144.419
$$

$$
(d_4^{20})_{cal} = \frac{7(12.01) + 16(1.0079)}{144.419} = 0.6938
$$

The experimental value for this isomer is 0.6901. In this manner, the densities of other isomers can also be calculated out one by one. Table 3-3 lists the experimental and all the calculated densities of C_7 isomers by Tatevskii's method.

Similar methods can be applied to other physical properties, such as boiling point temperature.

$$t\,^{\circ}bp = \frac{\lambda}{b - \log p} - 273.16 \qquad\qquad [3\text{-}35]$$

$$= \frac{\displaystyle\sum_{i \le j = 1}^{4} n_{ij}\,\lambda_{ij}}{\displaystyle\sum_{i \le j = 1}^{4} n_{ij}\,b_{ij} - \log P} - 273.16$$

where:

λ = latent heat of vaporization

P = vapor pressure

Table 3-3. Comparison of Experimental and Calculated Specific Gravities of Some Paraffins

Compound	d_4^{20}	
	Exp. (Rossini, 1953)	Calc.
n-Heptane	0.68376	0.6818
2-Methylhexane	0.67859	0.6773
3-Methylhexane	0.68713	0.6868
3-Ethylpentane	0.69816	0.6964
2,2-Dimethylpentane	0.67385	0.6789
2,3-Dimethylpentane	0.69508	0.6932
2,4-Dimethylpentane	0.67270	0.6730
3,3-Dimethylpentane	0.69327	0.6941
2,2,3-Trimethylbutane	0.69011	0.6937

3.2 FUNCTIONAL GROUPS IN ORGANIC COMPOUNDS

As discussed, hydrocarbons (the substitute sites of carbon are hydrogens) are the simplest class of organic compounds available; the typical example being the petroleum hydrocarbons. Basically, there are three major types of hydrocarbons. **Paraffins** are saturated chain-like molecules, typically of the form of C_nH_{2n+2} or RH, where R is C_nH_{2n+1}. If this saturated chain is folded into cyclic forms of 5- or

6- carbon-numbered rings, or the 6-numbered ring is further fused with other ring-forming methylene (CH_2) units to construct a condensed hexagonal system, then these hydrocarbons are termed **naphthenics**, represented by C_nH_{2n}; for example, cyclohexane n=3 or fused-ring naphthenic. A simple example is a polyacene-type naphthenic such as perhydronaphthalene and perhydroanthracene. This naphthenic can be represented by $C_nH_{3(n/2+1)}$. Another major type is the **aromatics**, which are the unsaturated or dehydrogenated naphthenics. Starting with benzene (consisting of alternative double and single bonds to build the fused-ring aromatics), for example, the **polyacenes**, the **polyphenylenes**, and the regular R[m,n] type are illustrated in Figure 3-4.

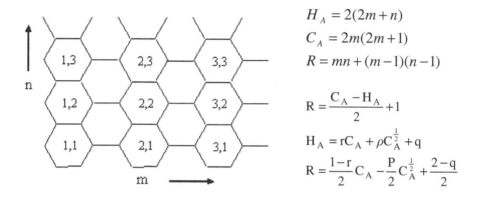

$$H_A = 2(2m + n)$$

$$C_A = 2m(2m + 1)$$

$$R = mn + (m-1)(n-1)$$

$$R = \frac{C_A - H_A}{2} + 1$$

$$H_A = rC_A + \rho C_A^{\frac{1}{2}} + q$$

$$R = \frac{1-r}{2}C_A - \frac{P}{2}C_A^{\frac{1}{2}} + \frac{2-q}{2}$$

Figure 3-4. The rectangular-type model R{m,n} represents polynuclear aromatic hydrocarbons. In this graph m or n is the index along the x- or y-axis. R{1,1} is benzene, R{1,2} is naphthalene, R{2,1} is biphenyl, R{2,2} is perylene, R{1,n}is polyacene, R{m,1} is polyphenylene, and R{m,n,} is a fused-ring aromatic hydrocarbon. (T. F. Yen, ref 2-7)

Polyacene $C_n H_{n/2+3}$

Polyphenylene $C_n H_{2n/3+2}$

Square fused aromatic $C_n H_{3n^{\frac{1}{2}} - \frac{3}{2}}$ (approximate value)   [3-36]

In a broad sense, all hydrocarbons can be represented by a general formula, $H_n C_m$, where the m values vary depending on the type.

The functional groups are derived from the substitution of one or more of the carbons or hydrogens of the $H_n C_m$ by **heterocyclic atoms**, such as N, O, S and X (X = F, Cl, Br, I). These compounds are sometimes referred to as **non-hydrocarbons**. The valence number of the heteroatoms is lower than four, which is the value of the carbon. Often these functional groups consist of unsaturated linkages. We will first discuss the oxygen groups.

3.2.1 Oxygen Functional Groups

As summarized in Table 3-4, the **alcohols** are important in that the hydrogens in the hydrocarbons (C_nH_{2n+2} or RH) are substituted by hydroxyl groups. ROH (where R is C_nH_{2n+1}) is the general formula for the paraffin series. The series begins with methanol (CH_3OH), ethanol (CH_3CH_2OH), propanol ($CH_3CH_2CH_2OH$), and so on. Similar to the bonding types, there is a primary alcohol, RCH_2OH, secondary alcohol, R_2CHOH, and tertiary alcohol, R_3COH, depending on the number of hydrogens remaining. The corresponding aromatic is phenol ArOH (Ar = C_6H_5), which exists in industrial wastewater derived from coal tar. Multiple OH groups substituted on one ring are also present, such as ortho- (catechol), meta- (resorcinol), and para- (hydroquinone) dihydroxybenzene. When hydrocarbons are partially oxidized, they can be converted into aldehydes such as those in automobile emissions.

$$RCH_3 \xrightarrow{\ [O]\ } RCHO$$

The condensation of two aldehydes, RC(H) = 0, will form a ketone, (R_2CO). The sharp odor of smog may originate from **aldehydes** and **ketones** formed by the oxidation of unsaturated hydrocarbons by ozone.

$$RCH{=}CHR' + O_3 \longrightarrow RCH \underset{\overset{\displaystyle |}{O}\!-\!\overset{\displaystyle |}{O}}{\overset{\displaystyle O}{\diagup \ \diagdown}} CHR' \xrightarrow{\ H_2O\ } RCHO + R'CHO + H_2O_2$$

Table 3-4. Substitution of Hydrocarbons (RH) by Functional Groups

Functional Groups	Name	Example
R-OH	alcohol	methanol ($R=CH_3$)
(secondary alcohol structure) R_1–C(H)(OH)–R_2	secondary alcohol	isopropanol ($R_1=R_2=CH_3$)
(tertiary alcohol structure) R_1, R_2, R_3–C–OH	tertiary alcohol	t-butyl alcohol ($R_1=R_2=R_3=CH_3$)
(benzene ring)–OH	hydroxyl-benzene	phenol
	aromatic diols	catechol (o)
		resocenol (m)
		hydroquinone (p)
R-COOH	carboxylic acid	acetic acid ($R=CH_3$)
		propionic acid ($R=C_2H_5$)
		butyric acid ($R=C_3H_7$)
RCOH	aldehyde	
$R_2C=O$	ketone	acetone, $R=CH_3$
ROR′	ether	diethyl ether, $R=C_2H_5=R'$
RX	halohydrocarbons	THM(chloroform, $CHCl_3$)
		(CFC, Freon, CH_2F_2)
$RCONH_2$	amide	acetamide, $R=CH_3$
RCN	nitrile	HAN (chloroacetonitrile)
RNO_2	nitro	TNT (Trinitrotoluene)
RSH	mercaptan	skunk
RSR	thioether	
$ArSO_3Na$	sulfonates	Anionic surfactant, Cationic exchanger

Table 3-5. Names of Mono- and Di-carboxylic Acids

No. of R	RH	R Abr.	R(C_sH_{2n+1})	Monocarboxylic Acid[*]	Dicarboxylic Acid[**]
1	Methane	Me	CH_3	Formic	—
2	Ethane	Et	CH_3CH_2	Acetic	Oxalic
3	Propane	Pr	$CH_3(CH)_2$	Propionic	Malonic
4	Butane	Bu	$CH_3(CH_2)_3$	Butyric	Succinic
5	Pentane	Pn	$CH_3(CH_2)_4$	Valeric	Glutaric
6	Hexane	Hx	$CH_3(CH_2)_5$	Caproic	Adipic
7	Heptane	Hp	$CH_3(CH_2)_6$	Enanthic	Pimelic
8	Octane	Oc	$CH_3(CH_2)_7$	Caprylic	Suberic
9	Nonane	No	$CH_3(CH_2)_8$	Pelargonic	Azelaic
10	Decane	De	$CH_3(CH_2)_9$	Capric	Sebacic

*Formic acid is HCOOH, Acetic is the 2-carbon acid CH_3COOH; the remaining acids follow the formula R'COOH with R' being one less than R(R' + 1).

** The first in the series is oxalic, $(COOH)_2$. The remaining ones are $(CH_2)_n(COOH)_2$ with n = R − 2

The **carboxylic acids** RCOOH and their corresponding esters RCOOR' are also abundant; e.g. in any landfill, cellulose is very easily converted to simple carboxylic acids before fermentation is initiated. Common monocarboxylic acids can be found in Table 3-5. Along with **dicarboxylic acids**, common names are also listed. These are intermediates after the oxidation of complex organics. Finally the **ethers** are those molecules in which oxygen serves as the bridge between two R groups, e.g. ROR'.

3.2.2 Nitrogen Functional Groups

Similar to the oxygen functional groups, an important group of compounds are **amines**, which bear the substituted ammonia functions: $-NH_2$ the amino, $=NH$ the imino, $=N-$ and the nitrilo. The primary amine is RNH_2, the secondary amine is R_2NH, and the tertiary amine is R_3N. The physiological properties of amines vary with the positions; for example,

The last structure is very strong in carcinogenicity. Nitrogen can also participate in cyclic 5- or 6-membered systems; for example,

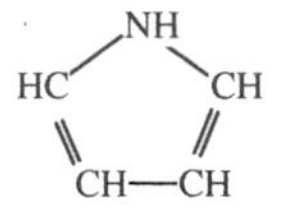

Pyrrole

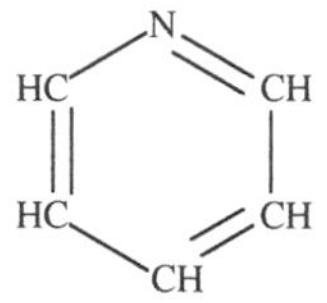

Pyridine

These are called heterocyclic compounds, which involve one or more heterocyclic elements such as N, O, S in the ring system; for example, indole and skatol have unpleasant odors from putrefaction in sludge digestion. **Amino acids** contain two different functional groups in a hydrocarbon skeleton; the simplest being glycine, H_2NCH_2COOH. Amino acids are the foundation of certain biological molecules. The amide is a condensation product of carboxylic acid and ammonia.

$$RCOOH + NH_3 = RCONH_2 + H_2O$$

Nitrile, RCN, is important in that potable water will yield haloacetonitrile (HAN) as an undesirable contaminant after chlorination. **Nitro compounds**, RNO_2, are important because certain structures will yield hazardous compounds such as TNT or picric acid.

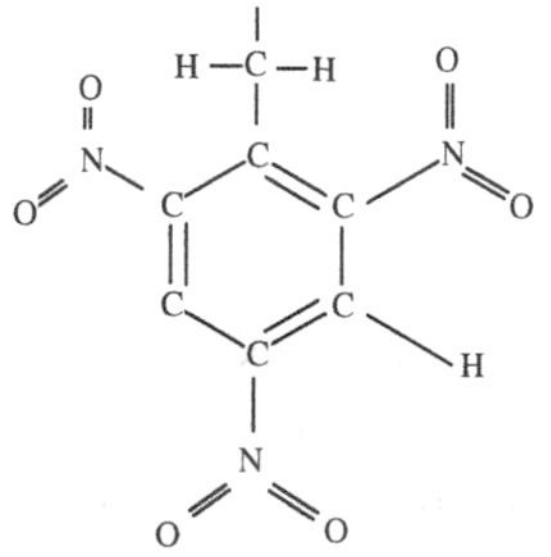

TNT

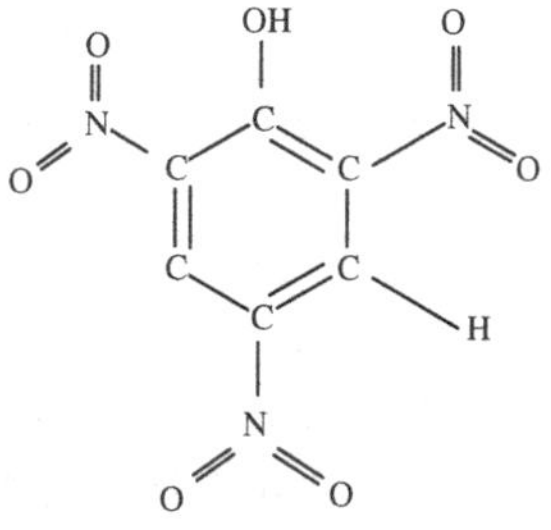

picric acid

These compounds can be found as soil contaminants for munitions operations.

3.2.3 Sulfur Functional Groups

Many sulfur compounds have the substituent of this group, -SH. These compounds are called **mercaptans**, and a well-known example is n-BuSH or n-$CH_3(CH_2)_3SH$ in the odor of skunk secretions, and nPrSH or $CH_3(CH_2)_2SH$ from freshly chopped onions. **Thioethers**, RSR′, are similar to ethers, the S taking the place of O as a bridge. An important class of sulfur compounds is the sulfonates, -SO_3Na; for example, linear alkylbenzene sulfonate (LAS) is an efficient **anionic surfactant**. A similar compound, polystyrene sulfonate, is an excellent **cationic ion exchanger**. Both classes contain the $ArSO_3^-Na^+$ functions, depending on the applications concerning the anionic or the cationic portions of the molecule.

3.2.4 Others — The Halides and Organometallics

Group VII elements of the periodic table, such as F, Cl, Br, and I, are termed halides and are represented by the letter X. These elements can replace the hydrogens of hydrocarbons. The **alkyl halides,** such as RX, are made from the corresponding alcohols.

$$ROH \xrightarrow{PCl_3} RCl + P(OH)_3$$

Trihalomethanes (THMs) are found in drinking water after chlorine disinfectant treatment; for example CHX_3 such as chloroform. The chlorofluorocarbons (CFC) such as freons, CCl_2F_2, are refrigerants or aerosol propellants easily released into the atmosphere. When metals interact with organics, they may form **organometallics**. For example, dimethyl mercury (CH_3HgCH_3) is more toxic than mercury (Hg) alone. Alkyl mercury halides, RHgX, have been used as preservatives for seeds in crops. Metals can form complexes with the functional groups of organic compounds. If there is more than one functional group, they can chelate with metals and form **metal chelates**. Some of the important chelates are listed in Figure 3-5. They can readily form the calcium or lead chelates. One way to remove lead or mercury from the body as a result of lead poisoning is by ingesting EDTA or BAL. Many other similar metal complexes and chelates, such as porphyrins, are essential in the biosphere.

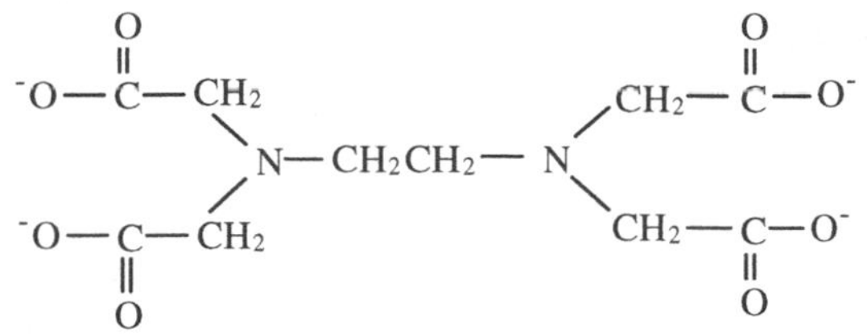

ethylenediaminetetraacetate

EDTA Anion

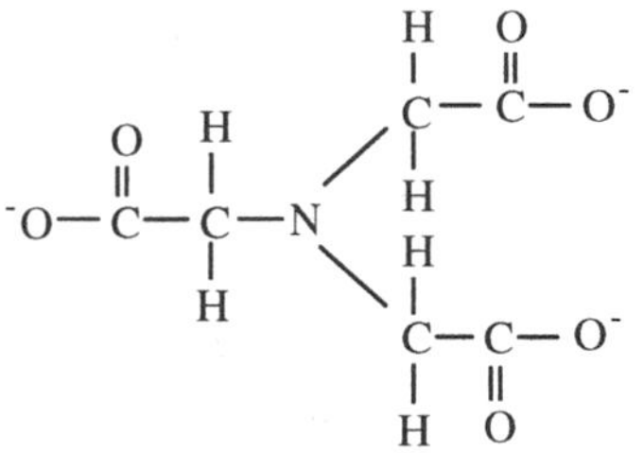

nitrilotriacetate anion

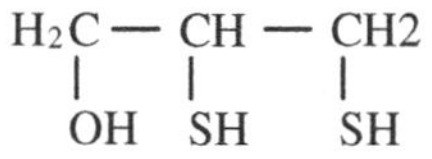

2,3-Mercaptopropanol
(BAL)
British anti lewisite

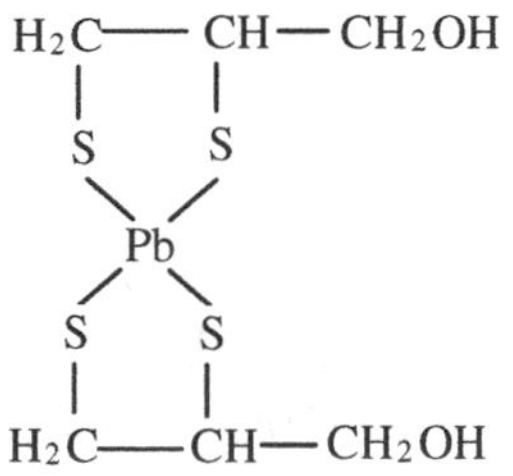

Pb bound to BAL 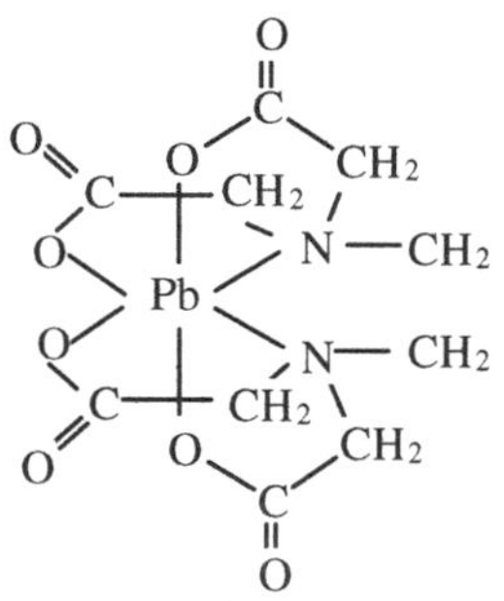   Pb bound to EDTA 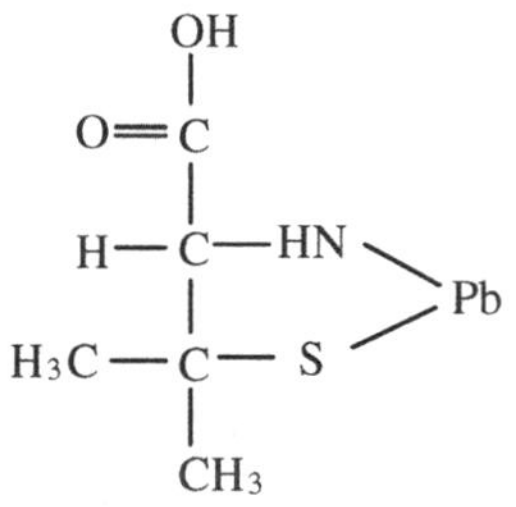   Pb bound to d-penicillamine

Ca^{2+} + NTA →

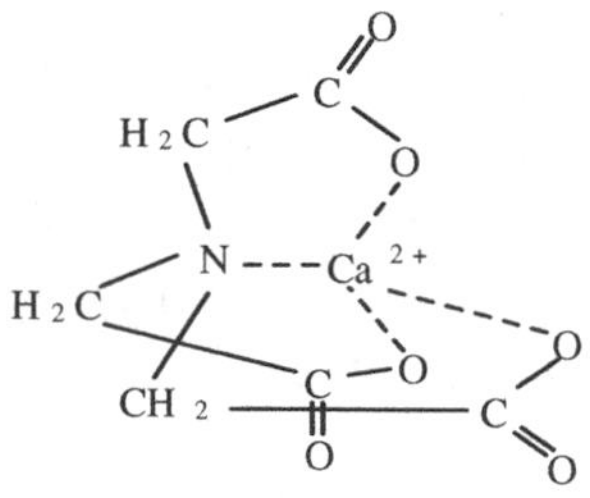

Figure 3-5. Chelating agents.

3.3 STRUCTURE ACTIVITY RELATIONSHIP

In the preceding section, we discussed the influence of molecular structure on physical properties. In addition to physical properties, other characteristics such as stability, reactivity, and toxicity can be described and explained in molecular terms. Much creative effort has been devoted to the development of methods that measure properties in relation to the chemical structure. It is too much effort to go through each method in detail. Instead, following this paragraph are some examples to illustrate structure activity relationships. From the **structure activity relationship** (SAR) studies, highly efficient specific chemical can be produced. This opens the door for the recent molecular modeling investigations for many dynamic properties.

[**Example 3-3**]   Predict the composition of chlorinated compounds that result from the reaction between chlorine and isopentane in vapor phase, given the reactivity ratios of the structures as primary: secondary: tertiary = 1 : 3.3 : 4.4.

For chlorination

$$\equiv CH + X_2 \rightarrow \equiv CX + HX$$

The chlorinated products are formed by free radical mechanism. The radical activity is

$$A_{k_3} : A_{k_2} : A_{k_1} = 4.4 : 3.3 : 1$$

The molecular structure of isopentane shows us that there are nine primary atoms, two secondary atoms and one tertiary hydrogen atom that could be substituted by chlorine (refer to Fig. 3-3, Structure B). The distribution can be written as

$$(4 - i)k_i A_{k_i}$$

in such a way that the % k_i can be calculated as

		%
k_1	$(3)(3)(1) = 9$	45
k_2	$(2)(3.3) = 6.6$	33
k_3	$(1)(4.4) = 4.4$	22

For k_1, which can result in two different isomers, carbon a and e result in one isomer ($k_1' = 2$), carbon d results in another isomer ($k_1'' = 1$) (refer to Fig. 3-3, Structure B). In this manner, there are 30% of the k_1' type and 15% of the k_1'' type. The preceding example illustrates how the difference in reactivity affects the products of a reaction.

Hammett studied the inductive effects of substituents in the phenyl ring on the hydrolysis of ethyl benzoate.

$$Y\phi COOEt \xrightarrow{\ OH^-\ } Y\phi COOH + EtOH$$

He correlated that

$$\log K_\sigma = \log K + \sigma \rho \qquad\qquad [3\text{-}37]$$

where

K_σ = reaction rate for substituted benzoate (with Y)

K = reaction rate for unsubstituted benzoate (without Y)

σ = Hammett constant

ρ = constant of 2.56 (slope)

The equation can be rearranged as

$$\log K_\sigma/K = \sigma \rho \qquad\qquad [3\text{-}38]$$

The results are also shown in Figure 3-6; for example, if it is ρ - OH (donating) $\sigma = -.357$; and if it is $\rho - NO_2$ (withdrawing) $\sigma = +.789$. A high value for the slope (+2.56) indicates that the reaction rate is enhanced by electron-withdrawing substituents.

Of the various parameters available for the study of SAR, those used to express electronic properties of steric factors, lipophilicity, are by far the most important. Quite often, therefore, a biological event will be described with the aid of the following **Hansch equation** to find the **quantitative structure activity relationship (QSAR)**:

$$\log BR = -a(\log P)^2 + b\log P + \sigma + cE_S + d \qquad [3\text{-}39]$$

where

BR = biological response

P = partition coefficient of the organic molecule examined

σ = summation of the Hammett constants of the various substituents

E_s = summation of their steric Taft parameters

and *a*, *b*, *p*, *c*, and *d* are constants. The preceding equation is for the use of DDT, as shown in Figure 3-7.

For another example, Hansch found a simple correlation in the study of narcosis

$$\log 1/C = 0.94 \log P + 0.87 \qquad [3\text{-}40]$$

where *C* is the molar concentration necessary to produce narcosis in case of α-carbolines. Table 3-6 shows the steric effects.

3.3.1 Polycyclic Aromatic Hydrocarbons

The next example shows the thermostability of a **polycyclic aromatic hydrocarbon** (PAH), which often refers to polynuclear aromatics (PNA), the general structure of which is shown in Figure 3-4.

The **number of unexcited forms of configuration**, N, is a qualitative measurement of either the boiling point or the thermal reactivity of these compounds. In general, the larger the N value, the higher the boiling point will be. Lower N values seem to indicate greater thermal reactivity. The number of unexcited configurations can be related to the number of rings in the system [m,n] and the type of bonding — that is, biphenyl or naphthalene — according to the classification scheme shown in Figure 3-4. In this model, m indicates the number of biphenyl-type linkages along the x-axis, and n indicates the number of

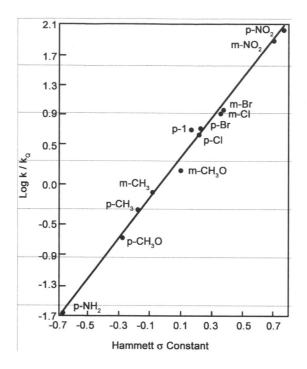

Figure 3-6. Hammett plot for the substituents on benzoate.

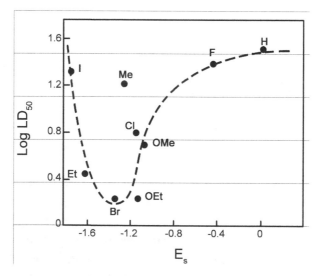

Figure 3-7. Relationship between toxicity and E_s for 1,1,1-trichloro-p-methyl-p'-x-diphenylethanes.

Table 3-6. Structure, Physical Properties, and Potency of Gamma Carbolines

X	Y	σR	Esp	Es8	Obs.	$^{pEd}50_{Eq.\ A}$
6CH$_3$	(CH$_2$)$_3$COC$_6$H$_4$-p-F	-0.14	-0.46	0.00	1.917	1.71
5CH$_3$, 8F	(CH$_2$)$_3$COC$_6$H$_4$-p-F	-0.32	-0.46	-0.46	1.745	1.50
8CN	(CH$_2$)$_3$COC$_6$H$_4$-p-F	1.00	-0.46	-0.55	2.167	2.16
H	(CH$_2$)$_3$COC$_6$H$_4$-p-F	0.00	-0.46	0.00	1.627	1.79
8CF$_3$	(CH$_2$)$_3$COC$_6$H$_4$-p-F	0.61	-0.46	-2.40	1.491	1.48
8OCH$_3$	(CH$_2$)$_3$COC$_6$H$_4$-p-F	-0.43	-0.46	-0.55	1.400	1.42
8Br	(CH$_2$)$_3$COC$_6$H$_4$-p-F	-0.16	-0.46	-1.10	1.377	1.42
8CH$_3$	(CH$_2$)$_3$COC$_6$H$_4$-p-F	-0.14	-0.46	-1.20	1.372	1.40
8F	(CH$_2$)$_3$COC$_6$H$_5$	-0.32	0.00	-0.46	1.441	1.10
8F	H	-0.32	0.00	-0.46	1.124	1.10
8F	CH$_2$C$_6$H$_5$	-0.32	0.00	-0.46	0.926	1.10
8Cl	H	-0.18	0.00	-0.98	1.134	1.04
8F	(CH$_2$)$_4$C$_6$H$_4$-p-F	-0.32	-0.46	-0.46	1.528	1.50
8F	(CH$_2$)$_4$COC$_6$H$_4$-p-F	-0.32	-0.46	-0.46	1.417	1.50
8F	(CH$_2$)$_3$CN	-0.32	0.00	-0.46	0.805	1.10
8F	a	-0.32	-0.55	-0.46	1.426	1.58
8F	(CH$_2$)$_3$CHOHC$_6$H$_4$-p-F	-0.32	-0.46	-0.46	1.449	1.50
8F	(CH$_2$)$_3$OC$_6$H$_4$-p-F	-0.32	-0.46	-0.46	1.589	1.50
8Cl	(CH$_2$)$_3$COC$_6$H$_4$-p-F	-0.18	-0.46	-0.98	0.90[b]	1.43
8F	(CH$_2$)$_3$COC$_6$H$_4$-p-F	-0.32	-0.46	-0.46	2.18[b]	1.50
8F	(CH$_2$)$_3$C$_6$H$_4$-p-NO$_2$	-0.32	-2.52	-0.46	<0.50	3.27
8F	(CH$_2$)$_3$COC$_6$H$_4$-p-C(CH$_3$)$_3$	-0.32	-2.78	-0.46	0.80	3.49
8F	(CH$_2$)$_3$COC$_6$H$_4$-p-NH$_2$	-0.32	-0.61	-0.46	1.09	1.63
8F	(CH$_2$)$_3$COC$_6$H$_4$-p-CH$_3$	-0.32	-1.24	-0.46	<0.50	2.17
8F	(CH$_2$)$_3$COC$_6$H$_4$-p-Cl	-0.32	-0.97	-0.46	<0.50	1.94
8F	(CH$_2$)$_3$COC$_6$H$_4$-p-NO$_2$	-0.32	-2.52	-0.46	<0.50	3.27

a $\log(1/ED_{50}) = 1.39 + 0.52\sigma R - 0.86\ Esp + 0.26\ Es8$; Eq. A $R^2 = 0.77$, s = 0.17, n = 18
b   Not included in calculation

naphthalene-type linkages along the y-axis. It can readily be seen that, for polya-cene, $R[1,n]$, $R = n$; and similarly for the polyphenylene, $R[m,1]$, $R = m$. In this model, the estimation of N can be obtained from

$$N = (n + 1)^m$$

Hence, for the polyacene series, N is equal to $n + 1$, and for the poly-phenylene series, N is equal to 2^m. For an equal number of rings, it can be shown that the polyphenylene has the highest value of N, while the polyacene has the least. Thus, it can be generalized that quinquephenyl is considerably more stable than pentacene, because the N value for the former is equal to 2^5 and for the latter it is only 6. In the polyphenylene family, N increases rapidly as it is equal to 2^R, and it is used as a nuclear coolant. Free rotation of a single bond is allowed in the longitudinal axis of the polyphenylene compound without a change of shape, and this molecular flexibility contributes to its stability. Boiling point of PAH can be related to stability. As shown in Figure 3-8 for the pentacyclic aromatic hydrocarbons, the N values correlate well with the stability.

In conclusion, we must point out the two isomeric 5-ring polycyclic aromatic hydrocarbons (PAH's), **benz(a)pyrene** and **benz(e)pyrene,** as shown in Figure 3-9. The former is a strong carcinogen and the latter is inactive (refer to Fig. 3-9). It is interesting to correlate the SAR, because there are a number of plausible mechanisms that will be found if one examines the K-region (kink) and the L-region (meso).

3.4 METHODS FOR STUDYING STRUCTURE ACTIVITY RELATIONSHIP

Three methods are generally used to study the structure activity relationship: molecular connectivity, factor analysis, and pattern recognition.

One of the topics in organic chemistry is reaction mechanism. We have intentionally omitted that here, as a number of specialized books are available; for example, R. A. Larson and E. J. Weber, Reaction Mechanisms in Environmental Organic Chemistry, Lewis Publishers, Boca Raton, Florida, 1994.

3.4.1 Molecular Connectivity

Molecular connectivity is an approach to the quantitative evaluation of molecular structure. This information is a nonempirical derivation of numerical values that encode within them sufficient information to relate them to many physico-chemical and biological properties. The basic idea assumes that significant properties of a molecule may be represented as bonds connecting the atoms in a molecule. Table 3-7 shows the detailed procedure for the calculation of the **connectivity index**, which will be used to predict the properties.

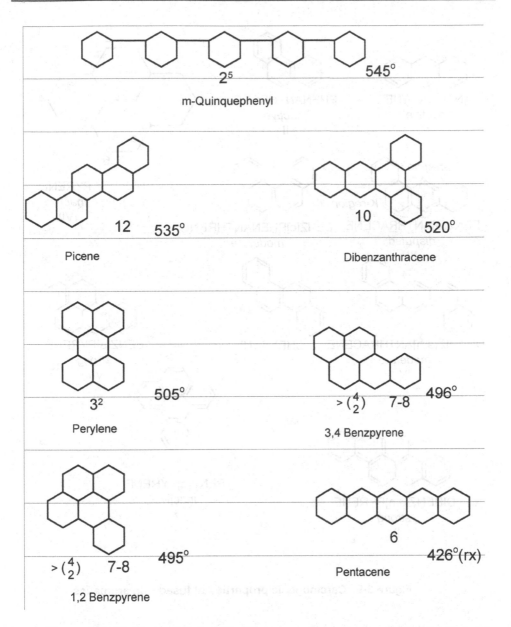

Figure 3-8. Boiling point (in °C) of some pentacyclic aromatic hydrocarbons. The N values are given under each structure.

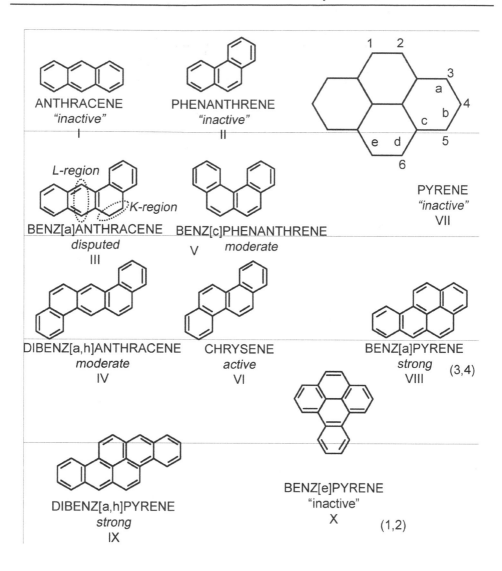

Figure 3-9. Carcinogenic properties of fused ring aromatics.

3.4.2 Factor Analysis

Factor analysis is a method that correlates experimental results with important parameters. The famous Hansch equation is a good example of this method. Figure 3-10 shows the stepwise procedure of factor analysis.

3.4.3 Pattern Recognition

Pattern recognition determines to which class a given pattern of chemical activity belongs. The decision rule is based on some statistical concepts of similarity between cases as determined by their properties. Patterns are merged into an existing class or cluster. The measure of similarity between patterns is their distance in property space as

$$d_{ij} = \left[\sum_{k=1}^{m} \left(x_{ik} - x_{jk} \right)^2 \right]^{\frac{1}{2}} \qquad [3\text{-}41]$$

$$\text{similarity} = s_{ij} = 1 - (d_{ij}/D_{ij})$$

where d_{ij} is the minimum distance between two patterns and D_{ij} is the maximum. Figure 3-11 shows the flow diagram of pattern recognition techniques.

Table 3-7. Procedure for Finding Sum of Edge Terms

Steps	2,2,3-trimethylbutane	2,4-dimethylpentane
Write structural formula		
Draw hydrogen-suppressed graph		
Write valence at each vertex		
Compute product of end point valences for each edge		
Compute each edge term as the reciprocal square root product		
Sum of edge terms:	2.943	3.126

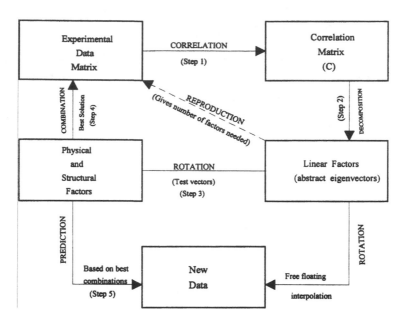

Figure 3-10. Factor analysis.

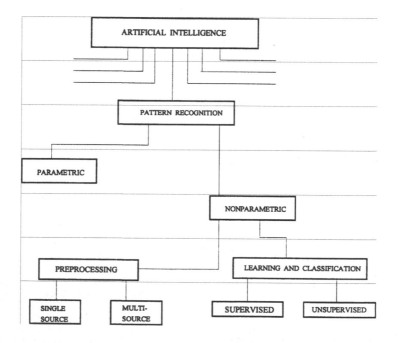

Figure 3-11. Flow diagram of pattern recognition techniques.

REFERENCES

3-1. J. A. K. Buisman, *Biological Activity and Chemical Structure*, Elsevier Science, New York, 1977.

3-2. L. B. Kier and L. H. Hall, *Molecular Connectivity in Chemistry and Drug Research*, Academic Press, New York, 1976.

3-3. D. F. Rossini, *Selected Values of Physical and Thermodynamic Properties of Hydrocarbons and Related Compounds*, Pittsburgh, Pennsylvania, Carnegie Press, pp. 1050, 1953.

3-4. V. M. Tatevskii, V. A. Benderskii, and S. S. Yarovoi, *Rules and Methods for Calculating the Physico-Chemical Properties of Paraffinic Hydrocarbons* (M. F. Mullins, translator), Pergamon Press, Elmsford, New Jersey, 1961.

3-5. T. F. Yen, and G. V. Chilingarian, "Note on Hydrocarbon 1. Saturated Hydrocarbons," *Energy Sources*, 9:71–74 (1987).

3-6. T. F. Yen, J. F. Kuo, and G. V. Chilingarian, "Note on Hydrocarbon 2. Naphthenic Hydrocarbons," *Energy Sources*, 9:125–132 (1987).

3-7. T. F. Yen, "Resonance topology of Polynuclear Aromatic Hydrocarbons," *Theoret. chim. Acta,* Berlin, *20*, 399–404 (1971).

3-8. T. F. Yen, "Terrestrial and Extraterrestrial Stable Organic Molecules," in *Chemistry in Space Research* (R. F. Landel and A. Rembaum, ed.), Am. Elsevier, New York, 1972, pp. 105–153.

3-9. R. A. Y. Jones, *Physical and Mechanistic Organic Chemistry*, 2nd ed., Cambridge University Press, London, 1984.

3-10. C. D. Johnson, *The Hammett Equation*, Cambridge University Press, London, 1973.

3-11. S. R. Hartshorn, *Aliphatic Nucleophilic Substitution*, Cambridge University Press, London, 1973.

3-12. W. H. Saunders and A.F. Cockerill, *Mechanisms of Elimination Reactions*, Wiley-Interscience, New York, 1973.

3-13. J. Miller, *Aromatic Nucleophilic Substitution,* Elsevier, Amsterdam, 1968.

3-14. R. A. Rossi and R.H. deRossi, *Aromatic Substitution by the $S_{RN}I$ Mechanism*, American Chemical Society, Washington, 1983.

3-15. H. J. Shine, *Aromatic Rearrangements*, American Elsevier, New York, 1967.

3-16. T. M. Lowry and K.S. Richardson, *Mechanism and Theory in Organic Chemistry*, 2nd ed., Harper and Row, New York, 1981.

3-17. R. A. Jackson, *Mechanism: An Introduction to the Study of Organic Reactions*, Clarendon, Oxford, 1972.

3-18. C. K. Ingold, *Structure and Mechanism in Organic Chemistry*, 2nd ed., Bell, London, 1964.

3-19. S. Ege, *Organic Chemistry, Structure and Reactivity*, 3rd ed., D.C. Heath, Lexington, Massachusetts, 1994.

3-20. G. Klopman, *Chemical Reactivity and Reaction Path*, Wiley-Interscience, New York, 1974.

3-21. N. Bondard, *Pattern Recognition*, Spartan-Macmillan, New York, 1970.

3-22. T. F. Yen, "A Scheme for Memorizing Thermodynamic Functions", *J. Chemic. Edu.* 31, 610 (1954).

3-23. K. G. Joreskog, J. E. Klovan and R. A. Regment, *Geological Factor Analysis*, Elsevier, Amsterdam, 1976.

PROBLEM SET

1. How much O_2 in mg/L is required to oxidize 100mg/L of benzene?

2. A chlorine-containing benzene derivative has the following elemental analysis C 66.4%, Cl 28%, H 5.53%. What is the structure?

3. In the following hydrocarbons identify how many primary, secondary, tertiary and quaternary structures are present.

$$
\begin{array}{ccc}
 & C\ C & \\
 & |\ \ | & \\
C\text{-}C\text{-}C\text{-}&C&\text{-}C \\
 & | & \\
 & C & \\
\end{array}
$$

Write out in bond matrix.

4. Using Tatevski's method, calculate the density of all heptane isomers.

5. For fused ring aromatic, prove $H_A = 3C_A^{1/2} - 3/2$

CHAPTER 4

ANALYTICAL CHEMISTRY

*Q*uantitative measurements of one sort or another serve as the keystone of engineering practices. Most engineering disciplines are demanding in this respect, for they require the use of not only the conventional measuring methods already employed by engineers, but also many of the techniques and methods of measurements used by chemists, physicists, and biologists.

Measurements play a key role in the sustainability of natural resources and the protection of the environment. They are needed to identify problems and to monitor the effectiveness of control and abatement technology. Decisions on vital questions such as the habitability of an area, the safety of drinking water, and the continued operations of an industrial plant are often based on measurement data. Equipment and technology for quantitative measurement are increasingly better developed these days, enabling us to identify new problems that may have existed for a long time. Consequently, stringent regulations can be set, and a cleaner and safer environment can be assured with suitable enforcement. In this chapter, acid-base chemistry, instrumental analysis, separation science, and the quality assurance of global measurements will be addressed.

4.1 ACID-BASE CHEMISTRY

A thorough examination of acid-base chemistry is important when studying aqueous chemistry in order to understand water and wastewater treatment processes. In this section, we will discuss some important acid-base chemistry concepts, such as pH, acidity, and alkalinity, which relate to engineering practices.

According to **Bronsted and Lowry's definition**, an **acid** is any substance that can donate a proton to any other substance. A **base**, then, is any substance that can accept a proton from an acid.

$$H^+ + OH^- = H_2O$$

An even broader theory of acids and bases has been proposed by **G.N. Lewis**. According to him, a base is any substance that donates a pair of electrons to the formation of a coordinate bond. In turn, an acid is any substance that accepts a pair of electrons to form such a bond.

$$R_2O : + BF_3 = R_3O : BF_3$$

Usanovich enlarged this concept to include the coordination, and examples are illustrated here:

$$B^- \quad + \quad HA \quad = \quad HB \quad + \quad A^-$$
$$(B1) \qquad\quad (A1) \qquad\quad (A2) \qquad (B2)$$

$$HCl \quad + \quad C_5H_5N \quad = \quad Cl^- \quad + \quad C_5H_5NH^+$$
$$(A1) \qquad\quad (B1) \qquad\quad (B2) \qquad\quad (A2)$$

$$HCl \quad + \quad H_2O \quad = \quad H_3O^+ \quad + \quad Cl^-$$
$$(A1) \qquad\quad (B1) \qquad\quad (A2) \qquad\quad (B2)$$

$$CO_3^{2-} \quad + \quad H_2O \quad = \quad OH^- \quad + \quad HCO_3^-$$
$$(B1) \qquad\quad (A1) \qquad\quad (B2) \qquad\quad (A2)$$

According to Usanovich, neutralization will form a secondary acid and base species. Water is described as amphoteric because it can be both an acid and a base; thus, water is an **ampholyte**.

The intensity of the acid or alkaline condition of a solution is universally expressed by the term **pH**. It is a way of expressing the hydrogen-ion activity, and it is important in almost every phase of engineering practice.

$$pH = -\log (H^+) = \log [1/(H^+)]$$ [4-1]

For water in the absence of other substances, $(H^+) = (OH^-)$ as required by electroneutrality, and at 25°C

$$H_2O = H^+ + OH^-$$

$$(H^+) (OH^-) = K_W = 10^{-14}$$ [4-2]

or

$$pH + pOH = 14 = pK_w$$

It is easily seen that, under neutral conditions,

$$pH = pOH = 7$$ [4-3]

A value of pH lower than 7 indicates that $(H^+) > (OH^-)$ and the water is acidic; when pH is greater than 7, the water is basic. If a strong acid such as nitric acid is added to the water with known concentration, C, then the pH of the water can be easily determined due to the neutrality requirement; for example,

$$(H^+) = (NO_3^-) + (OH^-)$$

or

$$(H^+) = C + K_W/(H^+)$$ [4-4]

and the equation can be expressed as

$$(H^+)^2 - C(H^+) - K_W = 0$$

and

$$(H^+) = 1/2 [C \pm (C^2 + 4 K_W)^{1/2}]$$ [4-5]

For an HNO_3 concentration of 10^{-7} M, the pH is calculated to be 6.79.

For weak acids such as acetic acid, which are not completely ionized like a strong acid, incomplete ionization will take place.

$$\frac{(H^+)(A^-)}{(HA)} = K_a = 1.8 \times 10^{-5} \qquad [4\text{-}6]$$

Again neutrality

$$(H^+) = (OH^-) + (A^-) \qquad [4\text{-}7]$$

and mass balance for a known concentration take place.

$$(HA) + (A^-) = C \qquad [4\text{-}8]$$

Using four equations (including K_w) (Eqs. [4-6], [4-7], [4-8], and [4-2]), we can solve four different species in the solution. By substituting the preceding equations, we get

$$(H^+)^3 + K_a(H^+)^2 - (K_aC + K_w)(H^+) - K_aK_w = 0 \qquad [4\text{-}9]$$

Given C is 10^{-2} M, the following are obtained

$$(H^+) = 4.15 \times 10^{-4}$$
$$(OH^-) = 2.42 \times 10^{-11}$$
$$(HA) = 9.59 \times 10^{-3}$$
$$(A^-) = 4.15 \times 10^{-4}$$

In many instances, an approximation can be made instead of the exact solution of the equation. In water supply, this is a factor that must be considered, whether in chemical coagulation, disinfection, water softening, or corrosion control. In wastewater treatment employing biological processes, pH must be controlled within a range favorable to the particular organisms involved. Chemical processes used to coagulate wastewater, dewater sludge, or oxidize certain substances such as cyanide ions, require that the pH be controlled within rather narrow limits.

Although pure water should have a pH of 7, the water in the atmosphere is not neutral. Most natural waters, domestic sewage, and many industrial wastes are buffered principally by a carbon dioxide-bicarbonate system. Carbon dioxide is a normal component of all natural waters. It may also enter surface waters by absorption from the atmosphere. Carbon dioxide may also be produced in water, particularly in polluted water, through the biological oxidation of organic matter.

The acidity of rain can be significantly increased by industrial pollutants, and acid rain is an environmental hazard. Acid waters are of concern because of their corrosive characteristics and the expense involved in removing or controlling these corrosion-producing substances. The corrosive factor in most waters is carbon dioxide, but in many industrial wastes it is mineral acidity. Carbon dioxide must be reckoned within water-softening problems where the lime or lime-soda ash method is used.

The **alkalinity** of water is the measurement of its capacity to neutralize acids. Bicarbonates represent the major form of alkalinity, because they are formed in considerable amounts from the action of carbon dioxide upon basic materials in the soil. A few organic acids are quite resistant to biological oxidation; for example, humic acid-formed salts that add to the alkalinity of natural water. In polluted or anaerobic waters, salts of weak acids such as acetic, propionic and hydrosulfuric acid may be produced and may also contribute to alkalinity. In other cases, ammonia or hydroxides may contribute to the total alkalinity of water.

In practice, the alkalinity can be conveniently expressed in terms of $CaCO_3$. In all cases, the multiplier is either 50 mg $CaCO_3$/meq, or 100 mg $CaCO_3$/mM. The following example can be illustrated. In general, where A_i^- is an anion and A_i^+ is a cation

$$T = \sum_i A_i^- - \sum_i A_i^+ \qquad [4\text{-}10]$$

[**Example 4-1**] A pH=10 natural water contains 100 mg/L carbonate and 75 mg/L bicarbonate. Compute the alkalinity as expressed in $CaCO_3$.

For CO_3^{2-}

$$(100/60) \times 100 = 167 \text{ mg/L as } CaCO_3$$

and for HCO_3^-

$$(75/61) \times 100/2 = 61 \text{ mg/L as } CaCO_3$$

At pH=10,

$$(H^+) = 10^{-10} \text{ M}$$

Thus expressed in mg/L

$$= 10^{-10}\,(10^3)(1) = 10^{-7}\,\text{mg/L}$$

or

$$10^{-7} \times 100/2 \times 1 = 5 \times 10^{-6}\,\text{mg/L as CaCO}_3$$

$$(OH^-) = K_w/H^+ = 10^{-14}/10^{-10} = 10^{-4}$$

Thus

$$(OH^-) = 10^{-4} 10^3 (17) \times 100/(17 \times 2)$$
$$= 5\,\text{mg/L as CaCO}_3$$

Thus

$$\text{alkalinity} = T = \Sigma\,A_i^- - \Sigma\,A_i^+$$
$$= 61 + 167 + 5 - (5 \times 10^{-6})$$
$$= 233\,\text{mg/L as CaCO}_3$$

4.1.1 Titration Curve

If a strong acid or base is titrated with a strong base or acid, a typical S-shape or inverse S-shape results as shown in Figure 4-1(a) and Figure 4-1(b). When a weak acid is titrated with a strong base, the character of the titration curve depends on whether the acid is monobasic (monoprotic) or polybasic (polyprotic); that is, whether the acid will yield one or more hydrogen ions (for example, the curves shown on Figure 4-1(c) and Figure 4-1(d) with its middle portion flattened out). Considering a weak monobasic acid being titrated with a strong base

$$HA = H^+ + A^-$$

$$(HA) = [C - (A^-)] \qquad\qquad\qquad [4\text{-}11]$$

$$K_a = \frac{(H^+)(A^-)}{(HA)} \qquad\qquad\qquad [4\text{-}12]$$

under neutrality condition

$$(H^+) + (B^+) = (A^-) + (OH^-) \qquad\qquad\qquad [4\text{-}13]$$

In the beginning stage of titration, $(B^+) = 0$, pH is low, and $(OH^-) \ll (H^+)$. Thus,

$$(H^+) \approx (A^-)$$

and after substituting into the K_a equation Eq. [4-12],

$$(H^+) \approx [K_a (HA)]^{1/2} = \{K_a[C-(A^-)]\}^{1/2} \qquad\qquad [4\text{-}14]$$

In many cases, at the beginning of titration $(A^-) \ll C$, so

$$(H^+) \approx (K_a C)^{1/2}$$

$$pH \approx 1/2 (pK_a - \log C) \qquad\qquad\qquad [4\text{-}15]$$

At the midpoint of titration (A^-) increases and (HA) decreases; when the neutralization is 50% completed, $(B^+) = (1/2)C$, both (B^+) and (A^-) become significant.

Since $(H^+) \ll (B^+)$ and $(OH^-) \ll (A^-)$,

$$(B^+) \approx (A^-) \approx (1/2)C$$

and

$$(HA) + (A^-) = C$$

Hence

$$(HA) \approx (1/2)C$$

Substituting into the K_a equation (Eq. [4-12]), then

$$(H^+) \approx K_a$$

or

$$pH \approx pK_a \qquad\qquad\qquad\qquad [4\text{-}16]$$

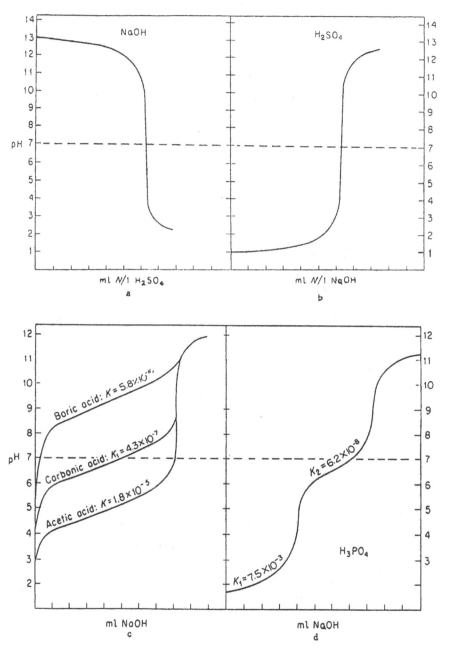

Figure 4-1. Titration curves for (a) strong base, (b) strong acid, (c) weak acids, and (d) weak bases.

At the stoichiometric endpoint of titration (also called the equivalence point), the equivalents of the bases added equal those of the acids, and as $(B^+)=C$, pH becomes high in such a way that $(H^+)<<(OH^-)$; the neutrality equation becomes

$$C \approx (A^-) + (OH^-)$$

Since

$$C-(A^-) = (HA)$$

then

$$(HA) \approx (OH^-) = K_w/(H^+)$$

Substituting into the K_a equation (Eq. [4-12]),

$$1/(H^+) = [(A^-)/K_wK_a]^{1/2}$$

At the end of titration

$$(A^-) \approx C$$

So

$$pH \approx 1/2 \ (\log C + pK_a + pK_w) \qquad [4\text{-}17]$$

For a weak base being titrated with a strong acid, such as

$$Ac^- + H_2O = HAc + OH^-$$

then

$$K_b = \frac{(OH^-)(HAc)}{(Ac^-)} \qquad [4\text{-}18]$$

In this instance

$$pK_a = pK_w - pK_b$$

Again, as before, in the beginning (H^+) is small, (OH^-) is important, $(OH^-) \approx (HAc)$, and $Ac^- \approx C$, so $(OH^-) \approx K_b^{1/2}C^{1/2}$ and

$$\frac{1}{H^+} = \frac{K_b^{\frac{1}{2}} C^{\frac{1}{2}}}{K_w}$$

Hence

Initial $pH \approx pK_w - 1/2\ pK_b + 1/2\ \log C$

Midpoint $pH \approx pK_w - pK_b$ [4-19]

Endpoint $pH \approx 1/2\ (pK_w - pK_b - \log C)$

The following is a summary:

pH	Weak Acid Versus Strong Base	Weak Base Versus Strong Acid
Initial	$1/2\ (pK_a + pC)$	$pK_w - 1/2\ (pK_b + pC)$
Midpoint	pK_a	$pK_w - pK_b$
Equivalence	$1/2\ pK_w + 1/2\ (pK_a - pC)$	$1/2\ pK_w - 1/2\ (pK_b - pC)$

4.1.2   Buffers and Buffer Index

The word **buffer** may be defined as substances introduced in a solution that offer resistance to change in pH as acids or bases are added to that solution. Buffer solutions usually contain a weak acid and its salt (conjugate base) or a weak base and its salt (conjugate acid). At some point, the smallest changes in pH occur, and consequently at that point the buffering capacity is the greatest.

Weak acids and bases and their salts are used as buffers at a pH near the pK value; that is, within ± 1 pH unit of the pK value. The effectiveness can be demonstrated as follows; for example, a weak acid

$$pH = pK_a + \log \frac{(salt)}{(acid)}$$

In a solution of 0.1 M of sodium acetate to 0.1 M of acetic acid

$$pH = 4.75 + \log \frac{0.1}{0.1} = 4.75$$

Now if a small amount of HCl is added, equivalent to 10% of the acetate present (i.e., 0.01 M), and the new salt concentration is 0.11 M, then

$$pH = 4.75 + \log\frac{0.09}{0.11} = 4.75 - 0.087 = 4.66$$

The pH has only decreased by 0.09 units, whereas if unbuffered, the amount of HCl added to the water (0.01 M) would change 5 units from pH=7 to pH=2. The reagent commonly used as a buffer near pH=7 is phosphoric acid, whose ionization constant is near pH=7. The salts, KH_2PO_4, and K_2HPO_4 are widely utilized for biochemical applications in nature.

The **buffering capacity** of a solution can be indicated quantitatively by the **buffering index**, β, which is defined as the slope of the titration curve pH versus the concentration of strong base added, C_B. On the other hand, the concentration of strong acid, C_A, can also be used.

$$\beta = \frac{dC_B}{dpH} = -\frac{dC_A}{dpH} \qquad\qquad [4\text{-}20]$$

If the concentration of the solution is given, this can be calculated. For a monoprotic acid

$$C_B + (H^+) = (OH^-) + (A^-)$$

$$(HA) + (A^-) = C$$

Substituting this into the K_a equation (Eq. [4-6]), then

$$(HA) = \frac{C(H^+)}{K_a + (H^+)}$$

and

$$(A^-) = \frac{CK_a}{K_a + (H^+)}$$

Finally, the charge-balance equation becomes

$$C_B = \frac{K_w}{(H^+)} - (H^+) + \frac{CK_a}{K_a + (H^+)}$$

Hence

$$\frac{dC_B}{dpH} = \frac{dC_B}{d(H^+)} \frac{d(H^+)}{dpH}$$

Because pH = $-\ln (H^+)/2.303$

$$\frac{d(H^+)}{dpH} = -2.303(H^+)$$

$$\beta = \frac{dC_B}{dpH} = -2.303(H^+) \left[\frac{dC_B}{d(H^+)} \right]$$

$$= 2.303 \left[\frac{K_w}{(H^+)} + (H^+) + \frac{CK_a(H^+)}{K_a + (H^+)^2} \right] \qquad [4\text{-}21]$$

This will numerically calculate the concentration of base needed to increase per pH unit, and usually β (increment of pH) will provide one with information of the concentration required to raise the pH value. Also β of a polyprotic acid or base is known.

4.2 INSTRUMENTAL ANALYSIS

A variety of sensitive instruments have been developed in recent years; these instruments have considerably increased the engineer's ability to measure and characterize pollutants with increasing complexity. An **instrumental analysis** is also called a physical method of analysis, since it is based on the differences between the physical properties of elements or **compounds**. Figure 4-2 is valuable for depicting the relationships between **electron voltage** E, **wavelength** λ, **frequency** v of the **radiant energy**, and also the operational ranges in which major analytical instruments function. At this time, we will describe some common instruments used in the field of engineering. Based on the electromagnetic series of energy,

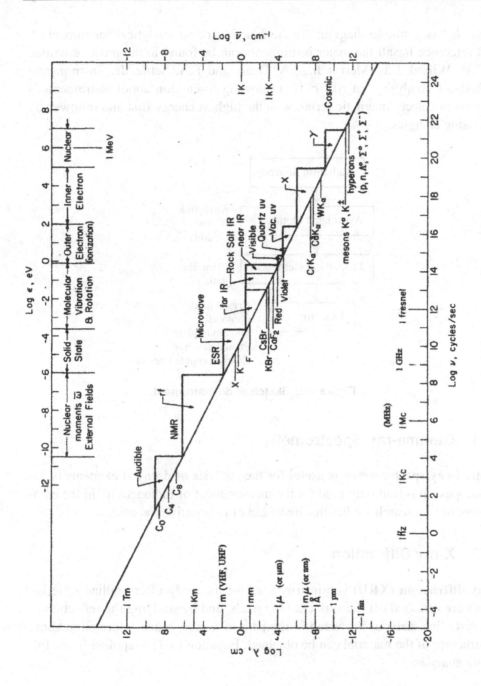

Figure 4-2. Generalized electromagnetic series.

Figure 4-3 is a simple diagram for the function of an analytical instrument. A good reference for all the major instruments can be found in the practice written by H.H. Wilard, L.L. Merritt Jr., J.A. Dean, and F.A. Settle, Jr., Instrumental Methods of Analysis, 6th ed., 1981. Following discussion about instruments is based on the electromagnetic series, with the highest energy first and followed by decreasing energies.

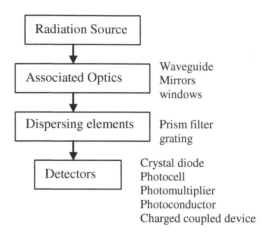

Figure 4-3. Sketch of an instrument.

4.2.1 Gamma-ray Spectrometry

Gamma-ray spectrometry is useful for the analysis of decayed elements or radioisotopes. It is routinely used for the measurement of radioactivity in the environment or in research studies that make use of radioactive tracers.

4.2.2 X-ray Diffraction

X-ray diffraction (XRD) is primarily used for the study of crystalline material. X-rays are reflected off the surfaces of crystals, and by studying the reflection as the crystalline material is rotated in the path of x-rays, much information about the structure of the material can be obtained. Equation [4-1] is applied in the following analysis:

$$n\,\lambda = 2d \sin \theta$$

[4-22]

where d is the range of interest that determines spacing between two centers and depends on the target material used. For very small angles, both small **angle x-ray diffraction** (SAXD) and **small angle neutron scattering** (SANS) are used to measure d-distance to 50–200 Å range. Another x-ray analysis commonly used is x-ray photoelectron spectroscopy (XPS), also called **Electron Spectroscopy for Chemical Analysis** (ESCA). To investigate the bonding pattern, an x-ray is applied to excite photons on a surface.

4.2.3 Mass Spectrometry

Mass spectrometry is an instrument that sorts out charged gas molecules or ions by bombardment with rapidly moving electrons. The ions formed are pulled from the gas stream by an electric field. A suitable detector can then record the particles of different mass either qualitatively, quantitatively, or both. Mass spectrometry when used with **gas chromatography (GC/MS)** can give positive identification and quantification for a large number of samples that are identified individually from the differences in their retention time when passed through a chromatographic column. Figures 4-4 to 4-6 show some typical mass spectra, and Figures 4-7 and 4-8 show typical gas chromatograms.

4.2.4 Ultraviolet/Visible Spectrophotometry

When a molecule absorbs radiant energy in either the **ultraviolet** (UV) or the **visible (vis) region**, **valence** or **bonding electrons** in the molecule are raised to high-energy orbits. The result is that fairly broad absorption bands are usually observed in both the ultraviolet and the visible region. The measurement of **UV absorption spectra** is primarily used to detect the presence of conjugated hydrocarbons. It is particularly suitable to have the selective measurement of low concentrations of organic compounds, such as benzene-ring-containing compounds or unsaturated straight-chain compounds containing a series of double bonds. **Beer's Law** is often used.

$$A = \log(I_o/I) = Ecb \qquad\qquad [4\text{-}23]$$

where

   A = absorbance = optical density
   E = molar absorptivity or extinction coefficient
   c = concentration in moles/L
   b = path length of sample cell in cm.
   I = resultant radiation
   I_o = incident radiation

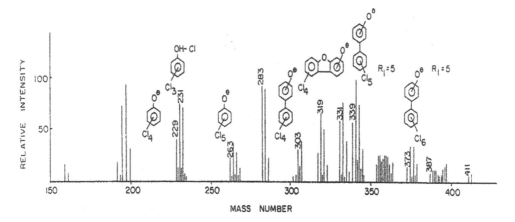

Figure 4-4. Negative chemical ionization mass spectra of a Baltic Sea seal extract.

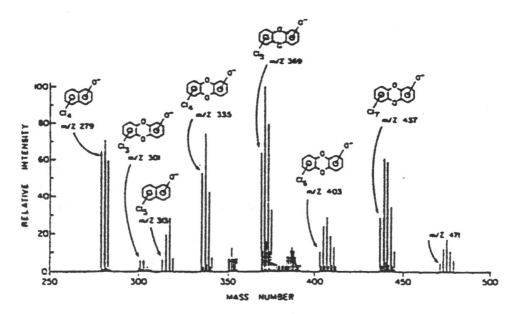

Figure 4-5. Negative chemical ionization mass spectra of an extract from Tittabawassee River bass caught near Dow Dam, ML.

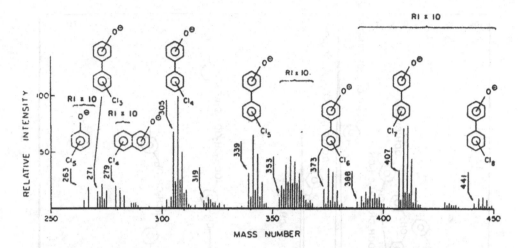

Figure 4-6. Negative chemical ionization mass spectra of a turtle caught in the Hudson River, NY.

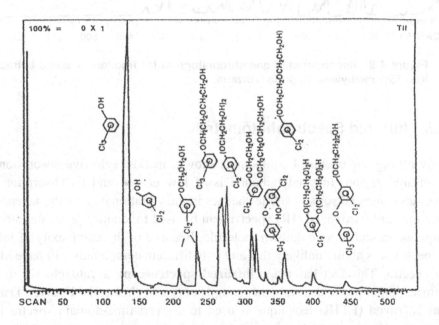

Figure 4-7. Reconstructed gas chromatogram for base-soluble fraction after extraction with 15% MeCl₂/PhH.

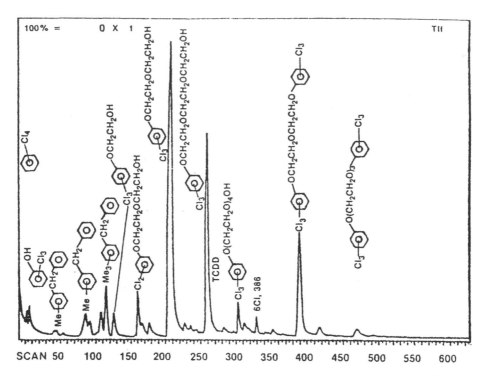

Figure 4-8. Reconstructed gas chromatogram for fraction of waste extracted into 15% methylene chloride/benzene.

4.2.5 Infrared Spectrophotometry

Nearly all organic chemical compounds show a marked selective absorption in the infrared region. Infrared radiation is of low energy and its absorption by molecules causes all sorts of subtle changes in the vibrational or rotational energy of the molecule. **Infrared (IR) spectra** can be used to identify particular atomic groupings present in an unknown molecule. Because of the complexity of infrared spectra, it is highly unlikely that any two different compounds will have identical spectra. This fact has made **infrared spectroscopy** a valuable aid in the identification of pesticides and other complex organic chemicals. **Fourier Transform Infrared (FTIR)** technique is used to convert time-domain spectra into frequency domain spectra with the aid of a computer to enhance the signal-to-noise ratio by performing signal averaging. Far IR can be used to determine the bonding of metal to carbon, whereas near IR can detect a large group oscillation.

4.2.6 Nuclear Magnetic Resonance

Nuclear Magnetic Resonance (NMR) is used to detect and distinguish between the nuclear particles present in a sample. It measures the changes in the nuclei of materials when placed in a fixed magnetic field and then subjected to an alternating magnetic field. A nucleus can be considered a spinning charged particle that has an associated magnetic moment. In the presence of a magnetic field, the magnetic moments can align themselves either with or against the field. The higher energy state, against the field, is somewhat less populated than the lower energy state, with the field, and the nuclei can be promoted from the lower to the higher state by the application of radio frequency energy. It is the absorption of this energy that we observed in the NMR experiment. Table 4-1 lists properties of common nuclei.

Table 4-1. Properties of Nuclei

Particle	Natural Abundance %	Relative Sensitivity for Equal Number of Nuclei (Constant Field)	Magnetic Moment (Nuclear Magnetons)[*]	Spin (Units of 2τ)
electron			1840	$\frac{1}{2}$
^{1}H	99.98	1.000	2.793	$\frac{1}{2}$
^{2}H	0.016	9.6×10^{-3}	0.857	1
^{12}C	99	0	0	0
^{13}C	1.11	1.59×10^{-2}	0.702	$\frac{1}{2}$
^{14}N	99.6	1.01×10^{-3}	0.404	1
^{16}O	99.996	0	0	0
^{17}O	3.7×10^{-2}	2.91×10^{-2}	-1.893	$\frac{5}{2}$
^{19}F	100	0.834	2.627	$\frac{1}{2}$
^{32}S	99.3	0	0	0
^{31}P	100	6.6×10^{-2}	1.130	$\frac{1}{2}$
^{35}Cl	75.4	4.7×10^{-3}	0.821	$\frac{3}{2}$
^{37}Cl	24.6	2.7×10^{-3}	0.683	$\frac{3}{2}$

[*] Nuclear Magneton — 5.049×10^{-22} erg/gauss

Many common nuclei have no magnetic moments and thus will not exhibit any magnetic resonance absorption. These include ^{12}C, ^{16}O, and ^{32}S because they have a nuclear spin of zero. But the spins of their isotopes are not zero and can be detected by NMR. e.g., ^{13}C **NMR** has been used routinely. Figure 4-9 shows typical NMR spectra. Similar to NMR, the **electron spin resonance (ESR)** in the microwave frequency range is used to detect free radicals or certain transition metals.

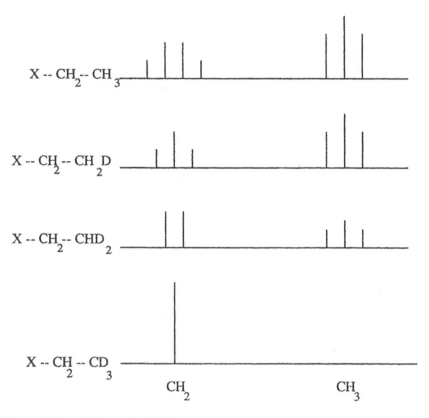

Figure 4-9. **Schematic NMR spectra of deuterated ethyl derivatives. The righthand set of lines is always a triplet when observable because of the two protons of the X-CH$_2$- group.**

Finally, it should be noted that for analysis of complex mixtures, usually more than one method is involved. A good example of the use of **multiple instruments** in identification of a complex hydrocarbon mixture in soil is illustrated by Figure 4-10. By a number of analytical approaches from puzzle-fitting, a final picture of the structure can be realized as shown in Figure 4-11.

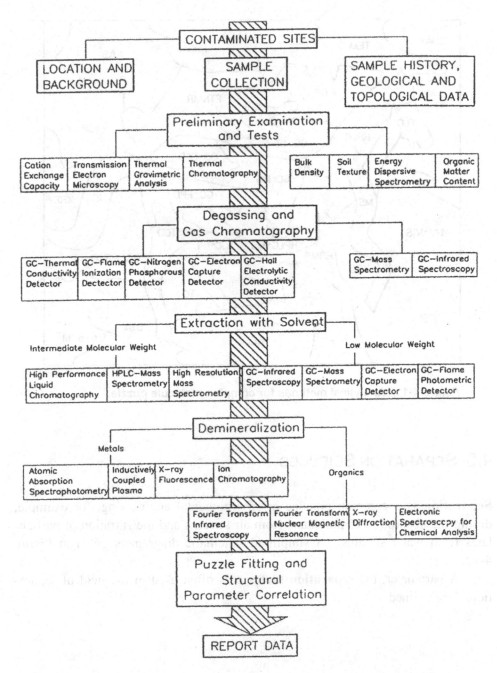

Figure 4-10. Analytical scheme for complex hydrocarbons mixtures in soil (Yen, 1988).

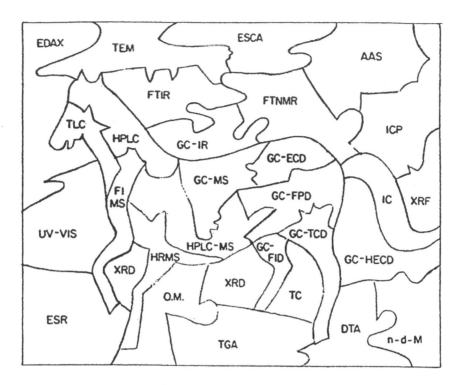

Figure 4-11. Analytical methods for complex molecule puzzle.

4.3 SEPARATION SCIENCES

Separation is a common practice in environmental engineering; for example, during the absorption of pollutants from air streams and the filtration of particulates from water streams. A simplified schematic diagram is given in Figure 4-12.

A parameter, the **separation factor,** α is often used in the field of separation. It is defined as

$$\alpha_{ij} = \frac{\dfrac{x_{i1}}{x_{j1}}}{\dfrac{x_{i2}}{x_{j2}}} \qquad\qquad [4\text{-}24]$$

where 1 and 2 denote the phases or products, while i and j denote the components. Therefore, X_{i1} represents the mole fraction of component i in phase 1 or product 1. If the separation factor $\alpha_{ij} = 1$, no separation will occur. If it is greater than unity, component i tends to be concentrated in product 1. On the other hand, if the separation factor is less than unity, component j tends to be concentrated in product 1.

For separation processes, as shown in Table 4-2, based upon the equilibration of immiscible phases, it is helpful to define the quantity

$$K_i = \left(\frac{X_{i1}}{X_{i2}} \right) \text{ at equilibrium}$$

K_i is called the equilibrium ratio for component i. The separation factor can then be written as

$$\alpha_{ij} = \frac{K_i}{K_j} \qquad [4\text{-}25]$$

For a vapor-liquid system, if the components of the mixture obey Raoult's and Dalton's Laws,

$$P_i = Py_i = P_i^o x_i \qquad [4\text{-}26]$$

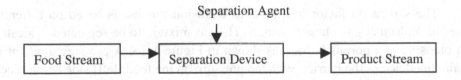

Figure 4-12. Schematics of a separation device.

In such a case

$$K_i = \frac{y_i}{x_i} = \frac{P_i^0}{P}$$

$$K_j = \frac{y_j}{x_j} \quad ,(1=y, 2=x) \qquad [4\text{-}27]$$

Table 4-2. Process Unit Operations and Principles

Type	Unit Operation	Principle
Equilibrium process	Distillation	Volatility
	Adsorption	Solubility
	Extraction	Solubility
	Crystallization	Crystal structure
	Gel filtration	Molecular size
Rate process	Gas diffusion	Diffusion rate
	Thermal diffusion	Diffusion rate
	Dialysis	Membrane transport
	Electrophoresis	Mobility
Mechanical process	Filtration	Size
	Settling	Density
	Centrifugation	Density
	Electrostatic	Charges
	Precipitation	Charges

So

$$\alpha_{ij} = \frac{K_i}{K_j} = \frac{P_i^0}{P_j^0}$$

[4-28]

The separation factor in a gaseous diffusion process is based on different rates of molecular gas-phase transport. The gas mixture to be separated is located on one side of a porous barrier, as shown in Figure 4-13. A pressure gradient is maintained across the barrier, with the pressure on the feed (left) side being much greater than that on the product (right) side. This pressure gradient causes a flux of molecules from the gaseous mixture to be separated across the barrier from left to right. If the barrier has sufficiently small pores and a sufficiently low gas pressure, the mean free path of the gas molecules will be large compared to the pore dimensions. As a result, the molecular flux will occur in what is called the **Knudsen flow**, as described by Equation [4-29].

$$N_i = \frac{a(P_1 y_{1i} - P_2 y_{2i})}{\sqrt{M_i T}}$$

[4-29]

where

N_i = flux of component i across the barrier
P_1 = pressure of the high-side
P_2 = pressure of the low-side
y_{1i} = mole fraction of component i on the high-pressure side
y_{2i} = mole fraction of component i on the low-pressure side
T = temperature
M_i = the molecular weight of component i
a = geometric factor depending only upon the structure of the barrier

In a continuum of particles diffusion, a dimensionless parameter known as **Knudsen number** is defined as

$$Kn = \frac{2\lambda}{d} \qquad\qquad [4\text{-}30a]$$

where λ is the mean free path of the particles in a medium and d is the aerosol particle diameter. If Kn << 1, a continum exists; if Kn >> 1, continuum does not exist. The mean free path is usually measured as

$$\lambda = \frac{\mu}{0.449 P \left(\dfrac{8M_a}{\pi RT}\right)^{\frac{1}{2}}} \qquad (in\ \mu m) \qquad [4\text{-}30b]$$

where

μ = viscosity of the medium (Pa-s)
P = pressure (Pa)
M_a = molar mass of medium (kg/ mol)

We shall now presume that the composition of the high-pressure side does not change appreciably through depletion of one of the gas species. The material on the low-pressure side has all arrived through the steady-state transport process so that

$$N_i = \frac{P_1 y_{1i}}{\sqrt{M_i T}}$$

$$N_j = \frac{P_1 y_{1i}}{\sqrt{M_j T}} \qquad\qquad [4\text{-}30c]$$

Porous barrier

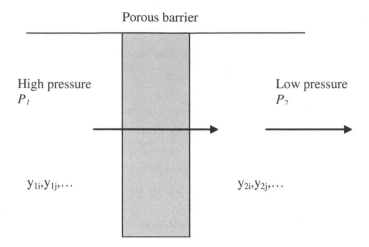

Figure 4-13. Simplified gaseous diffusion process.

$$\frac{N_i}{N_j} = \frac{y_{1i} M_j^{1/2}}{y_{1j} M_i^{1/2}} = \frac{y_{2i}}{y_{2j}}$$

We shall also presume for simplicity that $P_2 \ll P_1$. Combining Equations [4-24] and [4-30], we have

$$\alpha_{ij} = \frac{y_{2i} y_{1j}}{y_{2j} y_{1i}} = \frac{M_j^{\frac{1}{2}}}{M_i^{\frac{1}{2}}} \qquad [4\text{-}31]$$

which is independent of composition.

[Example 4-2]   Raw uranium is used for enrichment of U^{235} by gaseous diffusion process for a useful nuclear fuel. If the required U^{235} for the useful fuel is 2.1% and the natural abundance of U^{235} in ore is 0.7%, how many diffusion stages are required?

Raw uranium ore is converted first into the gaseous UF_6. Because the $U^{235}F_6$ is being separated from $U^{238}F_6$, the separation factor can be computed by Equation [4-31].

$$\alpha_{235-238} = [\frac{238 + 6(19)}{235 + 6(19)}]^{1/2} = (\frac{352}{349})^{1/2} = 1.0043$$

The improvement from 0.7% to 2.1% is three-fold, and this improvement can be evaluated by:

$$\alpha^n = \text{times of improvement}$$

and n is number of stages. Therefore,

$$(1.0043)^n = \frac{2.1}{0.7} = 3$$

Hence, $n = 263$, which requires 263 successive diffusion stages to perform this enrichment.

4.4 CHEMICAL MEASUREMENTS

Two of the essential principles that govern measurements are: the statistical methods for treating the measured numbers and confidence and acceptability of measured system.

4.4.1 Precision, Errors, and Data Reduction

Precision is a measure of agreement among individual measurements of the same property under prescribed similar conditions. **Accuracy** is a measure of the closeness of an individual measurement or the average of a number of measurements to the true value. Another term used is **bias**, which is a systematic error inherent in a method or caused by some artifact or idiosyncrasy of the measurement. Temperature effects are the examples of the first kind. Blanks, mechanical losses, and calibration errors are examples of the latter. There are various ways to analyze or reduce the data generated from measurements. The common method is through the use of frequency distribution curves, as shown in Figure 4-14.

An error function can be expressed in a probability function, P (ε), as shown here:

$$\int_{-\infty}^{\infty} P(\varepsilon)d\varepsilon = 1 \qquad\qquad [4\text{-}32]$$

where ε is the measured error and

$$P(\varepsilon) = \frac{1}{\sqrt{2\pi}\sigma} \exp(\frac{-\varepsilon^2}{2\sigma^2})$$ [4-33]

where σ is standard deviation

$$\sigma = \left(\frac{1}{n}\sum_{i=1}^{n}\varepsilon_i^2\right)^{1/2}$$ [4-34]

for a group of measurements of

$$x_i(i = 1, 2 \dots n)$$

and a frequency, f, as the frequency of the x_i value, is identical with the n population.

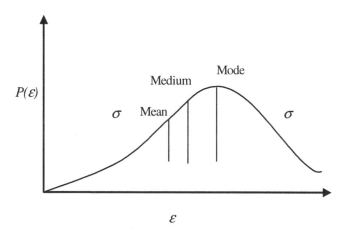

Figure 4-14. Error function.

Statistics commonly use measures of (a) central tendency, giving the location of typical value, (b) describing the shape of distribution which deals with symmetry and peak, and (c) of dispersion, showing the degree to which values vary and scatter.

For the central tendency, the following are important:

mode — most common measurement.

median — 50% of accumulated frequency or P_{50}, the 50th percentile.

arithmetic mean — $\overline{X} = \dfrac{\sum f_i X_i}{n}$, if frequency is not used, $\overline{X} = \dfrac{\sum X_i}{n}$.

geometric mean — $m_g = \left(\prod X_i^{f_i} \right)^{\frac{1}{2}}$, or frequency can be omitted.

For the shape category, the **moment** of the curve is determined by the formula; for example, the j is the moment:

$$m_j = \frac{\sum (x_i - \overline{x})^j}{n}$$
[4-35]

The **skewness** is defined as:

$$sk = m_3 / m_2^{3/2}$$
[4-36]

and the **kurtosis** is:

$$g_2 = (m_4 / m_2^2) - 3$$
[4-37]

For dispersion measures, the **variance** is defined as:

$$s^2 = \frac{1}{(n-1)} \sum (x_i - \overline{x})^2 f_i$$
[4-38]

the standard deviation is:

$$\sigma = \left(\frac{1}{n-1} \sum (x_i - \overline{x})^2 f_i \right)^{1/2}$$
[4-39]

the average deviation is:

$$d = \frac{1}{n} \sum f_i |x_i - \overline{x}|$$
[4-40]

and the **range** is:

$$n = x_n - x_1$$
[4-41]

A number of correlations using standard tables to compare two data sets or designs are also available, such as the t-tests. The **standard error** of the mean is defined as:

$$SE_m = S / n^{1/2} \qquad [4\text{-}42]$$

The **confidence limit** for a mean is:

$$CL = \pm 1.96 \ SE_m \qquad [4\text{-}43]$$

The **confidence interval** for a mean can be located from a statistics table of the t:

$$CI = \pm t \ SE_m \qquad [4\text{-}44]$$

where t is value from the table at a given level of confidence.

If only $n=1$ observation was available, this observation would be the sample mean and would give us some idea of the underlying population mean. Since there is no spread in the sample, however, it gives no idea about the population spread.

Only the extent that n exceeds one can we get information about the spread. That is, there are essentially only $(n-1)$ pieces of information for the spread, and this is the appropriate divisor for the variance.

For variance, there are $(n-1)$ d.f. (degree of freedom)

The statistics for chemical analysis are summarized in Table 4-3:

Table 4-3. Statistics for Chemical Analysis

	Population	Sample
Sample number	N (large)	n (small)
Mean	$\bar{\mu} = \sum (\dfrac{X_i}{N})$	$\bar{X} = \sum (\dfrac{X_i}{n})$
Variance	$\sigma^2 = \sum \left[\dfrac{(X_i - \bar{\mu})^2}{N} \right]$	$S^2 = \sum \left[\dfrac{(X_i - \bar{X})^2}{(n-1)} \right]$
Standard deviation (S.E.)	σ	S
Degrees of freedom (d.f.)	N	n-1
Confidence level 95%	$\mu = \bar{\mu} \pm 1.96 \left(\dfrac{\sigma}{N^{\frac{1}{2}}} \right)$	$\mu = \bar{X} \pm t_{0.025} \left(\dfrac{S}{n^{\frac{1}{2}}} \right)$

4.4.2   Quality Assurance and Quality Control

Quality Assurance (QA) may be defined as those operations and procedures that are undertaken to provide measurement data of a stated quality with a stated probability of being right. The quality of data should be (1) scientifically valid, (2) legally defensible, and (3) of a known and accepted precision as well as accuracy. In general, the goal is the use of data from a stable measurement system (statistical) to provide probable confidence of achievement (assurance) of a desired level of acceptability (quality).

The basic ingredients of a good measurement program are appropriate methodology, adequate calibration, and proper usage. Therefore, proper documentation of all operations and proper maintenance of facilities and equipment are important aspects of **quality control (QC)**. The relation of QC with QA is illustrated in Figure 4-15.

The quality of analytical measurements may be evaluated in terms of the precision and accuracy attained when compared with the requirements for the data. **Precision** is evaluated by repetitive measuring. Evaluation with respect to measurement bias and, hence, assessment of accuracy, is a more difficult process. **Accuracy** is infallible, always keeping an absolute standard. Measuring the same samples by several independent methods can identify measurement bias. Analyzing spikes, surrogates, or other internal standards can provide information on measurement bias. External evidence of measurement accuracy is usually desirable and is the easiest approach. Collaborative test exercises and proficiency tests using externally provided test samples are good ways to assess data quality and judge peer performance.

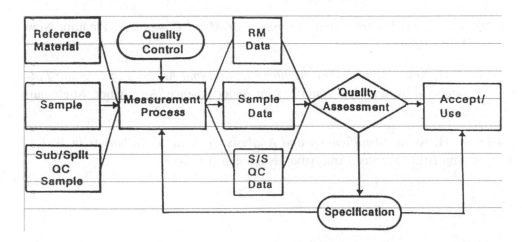

Figure 4-15. Measurement process quality assurance.

REFERENCES

4-1. C. N. Sawyer and P. L. McCarthy, *Chemistry for Environmental Engineering*, 3rd ed., McGraw-Hill, New York, 1978.

4-2. V. L. Snoeyink and D. Jenkins, *Water Chemistry*, Wiley, New York, 1980.

4-3. C. J. King, *Separation Processes*, McGraw-Hill, New York, 1971.

4-4. J. W. Cooper, *Spectroscopic Techniques for Organic Chemists*, Wiley, New York, 1980.

4-5. J. K. Taylor and T. W. Stanley, *Quality Assurance for Environmental Measurements*, American Society for Testing and Materials, Philadelphia, Pennsylvania, 1985.

4-6. J. S. Devinny, L. G. Everett, J. C. S. Lu and R. L. Stollar, *The Composition of Hazardous Wastes in Subsurface Migration of Hazardous Wastes*, Van Nostrand, 1990, pp. 15–39.

4-7. J. K. Taylor, *Quality Assurance of Chemical Measurements*, Lewis, Chelsea, Michigan, 1989.

4-8. L. L. Havilcek and R. D. Crain, *Practical Statistics for the Physical Sciences*, American Chemical Society, Washington DC.

4-9. H. L. Finston and A. C. Rychtman, *A New View of Current Acid-Base Theories*, Wiley, New York, 1982.

4-10. T. F. Yen, *Electron Spin Resonance of Metal Complexes*, Plenum, New York, 1969.

4-11. L. H. Keith, *Advances in the Identification and Analysis of Organic Pollutants in Water*, Vol. 1 and 2, Ann Arbor Science, Ann Arbor, Michigan, 1981.

4-12. L. H. Keith, *Identification and Analysis of Organic Pollutants in Water*, Ann Arbor Science, Ann Arbor, Michigan, 1976.

PROBLEM SET

1. Natural water contains 100 mg/L bicarbonate and 50 mg/L carboxylic acid. Compute the alkalinity as expressed in mg/L as $CaCO_3$. The pK_{a1} and pK_{a2} of H_2CO_3 are 6.3 and 10.3, respectively.

2. In a solution of 100 mg/L of NaF and 200 mg/L of HF, calculate the pH. The pK_a of HF is 3.2

3. A number of replicate measurements of moisture content have been made of a lignite sample. They are 15.2, 14.7, 15.1, 15.0, 15.3, 15.2, and 14.9. Calculate the mean, standard deviation, and degree of freedom.

CHAPTER 5

SURFACE CHEMISTRY

*T*he concepts of surface chemistry or colloid chemistry in environmental science and engineering are enormously important. The five major spheres in nature — atmosphere, hydrosphere, pedosphere, biosphere, and lithosphere — are each directly impacted by colloid and surface chemistry. A few obvious examples of this are the aerosol in air, the sedimentation process in water, the transport of nutrients across membranes, the erosion of soil, and many resource recovery processes.

This chapter will introduce some basic principles and definitions in colloid systems and especially in emulsions. Due to the increasing use of surfactants, we will group them with membrane-mimetic chemistry. Finally, we will address various commonly used techniques of surface analysis, in order to understand the molecular processes that occur at the solid-liquid interface.

5.1 COLLOIDAL SYSTEM

In Physical Chemistry, we briefly reviewed the three bulk phases: gas (g), liquid (l), and solid (s). Surface chemistry is a study of the chemical processes that occur at the interfaces between these phases. An interface is the boundary between two phases encompassing the complete zone where their properties, whether electric charge or chemical composition, differ from those exhibited in either of the bulk phases. A total of five such interfaces exist, that is, g-l, g-s, l-l, l-s, and s-s. Notice that it is difficult to form a g-g interface.

Usually there are three types of colloidal systems:

- **Macromolecules** — such as humic materials or proteins existing in a solution form.
- **Colloidal dispersion** — discrete particles (disperse phase) suspended in a dispersing medium to form a two-component dispersion. In general, there are the following eight categories of disperse phase–dispersing medium systems:

  - l-g, liquid aerosol, fogs
  - s-g, solid aerosol, smokes
  - g-l, foam
  - l-l, emulsion
  - s-l, sol, hydrosol (l = water)
  - g-s, solid foam
  - l-s, gel, solid emulsion
  - s-s, solid suspension, solid sol

- **Associate colloids** — formed when the concentration of a surface-active agent, either called a surfactant or an amphophile, exceeds a critical value and aggregation occurs.

There are many other classifications of colloids. For example, **hydrophilic colloids** have an affinity toward water and tend to form stable suspensions in water. Naturally occurring proteins or synthetic polymers bearing functional groups such as O^-, $-OH_2^+$, and $-NH_3^+$, will form large ionic molecules. Conversely, **hydrophobic colloids** do not form stable suspensions in water. They are all electrically charged, either positive or negative (electrokinetic property). The processing charge may result from charged groups within the protected surface, or it may be gained by the adsorption of a layer of ions from the surrounding medium.

A more general definition would be **lyophilic** and **lyophobic** colloids. The use of the prefix "lyo" instead of "hydro" includes other solvents and is not limited to water. In this manner, the colloids are termed protective colloids. Also, stability may give another classification to the unstable colloids that aggregate slowly, such as **diuturnal colloids**. The systems that aggregate rapidly are termed **cadicious colloids**.

5.2 COLLOIDAL RANGE AND ADSORPTION

The size of a colloidal particle is within the 1 nm–1 µm range. This is an arbitrary range that is located approximately between the dissolved solid in solution and the bacteria observed by light microscopy. Figure 5-1 shows airborne particles, and Figure 5-2 shows waterborne particles. An important feature of colloidal particles is the large surface area to volume ratio. Thus, the surface chemistry at the interface region controls the properties of the particle. The examples that follow indicate the extensive surface area of the small particles.

[**Example 5-1**]   What is the surface area of 1 cm^3 of a colloid that consists of packing each side with (a) cubic particles of 10 nm size, and (b) a thread of 10 nm in radius.

(a)   Because 10 nm = 10^{-6} cm, the cubic particles needed for packing are:

$$\frac{1}{\left(10^{-6}\right)^3} = 10^{18}$$

There are six sides; thus, the total area is

$$6 \times \left(10^{-6}\right)^2 \times 10^{18} \frac{1}{(30.48)^2} = 6500 \, \text{ft}^2$$

(b)   The total length of the thread is

$$l = \frac{1}{\pi r^2} = \frac{1}{\pi \left(10^{-6}\right)^2} = 0.34 \times 10^{12} \, \text{cm}$$

$$\text{Surface area} = 2 \, \pi r l = 0.34 \times 10^{12} \times 2\pi \times 10^{-6}$$

$$= 2 \times 10^6 \text{ cm}^2 = 2167 \text{ ft}^2$$

$$\approx 0.5 \text{ acre}$$

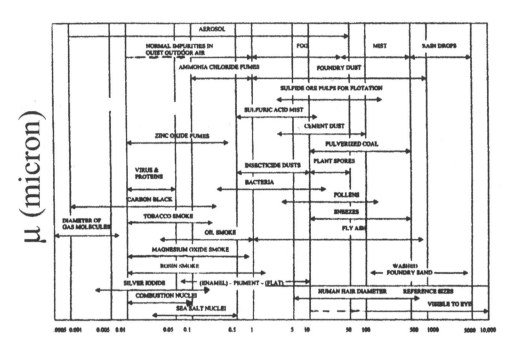

Figure 5-1.   Sizes of airborne particles (Redrawn, with permission, from L. Byers, "Controlling Atmospheric Particulates," *Technology Tutor* **1:43 (1971)).**

SIZE SPECTRUM OF WATERBORNE PARTICLES AND OF FILTER PORES

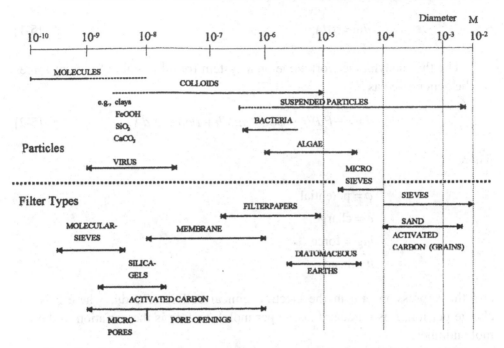

Figure 5-2. Suspended particles in natural and wastewaters vary in diameter from 0.005 to about 100 μm (5 x 10^{-9} to 10^{-4} m). For particles smaller than 10 μm, terminal gravitational settling will be less than about 10^{-2} cm sec^{-1}. Filter pores of sand filters, on the other hand, are typically larger than 500 μm. The smaller particles (colloids) can become separated either by settling if they aggregate or by filtration if they attach to filter grains. [From W. Stumm, *Environ. Sci.Technol.*, 11, 1066 (1977).]

The colloidal particle exhibits the characteristic light-scattering phenomena of the so-called **Tyndall effect** because its size approaches the wavelength of the light. The extensive surface area will give high interfacial energy and a high surface/charge density ratio, both of which are the requirements of a catalyst.

When a surface is created, there is work associated with the event. For a condensed phase, molecules in the interior will experience a spherical symmetry, attracting all adjacent molecules. Molecules at the interface will experience a net inward attraction around the surface.   This force is usually called the **interfacial tension**, σ, but for air-liquid interfaces it is usually referred to as **surface tension**, γ. This attraction increases the intermolecular distances for the intermolecules.   This process will eventually expand the surface area from the interior, A_s. The work will be

$$dw = \sigma \, dA_s \qquad\qquad [5\text{-}1]$$

The thermodynamic work done on a system for all possible forms of forces can be generalized as

$$dw = -PdV \;-\; \varphi\varphi d + mgdh + \mu dn + \sigma \, dA_s \qquad\qquad [5\text{-}2]$$

where

ϕ = potential

e = charge

mg = force

h = head

and the expression ϕde in the electrochemical cell is $zFE \, dn$, where z is the charge per ion, F is Faraday (charge per mole), and dn is the variation in the ion mole number.

Equation [5-2] is similar to various fluxes experienced in Equation [5-2a]. Likewise, the various fluxes for a given material in water may be written as

$$F = F_1 + F_2 + F_3 + F_4 = -D\Delta C - \kappa\Delta T - K\Delta h + k\Delta\phi \qquad\qquad [5\text{-}2a]$$

In this equation, F_1 is mass flux according to Fick's first law of diffusion, F_2 is heat flux according to Fourier's law, F_3 is fluid velocity according to Darcy's law, and F_4 is electric current per area according to Ohm's law. In Equation [5-2a], the following symbols are used.

D = diffusion coefficient

C = concentration

κ = thermal conductivity

T = temperature

K = hydraulic conductivity

h = hydraulic head

k = electric conductivity

ϕ = electric field

and the ∇ gradient function is the partial derivatives over spatial and temporal coordinates; for example, $\partial/\partial x$, etc. Details of the laws are beyond the scope of this chapter, but those laws have appeared later in the book.

Going back to the discussion of thermodynamics, the Gibb's free energy may be expanded to involve the surface deformation.

$$dG = VdP - SdT + \Sigma \mu_i d\, n_i + \sigma\, dA_s \qquad [5\text{-}3]$$

At constant T, P, and n_i, then

$$dG = \sigma\, dA_s \qquad [5\text{-}4]$$

The surface tension of water at 20°C is 72.75 mNm^{-1}; for seawater (salinity = 35%) it is 73.53 mNm^{-1}; and for soil water it is only 45 mNm^{-1}. (Other amounts of surface tension include 1erg/cm^2 = 10^{-3}J/m^2 = 10^{-3} N/m = 1mN/m; thus mNm^{-1} and erg cm^{-2} are interchangeable. Also because 1N=10^5 dyne and 1mNm^{-1}=10^2 dyne/m=dyne/cm, the units mNm^{-1} and dyne cm^{-1} are also interchangeable.) In general, electrolytes will increase, and surfactants (such as humic acid) will decrease the surface tension of a liquid.

The term **surface concentration,** or **surface excess,** Γ_i, is defined as the distribution of component i in the interface between the two adjacent phases α and β. If the total number of i moles is n_i, then the distribution is

$$n_i = n_{i,s} + n_{i,\alpha} + n_{i,\beta} \qquad [5\text{-}5]$$

In Equation [5-5], the first term is the concentration at the interface, whereas the second and third terms are the concentrations at phase α and phase β respectively. If concentration C and volume V are specified, then

$$n_{i,s} = n_i - C_{i,\alpha}V_\alpha - C_{i,\beta}V_\beta \qquad [5\text{-}6]$$

Moles at the interfaces can be taken as follows, assuming the interface is a plane:

$$\Gamma_i A_s = n_i - C_{i,\alpha}V_\alpha - C_{i,\beta}V_\beta \qquad [5\text{-}7]$$

Thus

$$\Gamma_i = \frac{n_{i,s}}{A_s}$$ [5-8]

The above equation indicates that when $\Gamma_i > 0$, component i will accumulate at the surface and adsorption is positive. Conversely, negative adsorption occurs when $\Gamma_i < 0$, and the solute will be excluded from the interface as in the case of salt in water. Either case will alter the surface tension and as exemplified by hydrocarbons in water, in the former case the attractive force between the solvent molecules will be greater than that between the solute and solvent molecules. At constant P and T, the surface concentration may be expressed as

$$\Gamma_i = \left(\frac{\mathrm{d}\gamma}{\mathrm{d}\mu_i} \right)_{T,P}$$ [5-9]

[Example 5-2]   Derive Equation 5-9.

From thermodynamics in Chapter 1, the internal energy is

$$E = TS - PV + \gamma A + \Sigma \mu_i n_i$$ [5-10]

Equation [5-10] can be written as

$$dE = TdS + SdT - PdV - VdP + \gamma dA + Ad\gamma + \Sigma \mu_i dn_i + \Sigma n_i d\mu_i$$ [5-11]

Also from the first and second laws of thermodynamics

$$dE = TdS - PdV + \Sigma \mu_i dn_i$$ [5-12]

Subtracting Equation [5-11] from Equation [5-12],

$$SdT - Vdp + Ad\gamma + \Sigma n_i d\mu_i = 0$$ [5-13]

Using T and P as constant,

$$dy = -\frac{\Sigma n_i}{A}\,d\mu_i = -\Gamma_i\,d\mu_i \qquad [5\text{-}14]$$

or,

$$\Gamma_i = -\frac{dy}{d\mu_i}$$

This chemical potential as Equation [5-9] may be expressed in terms of activity or called the **Gibb's adsorption isotherm**, because $\mu_i = \mu_o + RT\ln a_i$

$$\Gamma_i = -\frac{1}{RT}\left(\frac{dy}{d\ln a_i}\right) \qquad [5\text{-}15]$$

or

$$\Gamma_i = -\frac{a_i}{RT}\frac{dy}{da_i} \qquad [5\text{-}16]$$

This equation correlates adsorption, surface tension and concentration because concentration is proportional to activity.

For a two-component system, the preceding equation is

$$-dy = RT\,\Gamma_2\,d\ln a_2 \qquad [5\text{-}17]$$

Assuming ideal behavior, then

$$-dy = RT\,\Gamma_2\,d\ln x_2 \qquad [5\text{-}18]$$

Usually at sufficiently low concentrations of surface-active solute, a surface tension isotherm is linear at constant temperature, such that

$$\pi = \sigma_0 - \sigma = m\,x_2 \qquad [5\text{-}19]$$

where m is the slope of the isotherm, π is the spreading pressure, and σ is the same as γ. However, at high concentrations the isotherm is no longer linear, and an empirical equation is developed, which is called the **Szyszkowski equation**.

$$\pi = \sigma_0 - \sigma = RT\,\Gamma_m\,\ln\left(\frac{x_2}{a}+1\right) \qquad\qquad [5\text{-}20]$$

Now replacing σ with γ and differentiating Equation [5-20],

$$-\frac{d\gamma}{dx_2} = \frac{\Gamma_m\,RT}{x_2 + a} \qquad\qquad [5\text{-}21]$$

where

Γ_m = moles of component 2 per unit area at saturated concentration

a = empirical constant

Also, we can write the Gibb's equation from Equation [5-18]

$$-\frac{d\gamma}{dx_2} = \frac{RT\,\Gamma_2}{x_2} \qquad\qquad [5\text{-}22]$$

Combining the two equations, we have

$$\Gamma_2 = \frac{\Gamma_m\,x_2}{a + x_2} \qquad\qquad [5\text{-}23]$$

which is identical to the **Langmuir isotherm**,

$$\theta = \frac{x_2}{a + x_2} \quad , \quad \Gamma_m = \Gamma \qquad\qquad [5\text{-}24]$$

This is a general adsorption equation where θ usually is normalized to Γ_2/Γ, Γ being total adsorption. When Γ_2/Γ is plotted versus x_2, an isotherm results.

$$\frac{x_2}{\Gamma_2} = \frac{a}{\Gamma} + \frac{x_2}{\Gamma}$$

[5-25]

Equation [5-25] represents a Langmuir isotherm, and when x_2/Γ is plotted versus x_2, linear lines are obtained.

[Example 5-3]   The surface tension of an aqueous solution of valeric acid can be expressed by $\gamma = \gamma^*_{(H_2O)} - a\ln(1+bC_b)$, and a and b can be evaluated by experimental data as $a = 0.0131$ Nm^{-1}, $b = 0.01962$ m^3/mol. Calculate the surface concentration at $C_B = 200$m mol^{-3}. Then find the area of valeric acid adsorbed when the surface concentration is saturated.

$$\gamma = \gamma^*_{H_2O} - a\ln(1+bC_B)$$

$$\left(\frac{\partial\gamma}{\partial C_B}\right)_{T,P} = -\frac{ab}{1+bC_B}$$

(a)

From Equation [5-16] using concentration instead of activity,

$$\Gamma = -\frac{C_B}{RT}\left(\frac{\partial\gamma}{\partial C_B}\right)_{T,P} = \frac{abC_B}{RT(1+bC_B)}$$

(b)

When $C_B = 200$ mol/m^3

$$\Gamma = \frac{(0.0131)(0.01962)(200)}{(8.314)(2922)(1+0.01962\times200)}\ mol\big/m^2 = 4.30\times10^{-6}\ mol\big/m^2$$

From (b)

$$\Gamma = \frac{abC_B}{RT(1+bC_B)}$$

When C_B is large, $bC_B >> 1$ and $1 + bC_B \approx bC_B$

$$\Gamma_{max} = \frac{a}{RT} \tag{c}$$

which is the saturated condition.
    Thus

$$\Gamma = \frac{0.0131}{(8.314)(292.2)} = 5.39 \times 10^{-6} \text{ mol/m}^2$$

and the area of valeric acid is

$$[(5.39 \times 10^{-6})\ (6.023 \times 10^{23})]^{-1} = 3.08 \times 10^{-19} \text{ m}^2$$

Table 5-1. Surface-Active Agents

Anionic

Sodium stearate	$CH_3(CH_2)_{16}COO^-Na^+$
Sodium oleate	$CH_3(CH_2)_7CH=CH(CH_2)_7COO^-Na^+$
Sodium dodecyl sulphate	$CH_3(CH_2)_{11}SO_4^-Na^+$
Sodium dodecyl benzene sulphonate	$CH_3(CH_2)_{11}C_6H_4SO_3^-Na^+$

Cationic

Dodecylamine hysrochloride	$CH_3(CH_2)_{11}NH_3^+Cl^-$
Hexadecyltrimethyl ammonium bromide	$CH_3(CH_2)_{15}N(CH_3)_3^+Br^-$

Nonionic

Polyethylene oxides	$CH_3(CH_2)_7C_6H_4(OCH_2CH_2)_8OH$
Spans (sorbitan esters)	
Tweens (polyoxyethylene sorbitan esters)	

Ampholytic

Dodecyl betaine

$$C_{12}H_{25}N+\begin{array}{l}(CH_3)_2\\ CH_2COO^-\end{array}$$

5.3 SURFACTANTS

Surfactants are **surface-active agents**; they are also referred to as **detergents**. They are organics or organometallics that form association colloids or micelles in solution. They are also called **amphiphilic substances**, or **amphiphiles**, which signifies that they possess distinct regions of hydrophobic and hydrophilic character. Due to their polarity, the surfactants are also described as amphipathic, heteropolar, or polar-nonpolar molecules. In general, there are four types:

- **Cationic surfactants** — typically the "onium" structures such as ammonium (N), sulfonium (S), and phosphonium (P). They can be represented by RnX^+Y^-, where the "onium" portion is X^+. An example is dodecyltrimethyl ammonium bromide $(CH_3)_3N^+ (CH_2)_9CH_3 Br^-$, a long chain quaternary compound.

- **Anionic surfactants** — typically the alkali or alkaline earth salts of mono- or polybasic carboxylic (fatty) acids of the $RX^- Y^+$ type. The X^- denotes the carboxylic, sulfonic, or phosphoric portion of the acid. An example illustrated will be sodium dodecylbenzene sulfonate, CH_3 $(CH_2)_9 \phi SO_3Na$, which again is a long-chain salt.

- **Nonionic surfactants** — most of these can be represented by the polyoxyethylene or the polyoxypropylene derivatives, or the polyalcohols, carbohydrate esters or fatty alkanol amides. An example is polyoxyethylated t-octylphenol, which is commercially available as Triton X-100 t-Oc-ϕ-$(OCH_2$-$CH_2)_{10}$-OH. According to its name, this class of surfaces does not contain obvious ions or charges.

- **Ampholytic (zwitterionic) surfactants** — zwitterionic surfactants possess both cationic and anionic functional groups in the hydrophobic moiety. They can be anionic, cationic, or neutral, depending on the pH of the solution. They generally come from N-alkyl or C-alkyl sultaines (sulfonic), betaines (carboxylic), or phosphatidyl amino alcohols and acids (phosphoric).

These chemical classifications are based on the charges that are carried by the surface-active portion of the molecules. Most of the surfactants are synthetic.

The common types are listed in Table 5-1. Surfactants are also derived from naturally-occurring sources; these are called biosurfactants. The amphiphiles include lipids (e.g., carboxylic acid esters), complex lipids (e.g., simple lipids containing P⁻, N⁻ bases, and/or sugars), and bile acids such as cholic and deoxycholic acid. The biosurfactants play an important role in vivo transport and membrane processes. Some examples of biosurfactants obtained from bacteria, which may be commercially viable, are listed in Figure 5-3. In general, the biosurfactants may be more valuable when compared with their synthetic counterparts. The biosurfactants have the following features:

- They are polymeric by nature, and more stable under heat and weathering.

- They are either heterodisperse or polydisperse, and never monodisperse. For the former, broad diffuse distribution in randomness will be confronted with a broad range of targets. The latter will contain a Poisson distribution around a mean and, again, will be used for a probability distribution of targets.

- They contain polar heads and long tails, portions of which often consist of peptide linkages in helical configurations for the accommodation of various conformations during transport.

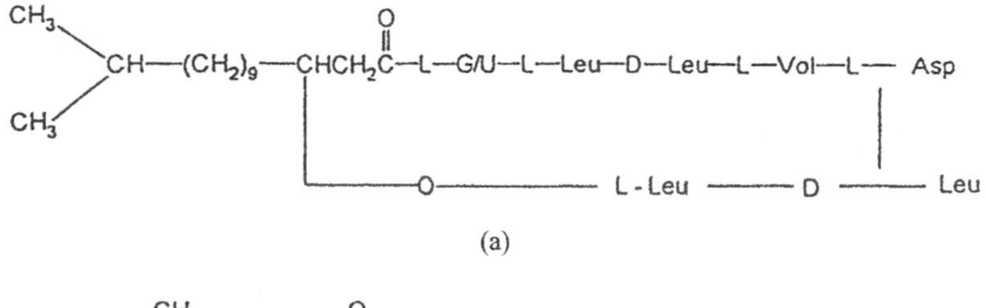

(a)

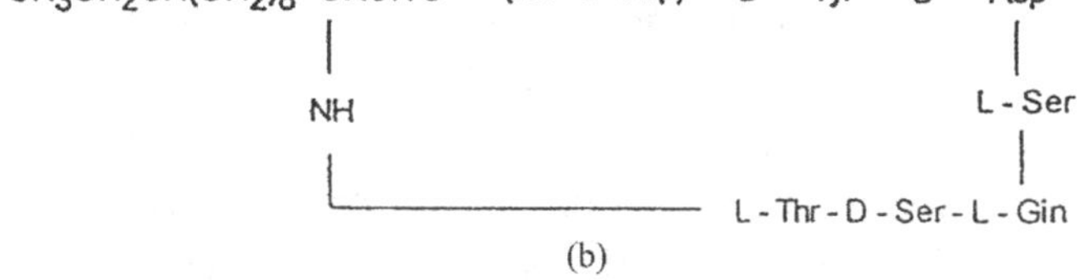

(b)

Figure 5-3. Two examples of biosurfactants. (a) The structure of surfactin or subtilysin, a lipopeptide isolated from *Bacillus subtilis*, (b) One of the surfactin analogs isolated from *Bacillus subtilis* containing a β-amino acid instead of a β-hydroxy fatty acid.

5.3.1 Emulsions

Next, we will discuss some phenomena related to the surfactants. An **emulsion** is defined as a mixture of particles of one liquid with some of a second liquid; and, because one of the liquids is often aqueous in nature, the two general types of emulsions are **oil-in-water** (O/W) and **water-in-oil** (W/O), the use of the word "oil" here denoting any water-insoluble fluid. In these emulsions, the outer phase is continuous, whereas the inner phase is not. Illustrations of the two types of emulsions are in Figure 5-4. In general,

$$\frac{A}{B} = \frac{inner\ phase}{outer\ phase}$$

Now let

$$\frac{V_A}{V_B} = \phi \qquad\qquad [5\text{-}25a]$$

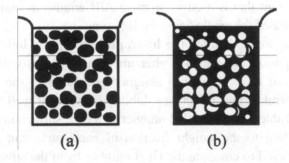

Figure 5-4. The two types of emulsion: (a) oil in water, O/W; and (b) water in oil, W/O.

In a dilute emulsion, the inner phase exists as spheres, and the **Einstein limiting law** is applicable,

$$\eta = \eta_0 (1 + 2.5\phi) \qquad\qquad [5\text{-}25b]$$

For a small ϕ value, the specific viscosity, η_{sp}, can be empirically related to ϕ by a power series

$$\eta_{sp} = \frac{\eta}{\eta_0} - 1 = a\phi + bd^2 + cd^3 + \dots \qquad [5\text{-}25c]$$

where a, b, and c are empirical coefficients.

For a stable emulsion, a surfactant is required. In this case, the surfactant added is called an **emulsifier** (emulsifying agent). The purpose of the surfactant is to lower the interfacial tension. For example, the interfacial tension for an oil-water emulsion is 41 dyne/cm, but if a trace amount of sodium oleate is introduced, the interfacial tension is reduced to 7.2 dyne/cm. If NaCl is also introduced, the interfacial tension can reach 0.01 or even 0.002 dyne/cm. This is the principle for tertiary oil recovery in enhanced oil recovery or the abatement process for oil-spilled sands and soils. The aging processes of an emulsion can result in **flocculation** (inner phase clustering together) or **creaming** (inner phase undergoing gravity separation), and eventually will break the emulsion to yield two liquid layers as shown in Figure 5-5. Another process is called **inversion**, where A/B becomes B/A; for example, a W/O emulsion is stabilized by a monovalent ion soap, whereas an O/W emulsion is stabilized by a polyvalent ion soap — the latter agent being called an antagonistic agent. A de-emulsifier or breaking agent is usually a surfactant that is used to promote A/B emulsions and can be used to break B/A emulsions if the emulsifier is the B/A type.

An empirical method called the **hydrophile-lipophile balance** (HLB) has been set up to quantify the emulsion phenomena as previously discussed. For a given surfactant, an HLB number is assigned according to the structure. The HLB scale can correlate the surfactant solubility in water as well as its functional use as shown in Table 5-2. The HLB number of a dual surfactant mixture will be calculated according to the weight fraction of each surfactant. An empirical method has been used to compute the HLB number from the structural groups of the surfactant.

$$HLB\ No. = 7 + n_h H + n_l L \qquad [5\text{-}26]$$

From the equation

n_h = number of hydrophilic groups

n_l = number of lipophilic groups

H = group contribution from hydrophilic

L = group contribution from lipophilic

The n_h, n_l, H, and L are listed in Table 5-3.

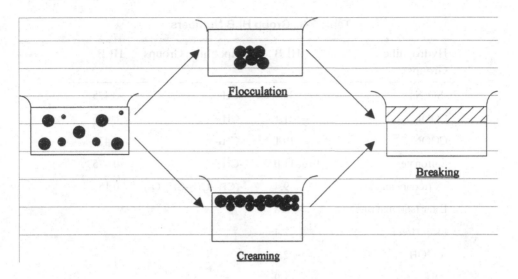

Figure 5-5. Types of emulsion instability.

Table 5-2. The HLB Scale

Surfactant Solubility Behavior in Water	HLB Number	Application
	0	
No dispersibility in water	2	W/O emulsifier
	4	W/O emulsifier
Poor dispersibility	6	
Milky dispersion; unstable	8	Wetting agent
Milky dispersion; stable	10	Wetting agent
Translucent to clear solution	12	Detergent, O/W emulsifier
	14	Detergent, O/W emulsifier
Clear solution	16	Solubilizer, O/W emulsifier
	18	Solubilizer, O/W emulsifier

Table 5-3. Group HLB Numbers

Hydrophilic Groups	HLB	Lipophilic Groups	HLB
-SO$_4$Na	38.7	-CH-	-0.475
COOK	21.1	-CH$_2$-	-0.475
COONa	19.1	-CH$_3$-	-0.475
Sulfonate	about 11.0	-CH=	-0.475
-N (tetiary amine)	9.4	-(CH$_2$-CH$_2$-CH$_2$-O-)	-0.15
Ester (sorbitan ring)	6.8		
Ester (free)	2.4		
-COOH	2.1		
-OH (free)	1.9		
-O-	1.3		
-OH (sorbitan ring)	0.5		

5.3.2 Membrane-Mimetic Chemistry

Membrane-mimetic chemistry describes the chemistry processes in simple media that mimic aspects of biomembranes. Any arrangement of amphiphilic (or surfactant) molecules forming monolayers, bilayers, multilayers, micelles, reversed micelles, unilamellar vesicles, multilamellar vesicles, or polymerized vesicles can be related to biomembranes. The emerging scientific discipline concerned with the development and utilization of membrane-moderated (and inspired) processes in organized surfactant assemblies have a great amount of potential engineering applications, especially in the field of chemical, biological and environmental engineering.

Most biological membranes consist of lipoprotein material. The classical model of lipid bimolecular layers has its foundation in hydrophobic bonding. The hydrophobic groups are always oriented toward the outside as shown in Figure 5-6. The importance of membranes to the maintenance of life is essential due to the necessity for the regulation of transport across them. One important characteristic of lipids or fatty acids is that they can form clusters or islands on a surface. The polar ends will form a film extending towards a polar media, similar to the spreading of monomolecular films over the surface of water. For example, a monolayer of cetyl alcohol can enhance the rate of evaporation by as much as 40%. A sizable reduction in terms of water conservation can be made in this manner in fresh water lakes and reservoirs.

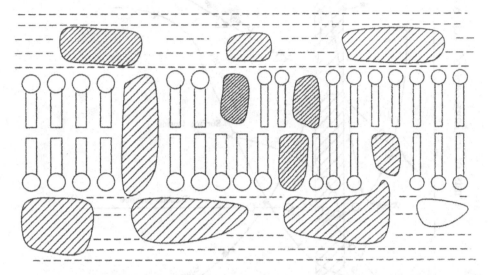

Figure 5-6. Schematic representation of a biological cell membrane. A bimolecular layer of phospholipid with hydrocarbon chains orientated to the interior and hydrophilic groups on the outside is penetrated by protein (shaded areas). Protein is also found adsorbed at the membrane surface.

Solutions of highly surface-active materials, such as colloidal electrolytes, exhibit unusual physical properties. For example, abrupt changes in osmotic pressure, turbidity, electrical conductivity, surface tension, and so on, take place at well-defined concentrations. A typical colloidal electrolyte — sodium dodecyl sulfate — is shown in Figure 5-7. These changes are interpreted as the idea that aggregation of the colloidal electrolyte must happen at that point of abrupt change. This type of aggregation is called a **micelle** and the concentration of these abrupt changes is termed the **critical micelle concentration** (CMC). Micelles are narrowly dispersed in size and may contain 50 to 100 monomer units. They may be either spherical or lamellar in geometry as shown in Figure 5-8.

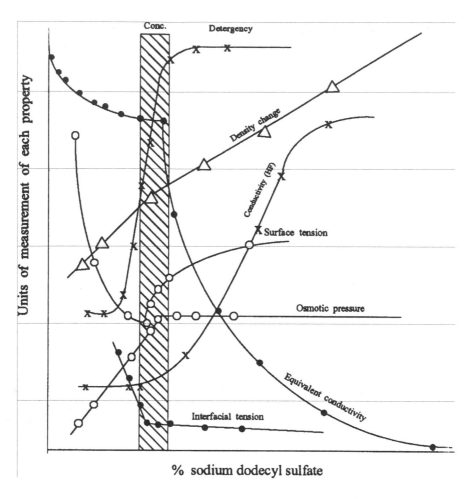

Figure 5-7. Properties of surfactant near CMC.

Many micelles found in non aqueous solvents are inverse (or reversed) in nature; in other words, the polar groups form the interior while the hydrophobic moieties are in contact with the nonpolar solvent. Asphaltene in oil behaves as a surfactant in a nonpolar medium and can be inverted by the use of ultrasound as shown in Figure 5-9. For a ternary system, normal micelles, inverse micelles, and liquid crystals may be formed, as shown in Figure 5-10. The vesicles can be single compartment or multicompartment lamellar; but most of them are liquid crystalline multilamellar vesicles that can be broken into unilamellar vesicles by

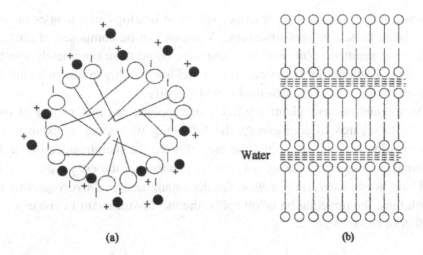

(a)

(b)

Water

Figure 5-8. Schematic representation of (a) a spherical micelle and (b) a lamellar micelle.

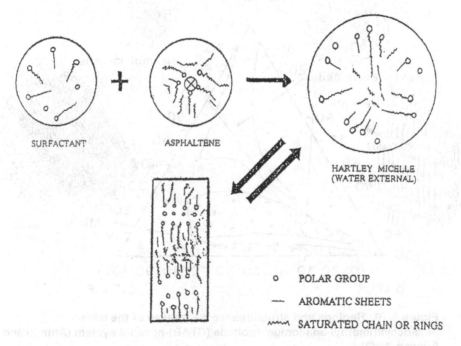

SURFACTANT

ASPHALTENE

HARTLEY MICELLE
(WATER EXTERNAL)

○   POLAR GROUP

—   AROMATIC SHEETS

∿∿   SATURATED CHAIN OR RINGS

Figure 5-9. Interaction of surfactant with asphaltene to form Hartley micelle and the consequent formation of liquid crystals.

ultrasound. The term vesicle describes spherical or ellipsoidal single- or multi-compartment closed bilayer structures. Vesicles can be composed of naturally-occurring or synthetic phospholipids, and may come from completely synthetic surfactants. Exploitation of vesicles for mimicking membrane functions has been prompted by the ease of their formation and stability.

Membrane-mimetic chemistry has many applications in a variety of industries. With regards to engineering, the following will serve as examples: the desulfurization of coal by multiphase biocatalysis, the maintenance of masking and demasking reactions during wastewater processing, the enzymatic treatment of industry wastewater via micelles, the decontamination of hydrocarbons from soil columns, the remediation of oil spills, the use of surfactant in enhanced bio-remediaton, and so on.

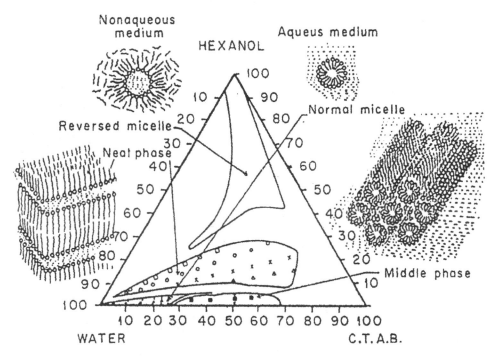

Figure 5-10. Regions and structures for the phases of the water-hexadecyltrimethyl-ammonium bromide (CTAB)-hexanol system (Ahmad and Friberg, 1972).

5.4 SURFACE ANALYTICAL TECHNIQUES

In order to understand the molecular processes that occur at the solid-liquid interfaces, different methods of surface analysis have been developed to obtain information about changes in chemical bonding, lattice structure, and surface topography during reactions involving natural materials. Following are some examples:

5.4.1 Scanning Electron Microscopy (SEM)

Scanning Electron Microscopy (SEM) is a technique that can provide information about surface features, morphology, compositions, and crystallographic information of the target sample. This technique provides high-resolution images of the sample surface on a nanometer scale. SEM's large depth of field allows for the examination and analysis of bulk specimens; it also permits the observation of *in situ* solid-liquid interface reactions.

Figure 5-11 illustrates the scheme of SEM. A beam of electron first sweeps across the surface of the sample and then generates **secondary electrons** (SE), **backscattered electrons** (BE), and characteristic X-rays. These signals are captured and transformed into images by detectors and then showed on a viewing screen.

One application of SEM in environmental engineering is the scale study. The occurrence of barium sulfate scale is a severe problem in oil and gas production concerned by environmental engineers. An example is the study of the dissolution rate of barium sulfate scale by chelating agents using SEM examination. The SEM images provided the micrography of the barite before and after dissolution process. Figure 5-12 shows the formation of etch pits on the sample surface. The chelating agent is diethylenetrinitilopentaacetic acid (DPTA).

Using SEM, a kinetic model was developed to analyze the barite dissolution rate

$$r_{Ba^{2+}} = kcCs = k_c(C_\infty - C_t) \qquad [5-27]$$

where

r_{Ba2+} = reaction rate of $BaSO_4$ dissolution at steady state (mol h^{-1})

k_c = reaction rate constant(h^{-1})

C_s = mass concentration of $BaSO_4$ (s) remaining in the solution (mg/L)

C_∞ = ultimate concentration of barium (Ba^{2+}) in solution (mg/L)

C_t = concentration of barium (Ba^{2+}) in solution at time t (mg/L)

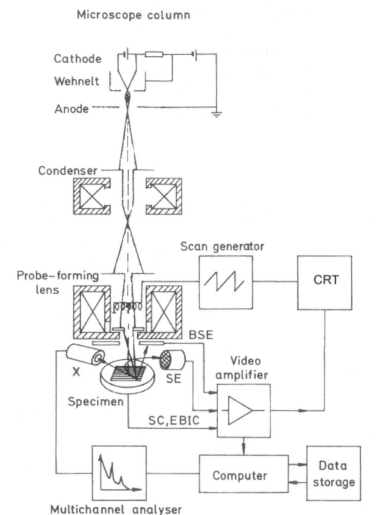

Figure 5-11.   Diagram of the scanning electron microscope (BSE=backscattered electrons, SE=secondary electrons, SC=specimen current, EBIC=electron-beam-induced current, X=x-rays, CRT=cathode-ray tube) (after L. Reimer, Scanning Electron Microscopy: Physics of Image Formation and Microanalysis, New York,1998).

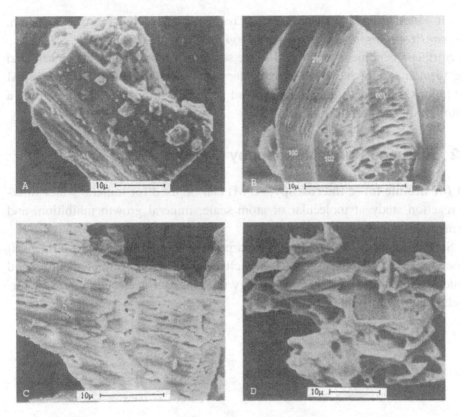

Figure 5-12. SEM investigations of barite dissolution by DTPA: (A) before dissolution process, (B) after 0.5 h dissolution, (C) after 1 h dissolution, (D) after 7 h dissolution. (After K. Dunn and T.F.Yen, Env. Sci.Tech, _33_, 1999,2821)

Since

$$r_{Ba^{2+}} = dC_t / dt \qquad\qquad [5\text{-}28]$$

integration yields

$$\int_0^{C_t} \frac{dC_t}{C_\infty - C_t} = k \int_0^t dt$$

and

$$\frac{C_\infty - C_t}{C_\infty} = \exp(-k_c t) \qquad\qquad [5\text{-}29]$$

Pit kinetics is also discussed in Chapter 12 Mechanisms of Corrosion, *Chemical Processes for Environmental Engineering* by Teh Fu Yen.

Surface phenomena are crucial to the scale dissolution study. They can lead to the formation of etch pits as defect centers on a specific surface. In these SEM observations, the morphology changes and scale dissolution are considered as a function of dissolution time.

5.4.2 Scanning Force Microscopy (SFM)

SFM (**or atomic force microscopy**, AFM) can be used for mineral-water interface reaction study at molecular or atom scale, mineral growth inhibition, and mineral etching in solution.

SFM is an instrument that uses a sharp probe to scan the sample surface. It can be applied to all types of surfaces including insulators, semiconductors and conductors so that the surface morphology imaging, surface roughness, and surface characterization can be obtained.

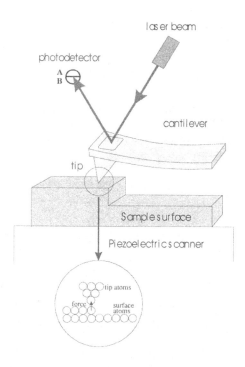

Figure 5-13. Schematic illustration of the scanning force microscopy: The essential elements are the tip, cantilever, photodetector, piezoelectric scanner, and sample. (K.-S. Wang, 2001, PhD Dissertation USC)

As Figure 5-13 shows, the probe is a tip on the end of a cantilever that bends in response to the force between the tip and the sample. As the cantilever flexes, the light from the laser is reflected onto the split photo-diode. By measuring the difference in signals (A-B), changes in the bending of the displacements, the interaction force between the tip and the sample can be found. The movement of the sample is performed by a precise positioning device made from a piezoelectric tube. There is a fluid cell attached to SFM that can be used to directly observe the mineral-liquid interface during the course of growth, precipitation, or dissolution processes of minerals in real time.

Permanent contact (contact-mode atomic force microscopy, CM-AFM) between the tip and the sample surface is a widely used imaging procedure for SFM. However, due to the lateral shear forces, the scan process of this mode of SFM can cause damage or modification on the sample surface. Another technique called noncontact-mode AFM (NC-AFM) can avoid such damages or intermittent contact by using oscillating cantilever with a tip located a few nanometers above the sample surface.

Figure 5-14. A three-dimensional SFM image of a barite surface after dissolution CDTA. The information of a layer-by-layer of depth etch pits can be seen. (K.-S. Wang, 2001, PhD Dissertation USC)

SFM can also be used in the study of barite dissolution process. Figure 5-14 shows a three-dimensional SFM image of a barite surface after dissolution with trans-1,2 cyclohexylenediaminetetraacetic acid (CDTA). The etch pits on the surface grow from shallow triangular pits as dissolution time increased, and the dissolution proceeds in a layer-by-layer process. Figure 5-15 shows the height profile of a two dimensional SFM image of one pits on the sample surface along the white marked axis.

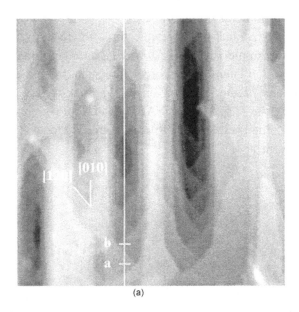

(a)

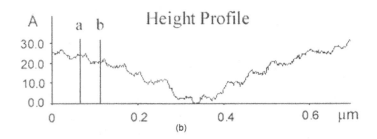

(b)

Figure 5-15. (a) SFM image of a barite surface after dissolution in 0.18 M CDTA for 10 min. (b) A line profile taken across the pit along b axis. The morphology of etch pits was defined by monolayer steps parallel to the [010] and [120] directions. Flat terraces at the bottoms of the etch pits were observed. The height difference between the two terraces denoted as "a" and "b" is 3.64Å. Image size: 0.7 μm × 0.7 μm. Depth scale from black to white:6nm.(K.-S. Wang et al, 2000, Langmuir, 16, 649)

5.4.3 X-ray Photoelectron Spectroscopy (XPS)

XPS (or Electron spectroscopy for chemical analysis, **ESCA**) is a surface chemical analysis technique, which can be used for most solids: insulators, conductors, organics, or powders. The XPS technique can analyze a variety of surfaces with minimal sample damage.

As Figure 5-16 illustrates, the target atom adsorbs the energy carried by an incoming X-ray photoelectron and is raised into an excited state from which it relaxes by the emission of a photoelectron. The electron energy spectrum is characteristic of the emitting atom type. Lines in the spectrum are labeled according to the energy level from which they originate. An XPS spectrum for copper is shown in Figure 5-17 as an example.

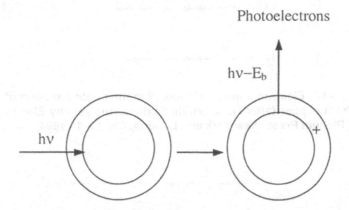

Figure 5-16. Achematic illustration of the physical basis of X-ray adsorption (After W. M. Riggs and M. J. Parker, Surface Analysis by X-ray photoelectron spectroscopy in Method of Surface Analysis, (Ed. Czandera, A.W.), Elsevier Science Publishing Co., Amesterdam, 1975, Ch.4, 103-158).

XPS can be used in combination with SFM or SEM with or without some treatment such as exposure to reactive gases or solutions, etc. It can provide chemical bonding identification of surface species, elemental surface analysis (except hydrogen and helium), and compositional depth profiles

An example of the application of XPS is the identification of carbon steel surface species. Corrosion of iron and steel equipment is a major problem of many industries. Hence, the corrosion inhibitors are being developed to increase

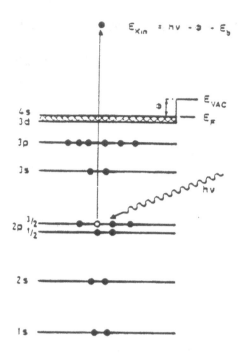

Figure 5-17. Electron transition involved in the photo emission of a 2P$_{3/2}$ electron from copper (after G. C. Smith, Surface Analysis by Electron Spectroscopy. Plenum Press, New York and London, Ch. 1,3-14,1994).

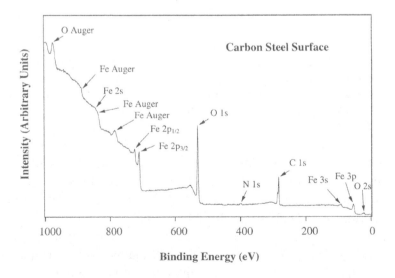

Figure 5-18. XPS survey spectrum of polished carbon steel.(K.-S. Wang, loc. cit)

equipment lifetime. Figure 5-18 shows a survey scan of XPS spectrum for a clean polished carbon steel sample. The spectrum consists of carbon, oxygen, and iron photoelectron and Auger peaks. The assignments of binding energy for the chemical composition of the analyzed surfaces are shown in Table 5-4.

Table 5-4. XPS Binding Energies of Various Elements/Transitions (K.-S. Wang, loc. cit)

Element/Transition	Binding Energy (eV)	Peak Assignment
C1s	284.8-285.2	C-C, C-H
	286.0-286.4	C-N
	287.9-289.4	C-O
O1s	529.9-530.2	Fe-O-Fe
	531.8-532.0	Fe-OH
	533.0-534.6	C-O, H_2O
N1s	398.0-398.1	=N-
	398.7-400.1	-NH-
	400.1-400.6	Fe-N=
	401.7-402.0	$-N^+$
Fe $2P_{3/2}$	706.7-706.9	Fe^0
	709.9-711.6	Fe(II)Ox, FeOOH
	712.0-712.9	Fe(III)Ox
	713.0-714.0	Fe-org

The surface of the sample after reaction with a corrosion inhibitor OI (oleic imidazoline) shows the fingerprint of the adsorbed compounds. High-resolution XPS spectra Figure 5-19 shows the high resolution spectra by XPS of N(1s), and C(1s) for the polished carbon steel absorbed with OI.

5.4.4 Vertical Scanning Interferometry (VSI)

Another updated method of surface analysis techniques is the **vertical scanning interferometry** system, which can provide three-dimensional topography data of a surface. VSI is a non-invasive technique for quantifying the surface topography of solids for a wide range application from study of mineral dissolution, microbial cells to glass dissolution. It is capable of absolute rates measuring for the dissolution reactions with its large field of view.

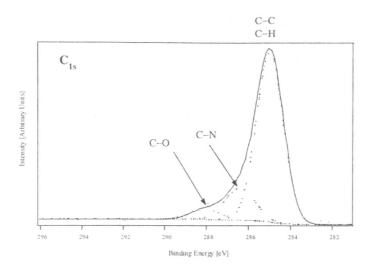

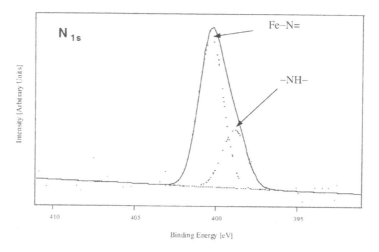

Figure 5-19. C(1s) and N(1s) spectra of polished carbon steel adsorbed with Ol.(K.-S. Wang, loc. cit)

As shown in Figure 5-20, the interference of two beams of lights—a reference beam and a sample beam are used to form an interference pattern of light and dark fringes, known as an interferogram. In combination with the wavelength of the light, the light and dark fringes are used to determine height differences between each fringe. The sample is moved by a piezoelectric stage vertically with nanometer precision. A charge-coupled device (CCD) camera is installed to capture the interferograms and quantify the topography of the surface.

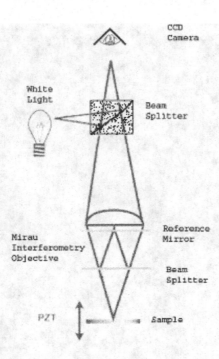

Figure 5-20. Diagram of vertical scanning interferometry (VSI). The VSI consists of a white light source, a double-beam Mirau interferometer with CCD camera and a piezoelectric stage. (After A. Lüttge et al. Am J Sci, 1999 299 652)

A good example of a VSI application is the surface study of an alumna membrane that is used as a template for a carbon nanotube (CNT) layer. VSI is a helpful tool for roughness measurements. The information can be used for optimizing the finish surface before and after applying the CNT layer. Figure 5-21 shows an image of an alumina membrane (Whatman Anodisc) with an exaggerated vertical axis for better image of roughness. Another interesting aspect of VSI is that by transferring the beam to any corner of the sample, thickness of the sample can be measured. Figure 5-22 and 5-23 show the thickness and the vertical profile of the sample.

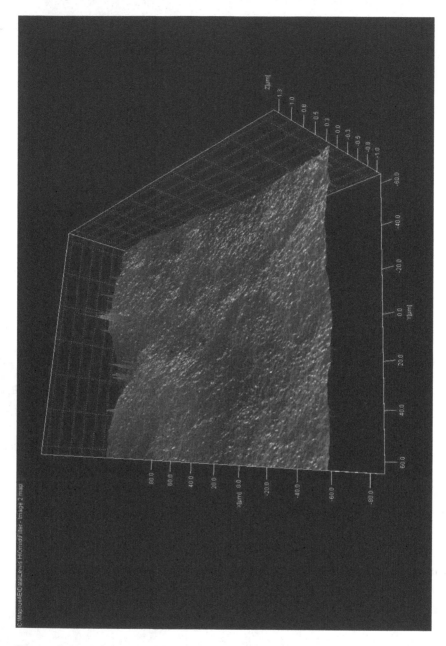

Figure 5-21. VSI image of 20nm pore-size alumina membrane (Whatman Ano-disc). (After O. Etemadi, PhD Dissertation, USC, 2006)

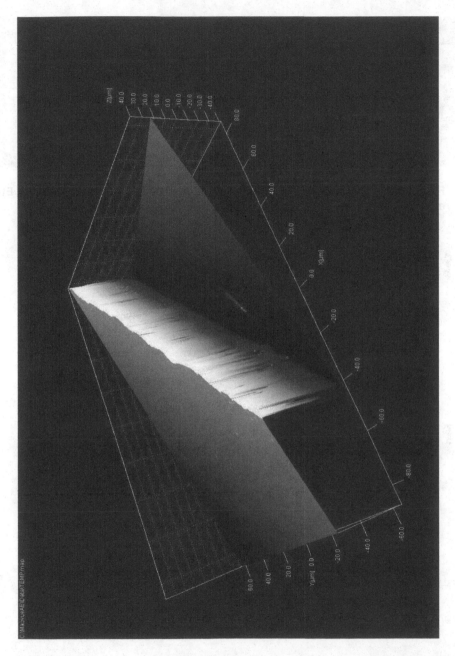

Figure 5-22. 3-dimensional VSI image of the alumina membrane thickness.
(After O. Etemadi, PhD Dissertation, USC, 2006)

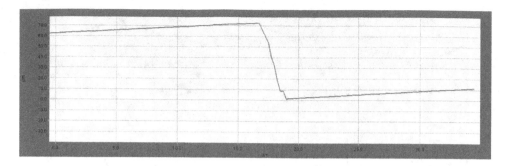

Figure 5-23. Vertical profile of an alumina membrane (≈60µm). (After O. Etemadi loc. cit)

REFERENCES

5-1. A. W. Adamson, *Physical Chemistry of Surfaces*, 4th ed., Wiley, New York, 1953.

5-2. D. J. Shaw, *Introduction to Colloid and Surface Chemistry*, 3rd ed., Butterworths, London, 1990.

5-3. D. A. Sabatini and R. C. Knox, *Transport and Remediation of Subsurface Contaminants*, American Chemical Society, Washington DC, 1992.

5-4. J. H. Fendler and E. J. Fendler, *Catalysis in Micellar and Macromolecular Systems, Academic Press*, New York, 1975.

5-5. J. H. Fendler, "Membrane Mimetic Chemistry," *Chem in Britain* 1098–1103 (1984).

5-6. S. Ross and I. D. Morrison, *Colloidal Systems and Interfaces*, John Wiley, New York, 1988.

5-7. J. H. Fendler, "Surfactant Vesicles as Membrane Mimetic Agents," *Acc. Chem. Res. 13*, 7–13 (1980).

5-8. M.-A. Sadeghi, K. M. Sedeghi, D. Momeni, and T. F. Yen, "Microscopic Studies of Surfactant Vesicles Formed during Tar Sand Recovery," *ACS Symp*. Ser. *396*, 391–407 (1989).

5-9. J. H. Fendler, *Membrane-Mimetic Chemistry*, Wiley, New York, 1982.

5-10. T. F. Yen, R. D. Gilbert and J. H. Fendler, *Membrane-Mimetic Chemistry*, Plenum Press, New York, 1994.

5-11. E. J. W. Verwey and J. Th. G. Overbeek, *Theory of the Stability of Lyophobic Colloids*, Elsevier, Amsterdam, 1948.

5-12. B. V. Derjaguin, L. Landau, "Theory of the Stability of Strongly Charged Lyophobic sols and of the Adhesion of Strongly Charged Particles in Solutions of Electrolyte," *Acta Physicchim. URSS, 14*, 733–762 (1941).

5-13. Y. Moroi, *Micelles, Theoretical and Applied Aspects*, Plenum, New York, 1992.

5-14. R.Zana and E.W.Kaler, *Giant Micelles, Properties and Applications,* CRC Press, 2007.

PROBLEM SET

1. Approximate the HLB (hydrophilic-lipophile balance) number for cetyl alcohol $CH_3(CH_2)_{15}OH$.

2. Suppose we have a gram of crystalline silica. For simplicity, the crystals with the density of 2.30 g/cm^3 are considered to be small cubes. Calculate the surface area in cm^2/g at various particle sizes with the cube sides as 10^{-1}, 10^{-3}, 10^{-5}, 10^{-6}, 10^{-7} cm.

3. For the following table,
    (a)   Suggest a case of stable mixture for water in oil emulsion.
    (b)   Suggest a case of stable mixture for oil in water emulsion.
    (c)   Also, suggest a mixture to break the water in oil emulsion.

Surfactant Name	HLB
Span 65 (sorbital tristearate)	2.1
Span 85 (sorbital tristearate)	1.8
Tween 20 (polysoxyethylene sorbitan monolamate)	16.7
Tween 60 (polysoxyethylene sorbitan monolamate)	14.9

CHAPTER 6

BIOCHEMISTRY

*T*his chapter aims to introduce the molecular basis of life. First, the fundamental chemistry of the molecules is essential to the cells. The molecules covered include amino acids and sugar phosphate derivatives and the macromolecules derived from these, i.e., proteins and nucleic acids. Steric hindrances limit the conformation opportunities open to nest amino acid residues in a protein and give rise to the requirement of both α helix and β sheet. These secondary structure units exploit hydrogen bonding arrangements to fold into globular or fiber structures. Enzyme kinetics such as Michaelis-Menten is discussed. The basis of biological membrane as well as clathrin, tubulin and virus geometry are examined. Finally genetic engineering is explored.

6.1 CELL CHEMISTRY

Cells in living organisms consist of a semipermeable membrane enclosing an aqueous solution rich in a diverse range of chemicals. Cells are engineered machines that undertake sophisticated, organized complex chemistry that can be

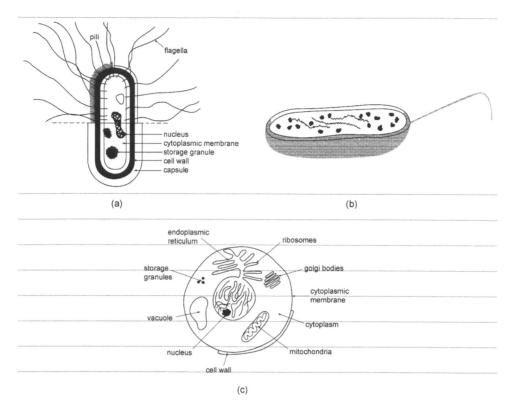

Figure 6-1. Typical cell structures: (a) prokaryotic cell, (b) a schematic view of a bacterial cell, (c) eukaryotic cell of multicellular organisms.

handed down to the daughter cell. A **prokaryotic cell** (Fig. 6-1) contains the gross features, which are not vastly different from those of the **eukaryotic cell** of multicellular organisms such as plants and animals. Some of these features are:

- **Cytosol** – an aqueous solution of water-soluble inorganic ions such as K^+ and Cl^- and a small amount of water-soluble organic molecules such as sugars and amino acids and organic polymers such as proteins and nucleic acids.

- **Cell Membrane** – lipids of low solubility in water, mixed with water-insoluble proteins, provide a semipermeable barrier to the outside.

- **Cell Wall** – crosslinked polymers with mechanical strength for support.

• **Ribosome** – an assembly of polymers of protein and RNA nature which can catalyze the production of proteins essential to life.

• **Flagellum** - molecular machine built from fibrous proteins engineered to propel or move.

• **DNA** – polymer acting as the genetic information store.

As biosphere materials are derived from seawater and seawater is squeezed out from the lithosphere, there is a one-to-one relationship between them as shown in Figure 6-2. Furthermore, water is an essential part of the biosphere. All the chemicals in the cell are essential to the energy and activity of particular cells.

The molecular reactions carried out in the cells can be summarized as **metabolism**. The chemicals that perform the reactions are collectively known as **metabolites**.

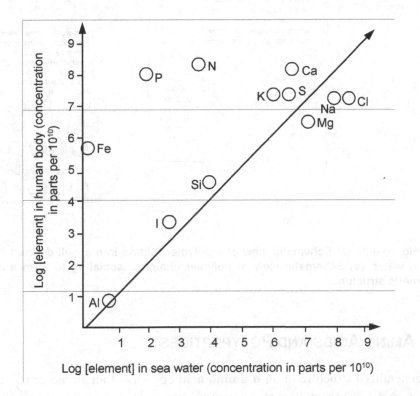

Figure 6-2. Comparison of the concentrations of elements in the human body and in sea water.

The polarity of water molecules and the resultant restoration of the hydrophilic and hydrophobic nature can affect the ordered structure in biology. This is illustrated in the monomeric (e.g., amino acids), oligomeric (e.g., small polypeptides), and high molecular weight polymers (e.g., proteins). Fig. 6-3 illustrates two different ways by which ordered biomolecular structures can be formed. This suggests that **multimeric structure** can also be formed.

Chains can fold into an isolated center (see Fig. 6-3a), or chains can form ordered **double helixes** (see Fig. 6-3b). These are explained in the next section.

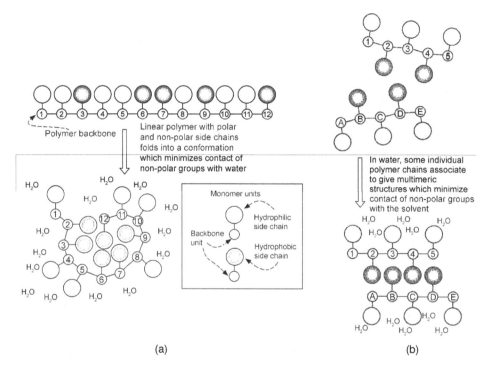

(a)   (b)

Figure 6-3. (a) Schematic view of a polymer folding into a well-defined shape in water. (b) Schematic view of polymer chains associating to form a multimeric structure.

6.2 AMINO ACIDS AND POLYPEPTIDES

The generalized structure of an **α-amino acid** consists of an amino group, a hydrogen, and substituent R at the α-carbon, linked by a carbonyl group. Those amino acids can form oligomers and polymers through the peptide linkages.

Depending on the difference of the R group, there are twenty different α-amino acids for protein buildup. These are **3-letter** or **1-letter** abbreviations (see Table 6-1).

Table 6-1. One- and Three-Letter Symbols for the Amino Acids

A	Ala	Alanine	M	Met	Methionine
B	Asx	Asparagine or aspartic acid	N	Asn	Asparagine
C	Cys	Cysteine	P	Pro	Proline
D	Asp	Aspartic acid	Q	Gln	Glutamine
E	Glu	Glutamic acid	R	Arg	Arginine
F	Phe	Phenylalanine	S	Ser	Serine
G	Gly	Glycine	T	Thr	Threonine
H	His	Histidine	V	Val	Valine
I	Ile	Isoleucine	W	Trp	Tryptophan
K	Lys	Lysine	Y*	Tyr	Tyrosine
L	Leu	Leucine	Z	Glx	Glutamine or glutamic acid

Generalized protein structures are illustrated by Fig. 6-4, consisting of different amino acids residues, which amount to 30,000 within the human body. A protein is described by its sequence, indicating the order of amino acid residues in the chain (See Figure 6-5).

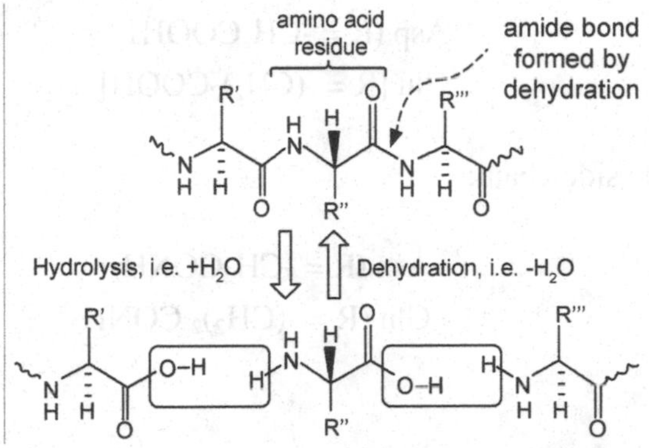

Figure 6-4. Generalized protein structure: amino acids are joined in a linear chain.

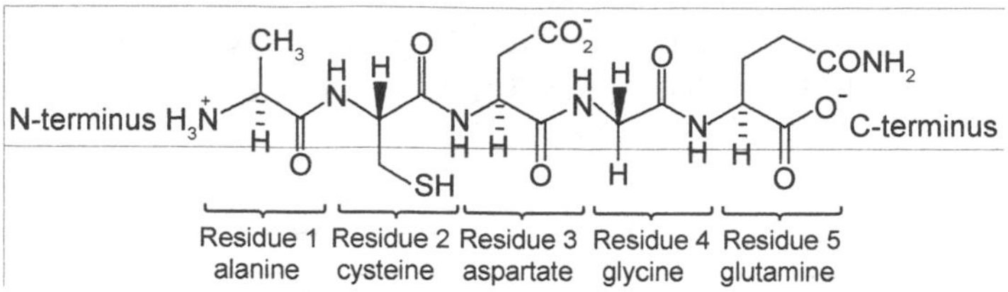

Figure 6-5. A structure of a representative peptide.

Amino acid residue can be classified according to the following:

- Basic skeleton of amino acid is **H₂N-CH(R)-COOH**,

    Gly(R=H)   (a)*

- With hydrocarbon side chain:

    Ala (R = Me)   (b)

    Val (R = Me₂)   (c)

    Ile (R = Me, Ee) Leu (R = iPr)   (d)

    Ile(R=Me, Ee)   (e)

    Phe (R = Bz)   (g)

- With carboxylic acid side chains:

    Asp (R = -CH₂COOH)   (n)

    Glu [R = -(CH₂)₂COOH]   (o)

- With amide side chain:

    Asx (R = -CH₂CONH₂)   (q)

    Gln (R = -(CH₂)₂ CONH₂   (r)

---

* For structures, see Fig. 6-6.

- With basic nitrogen-containing side:

$$\text{Lys } [R = -(CH_2)_4NH_2] \qquad\qquad\qquad\qquad (s)$$
$$\text{Arg } [R = -(CH_2)_3NH - C(NH_2)_2] \qquad\qquad (t)$$

- With hydroxyl functional groups:

$$\text{Ser } (R = -CH_2OH) \qquad\qquad\qquad\qquad (g)$$
$$\text{Thr } (R = -CH(Me)OH \qquad\qquad\qquad\qquad (l)$$
$$\text{Tyr } [R = CH_2 - OH)\text{- Bz }] \qquad\qquad\qquad (m)$$

- Sulfur-containing:

$$\text{Cys } (R = -CH_2SH) \qquad\qquad\qquad\qquad (k)$$
$$\text{Met } [R = -(CH_2)_2SMe] \qquad\qquad\qquad\qquad (i)$$

- Nitrogen-heterocyalics containing:

$$\text{Trp } (-CH_2\, \alpha\text{-indole}) \qquad\qquad\qquad\qquad (h)$$
$$\text{His } (-CH_2\, \alpha\text{-imidazole}) \qquad\qquad\qquad\qquad (p)$$

- An unusual structure is Pro (proline) which has a 3-carbon unit chain fused on the α-amino nitrogen side to become a 5-membered heterocyclic ring.

This unusual structure

$$(f)$$

has significant effect on the 3-dimensional structure of protein. Proline often acts to terminate α-helices and is called a **helix breaker**.

Twenty amino acids are listed in Fig. 6-6 together with a peptide formed by them. These amino acids can form the nucleotide sequence of DNA as can be seen later.

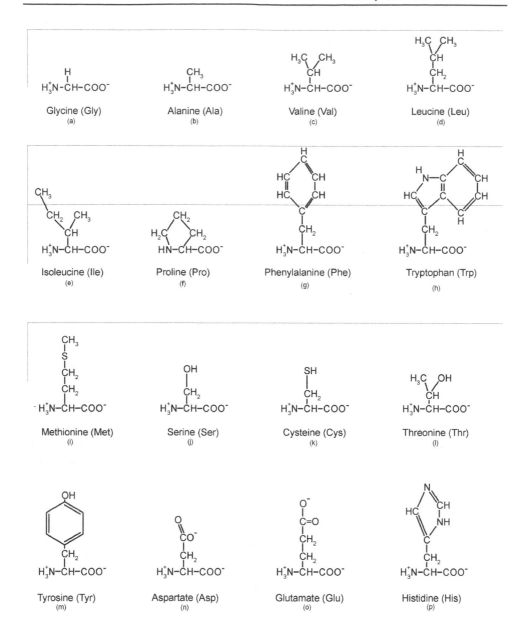

Figure 6-6. The structure of the 20 amino acids, which can be coded for by the nucleotide sequence of DNA, and the structure of a peptide of some of these amino acids (the last entry). (Continued on next page)

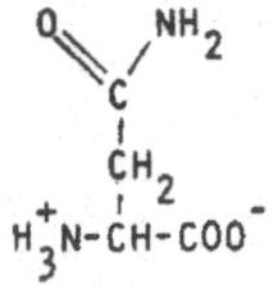

Asparagine (Asx) Glutamine (Gln) Lysine (Lys) Arginine (Arg)
     (q)   (r)   (s)   (t)

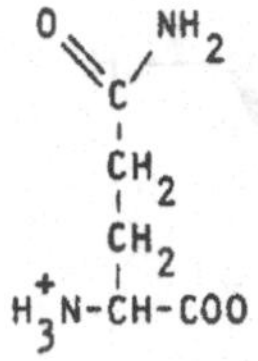

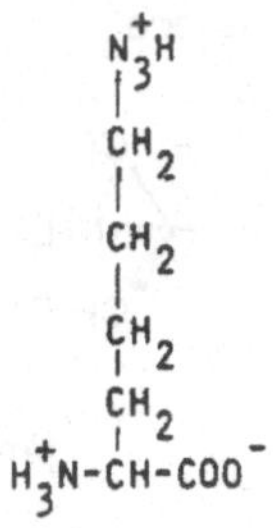

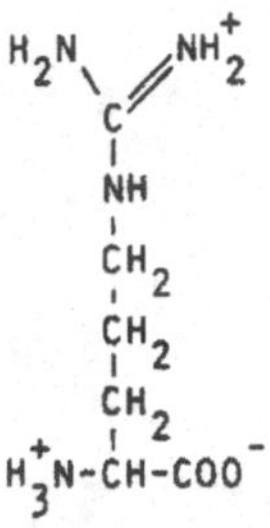

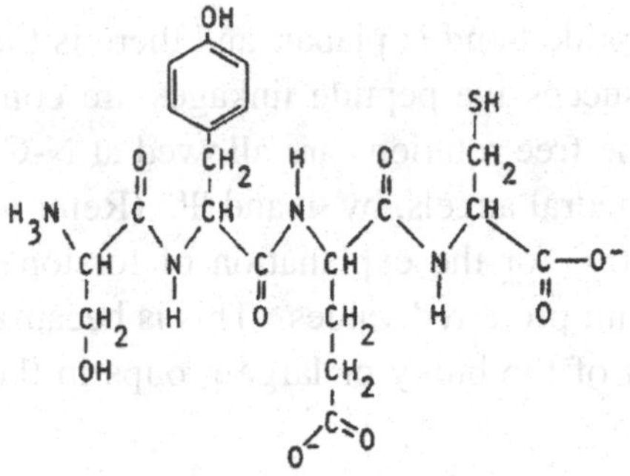

Serylphenylalanylglutamylcysteine
              (u)

Figure 6-6. Continued.

6.2.1 Stereochemistry

All twenty amino acids except glycine have four different substituents attached to the α-carbon. As indicated by the carbon tetrahedron arrangement, there are two different configurations as seen in the stereochemical view of mirror images as from Fig. 6-7.

Different configurations related to the mirror image are known as **enantiomers**, and the α-carbon of the 19 amino acids is called a **chiral center**. (Naturally occurring amino acids usually have L-configuration.) To describe the absolute configuration about the chiral center, a notation known as the **Cahn-Ingald-Prolog (CIP) rule** is used. As indicated by Fig. 6-8, there are assigned priorities to the R-side of the groups. In anticlockwise fashion, N > C (carbonyl) > C (sidechain) > H.

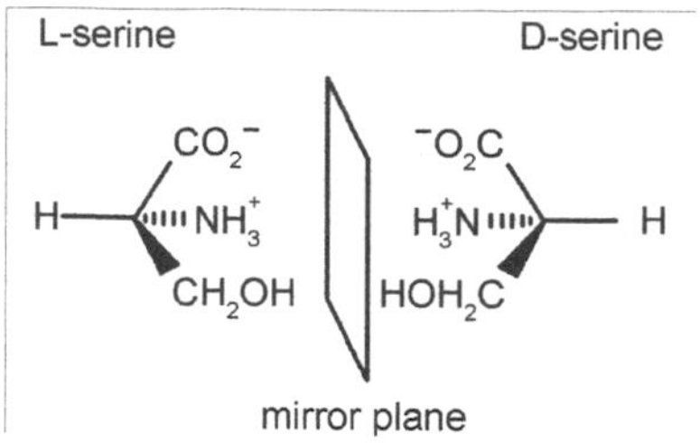

Figure 6-7. The enantiomers of serine.

Actually, the stereochemistry of the peptide bond is planar, and there is the possibility of trans and cis forms. If two successive peptide linkages are connected as shown in Fig. 6-9 in trans form, the free rotations are allowed at N-C^{α} and C^{α}-C bonds. These are referred to as dihedral angels, by Φ and Ψ. (Refer to Chapter 3 of this volume, Organic Chemistry, for the explanation of torsional angle.) Both Ψ and Φ angles have only certain preferred values. This is because free rotation is restricted by the steric effect of the bulky or large groups in the available space.

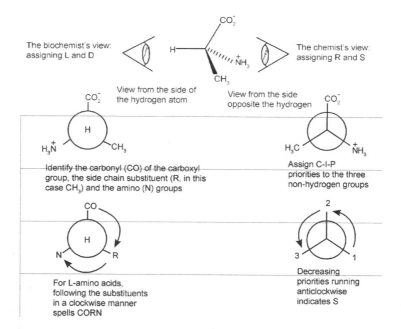

Figure 6-8. Chital characteristics identifying the configuration of amino acids: alanine as an example. View for the left-hand side is L- and D- configuration. View for the right hand side is Cahn–Ingold–Prelog rule for the configuration S- and R-.

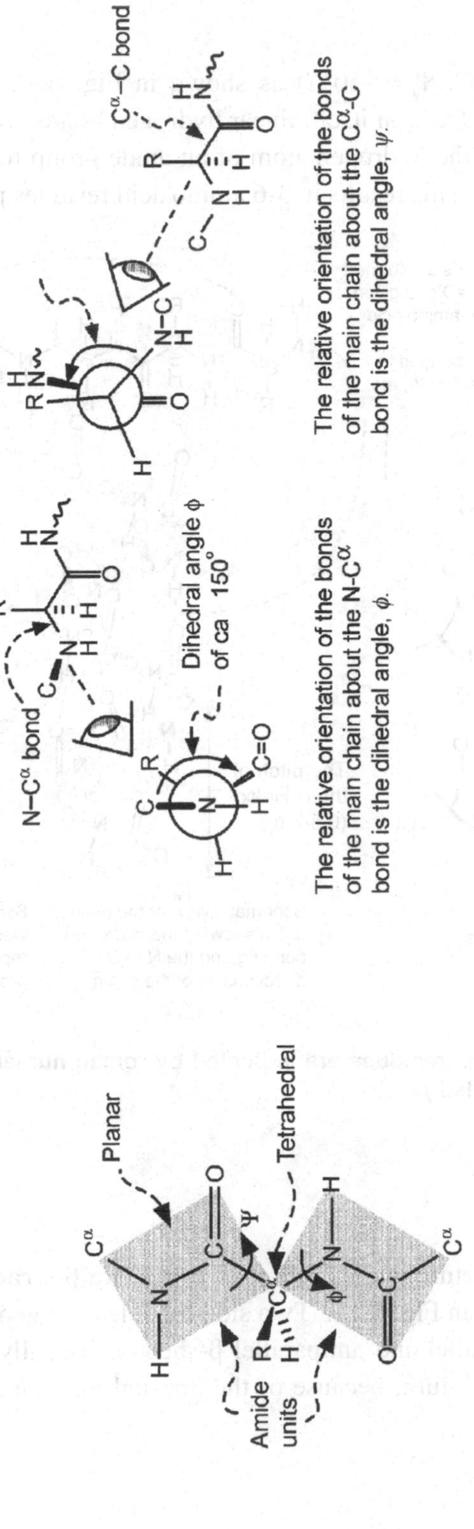

C^α–C bond

The relative orientation of the bonds of the main chain about the C^α-C bond is the dihedral angle, ψ.

N–C^α bond

Dihedral angle ϕ of ca 150°

The relative orientation of the bonds of the main chain about the N-C^α bond is the dihedral angle, ϕ.

Figure 6-9. The dihedral angles Ψ and ϕ.

Planar

Tetrahedral

C^α

Ψ

ϕ

C^α

Amide units

6.2.2 α-Helix

The α-helix (Φ = -60°C, Ψ = -50°C) as shown in Fig. 6-10 is quite compact whereas the chain turns back on itself, linear hydrogen bonds are formed between a carbonyl oxygen and the hydrogen atom of an anide group four residues down the chain (from I to V). This results in 3-6 amino acid residues per turn.

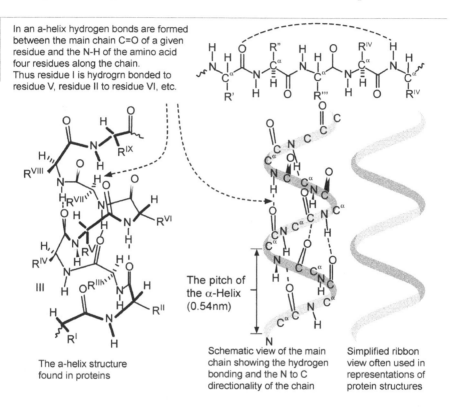

In an a-helix hydrogen bonds are formed between the main chain C=O of a given residue and the N-H of the amino acid four residues along the chain. Thus residue I is hydrogrn bonded to residue V, residue II to residue VI, etc.

The a-helix structure found in proteins

The pitch of the α-Helix (0.54nm)

Schematic view of the main chain showing the hydrogen bonding and the N to C directionality of the chain

Simplified ribbon view often used in representations of protein structures

Figure 6-10. α-Helix (residues are indicated by roman numerals from Rᴵ to Rᴵˣ, again the Cᵅ is labeled).

6.2.3 β-Sheets

Another secondary structure due to hydrogen bonds are **β-strands** (Φ = -120°C, Ψ = +120°C), as shown in Fig. 6-11. Two side-by-side arrangements are possible for the orientation (parallel and antiparallel β-sheets). Usually Pro-Gly is a favorable sequence for a β-turn, because of the unusual individual conformational preferences.

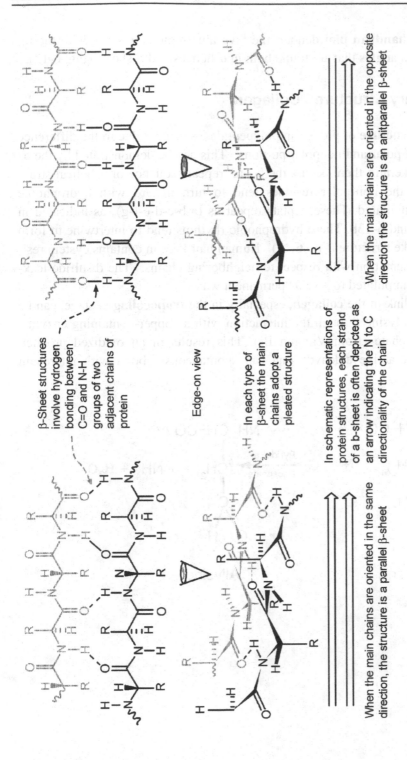

β-Sheet structures involve hydrogen bonding between C=O and N-H groups of two adjacent chains of protein

Edge-on view

In each type of β-sheet the main chains adopt a pleated structure

In schematic representations of protein structures, each strand of a b-sheet is often depicted as an arrow indicating the N to C directionality of the chain

When the main chains are oriented in the opposite direction the structure is an anitparallel β-sheet

When the main chains are oriented in the same direction, the structure is a parallel β-sheet

Figure 6-11. The two types of β-sheet structures.

A **Ramachandran plot** denotes the favorable regions α and β with the favorable dihedral angles that are found both in α-helices and β-sheets (Fig. 6-12).

6.2.4 Tertiary Structure: Collagen

The **tertiary structure** of protein in polypeptides is usually due to the difference in the region of polar and nonpolar portions. This can be demonstrated by the **α-keratin**. In α-keratin there occurs the heptat repeat sequence of α-helical structure in which the non-polar residues tend to turn inward with hydrophobic sidechain interdigitated. These heptats repeat as $(a\text{-}b\text{-}c\text{-}d\text{-}e\text{-}f\text{-}g)_n$ as indicated in Figures 6-13a and 6-13b. These hydrophobic rhythms tend to intertwine to form a coiled rope-like structure (Fig. 6-13). Human hair keratin contains cystein residue in fixed orientation with respect to neighboring chains. The disulfide leakages may be manipulated to give a "permanent wave."

The crosslink in the **collagen**, especially in the **tropocollagen** stage, can be originated from lysine through the interaction with a copper-containing enzyme, pyridoxal phosphate (from Vitamin B_6). This results in an oxidized product, allysine, a very active aldehyde that can spontaneously bond with interchain bonds.

Lysine   Allysine

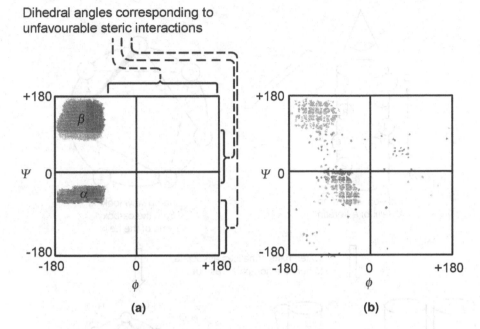

Figure 6-12. (a) Stylized Ramachandran plot for an L-amino acid. (b) Ramachandran diagram in which the experimental Φ, ψ angles for a range of residues other than glycine are shown for a representative set of proteins.

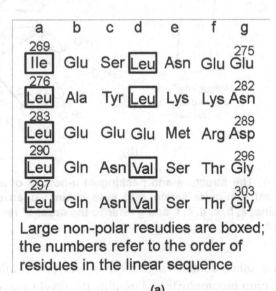

(a)

Figure 6-13. (a) A portion of the sequence of mouse α-keratin illustrating the heptat repeat.

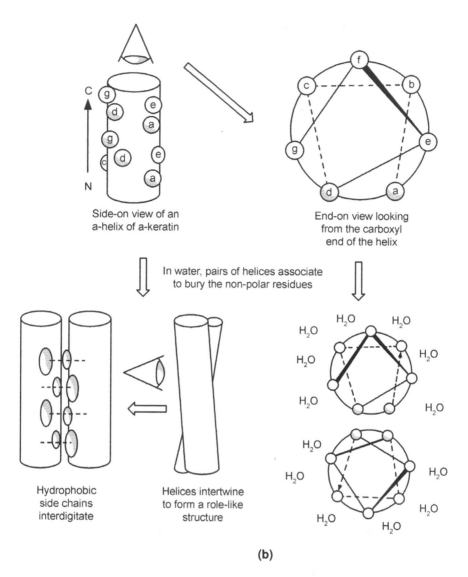

Side-on view of an
a-helix of a-keratin

End-on view looking
from the carboxyl
end of the helix

In water, pairs of helices associate
to bury the non-polar residues

Hydrophobic
side chains
interdigitate

Helices intertwine
to form a role-like
structure

(b)

Figure 6-13. (b) The structure and packing of α-helices of α-keratin. Shaded circles represent non-polar amino acid side chains; open circles represent other side chains; a, b, c, d, e, f, and g refer to the order of residues in the heptad repeat as shown.

If humans and other animals ingest the sweet pea (*Lathyrus odoratus*), which contains β-amino propionitrile and inhibits the lysyl oxidase, they can contract the disease **lathyrism**. One collagen deficiency disease is **Ehlers-Damlos**

syndrome type V, causing hypermobile joints and hyper extensibility of the skin. Nucleation for bone formation in collagen fiber is illustrated by Fig. 6-14.

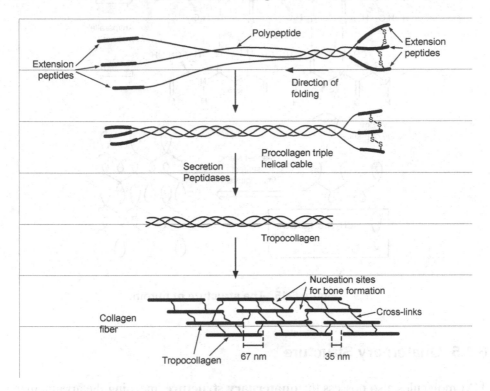

Figure 6-14. Role of the extension peptides in the folding and secretion of pro-collagen. Once secreted out of the cell, the extension peptides are removed and the resulting tropocollagen molecules aggregate and are cross-linked to form a microfibril.

Similarly, the crystalline β-sheet regions of cocoon silk of the **silkworm** *Bombyx mori* are responsible for mechanical strength. The region consists of sextets $(Gly-Ala-Gly-Ala-Gly-Ser)_n$ as shown in Fig. 6-15. Collagen has a triple-helical structure, which has strong mechanical strength.

The tertiary structure of **triose phosphate isomerase (TIM)** comprises alternate β strands and α-helices linked by turns. The overall shape is like a barrel (Fig. 6-16). The amphophilic helices consist of a non-polar face polarity toward the protein core and a predominately polar surface pointing into the external solution. There are eight β-strands arranged in a parallel β-sheet.

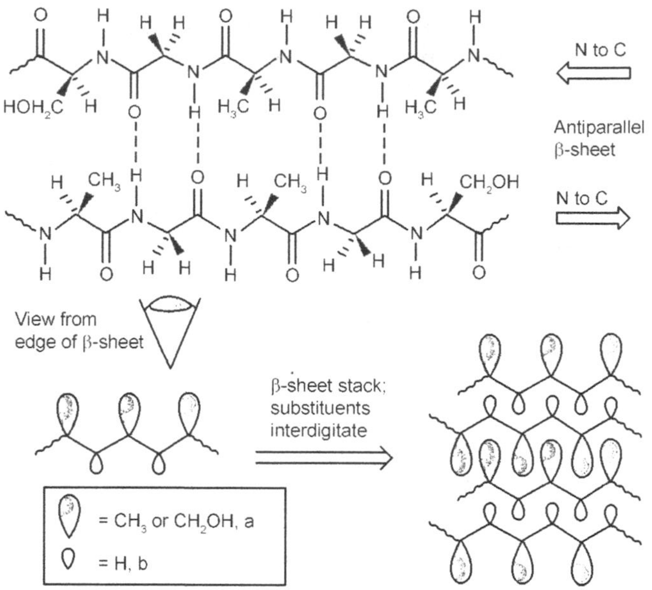

Figure 6-15. The structure of fibroin.

6.2.5 Quaternary Structure

TIM molecules also possess the **quaternary structure**, meaning the arrangement of the particular part of the surface of each protein chain will demonstrate the catalytic activity of the enzyme. Actually the exhibition of the active site is the nature of the quaternary structure of TIM. For the catalytic mechanism, TIM can catalyze the interconversion of the phosphorylated sugars, D-glyceraldehyde phosphate(D-GAP) and dihydroxyacetone phosphate (DHAP). A significant amount of experiments and evidence have shown that the residue Glu-165 of the β6 chain of TIM (see Fig. 6-16) is responsible for the deprotonation and reprotonation by the catalyst. To assist the catalysis, nearby positions of Lys-13 and His-95 also participate. This actually presents a fixed space conformation for the quaternary structure. Fig. 6-17 illustrates a possible catalytic mechanism. The overall reaction in the presence of TIM proceeds a billion times faster than the process in solution catalyzed by acetate im. Often, this is referred to as **turnover frequency**.

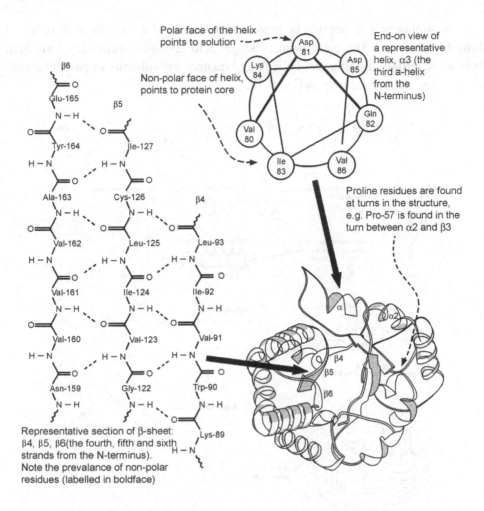

Polar face of the helix points to solution

Non-polar face of helix, points to protein core

End-on view of a representative helix, α3 (the third a-helix from the N-terminus)

Proline residues are found at turns in the structure, e.g. Pro-57 is found in the turn between α2 and β3

Representative section of β-sheet: β4, β5, β6(the fourth, fifth and sixth strands from the N-terminus). Note the prevalance of non-polar residues (labelled in boldface)

Figure 6-16. The tertiary structure of triose phosphate isomerase (TIM).

Figure 6-17. A mechanism for the interconversion of D-GAP catalysed by TIM.

In summary, polypeptide or protein structure can be thought of in terms of **four hierarchies**. There are primary single acid residues; secondary, hydrogen bondings; tertiary, folding and helical; and quaternary, subunits in particular spatial arrangements, as shown in Fig. 6-18.

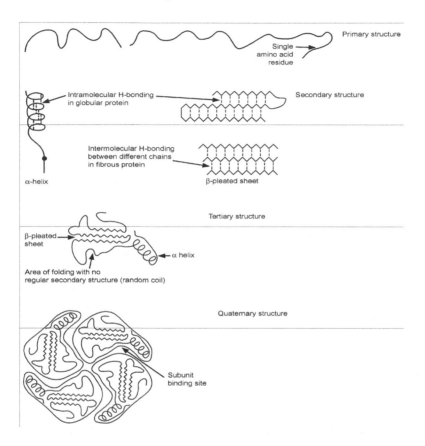

Figure 6-18. Protein structure can be thought of in terms of four hierarchies.

6.3   ENZYME KINETICS

Enzyme catalysts as well as enzyme inhibitors are important to humans. Let us cite one of the inhibitor examples. Penicillin, a well-known antibiotic, inhibits a family of bacterial enzymes, which catalyze the production of strong outer cell walls, since we do not have the equivalent enzymes or cell walls. Penicillin can kill bacteria, but is innocuous to humans. Drug therapy is based on this competitive theory as shown in Fig. 6-19.

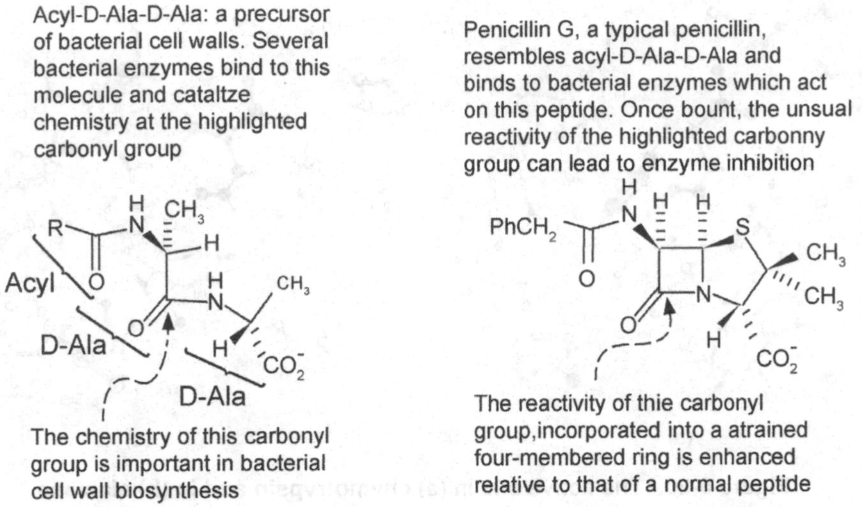

Acyl-D-Ala-D-Ala: a precursor of bacterial cell walls. Several bacterial enzymes bind to this molecule and cataltze chemistry at the highlighted carbonyl group

Penicillin G, a typical penicillin, resembles acyl-D-Ala-D-Ala and binds to bacterial enzymes which act on this peptide. Once bount, the unsual reactivity of the highlighted carbonny group can lead to enzyme inhibition

The chemistry of this carbonyl group is important in bacterial cell wall biosynthesis

The reactivity of thie carbonyl group, incorporated into a atrained four-membered ring is enhanced relative to that of a normal peptide

Figure 6-19. Penicillins, important antibiotics, are inhibitors of bacterial enzymes.

One important consideration of enzyme kinetics of the 3-dimensional complex folding geometry is the so-called **charge relay system**, as seen in the following example:

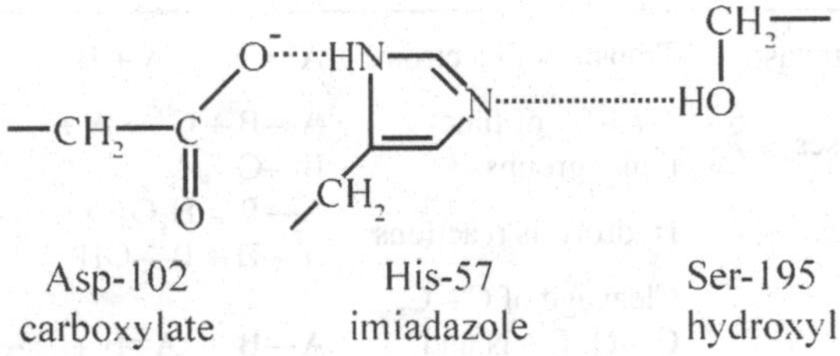

Asp-102	His-57	Ser-195
carboxylate	imiadazole	hydroxyl

For most proteolytic enzymes, the hydrolysis of the bond will take particular geometry to the **charge relay system**. For example, the case of chymotrypsin and elastase are similar; His-57 is from chain B and Ser-195 is from chain C. Asp-102 is from another chain (Fig. 6-20a and 6-20b depict the active site of chymotrypsin and elastase respectively). The combined cooperative bio-catalytic effect can be understood. Usually the carboxylate oxygen is highly nucleophilic (see the scheme) and can be easily inhibited and blocked by disopropyl-fluorophosphate structure.

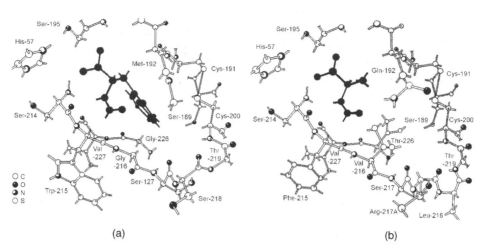

(a)   (b)

Figure 6-20. The active site in (a) chymotrypsin and in (b) elastase.

Classification of enzymes is summarized in Table 6-2, and certain examples are given.

Table 6-2. International Cassification of Eymes

Class	Name	Type of reaction catalyzed		Example
1	Oxidoreductases	Transfer of electrons	$A^- + B \rightarrow A + B^-$	Alcohol dehydrogenase
2	Transferases	Transfer of functional groups	$A{-}B + C \rightarrow A + B{-}C$	Hexokinase
3	Hydrolases	Hydrolysis reactions	$A{-}B = H_2O \rightarrow A{-}H + B{-}OH$	Trypsin
4	Lyases	Cleavage of C – C, C – O, C – N and other bonds, often forming a double bond	$A{-}B \rightarrow A{=}B + X{-}Y$ X Y	Pyrivate decarboxylase
5	Isomerases	Transfer of groups within a molecule	$A{-}B \rightarrow A{-}B$ X Y Y X	Maleate isomerase
6	Ligases (or synthases)	Bond formation coupled to ATP hydrolysis	$A + B \rightarrow A{-}B$	Pyrivate carboxylase

Some enzymes require the presence of **cofactors**, small protein units, to function. Cofactors may be inorganic ions or complex organic molecules called **coenzymes**. A cofactor that is conveniently attached to the enzyme is called a prosthetic group. A **haloenzyme** is the catalytically-active form of the enzyme with its cofactor. An **apoenzyme** is the protein-part of the coenzyme. Many co-enzymes are derived from the dietary vitamin precursors, and deficiencies in them can lead to certain diseases. Nicotine made adenine dinucleotide phosphate (NADP$^+$) and flavin adenine dinucleotide (FAD) as shown in Table 6-3 are widely occurring coenzymes involved in redox reactions.

Table 6-3. Some Common Coenzymes, Their Vitamin Precursors and Deficiency Diseases

Coenzyme	Precursor	Deficiency disease
Coenzyme A	Pantothenic acid	Dermatitis
FAD, FMN	Riboflavin (vitamin B$_2$)	Growth retardation
NAD$^+$, NADP$^+$	Niacin	Pellagra
Thamine pyrophosphate	Thiamine (vitamin B$_1$)	Beriberi
Tetrahydrofolate	Folic acid	Anernia
Deoxyadenosyl cobalamin	Cobalamin (vitamin B$_{12}$)	Pernicious anemia
Co-substrate in the hydroxylation of proline in collagen	Vitamin C (ascorbic acid)	Scurvy
Pyridocxal phosphate	Pyridoxine (vitamin B$_6$)	Dermatitis

6.3.1 Michaelis-Menten Equation

Enzyme catalysis depends on the particular conformation, which was situated and located in such a fashion that the **turnover frequency** is enhanced. If E, S, and P represent the enzyme, the substrate, and the product respectively, the reaction scheme can be written as

$$E + S \underset{k_{-1}}{\overset{k_1}{\rightleftharpoons}} ES \underset{k_{-2}}{\overset{k_2}{\rightleftharpoons}} P + E \qquad [6\text{-}1]$$

The kinetic constants can be written as k; at steady state the rate of ES formation is equal to rate of ES removal, ES being the complex intermediate.

$$\frac{d(ES)}{dt} = 0 \tag{6-2}$$

The usual way of kinetics can be written as

$$-\frac{d(ES)}{dt} = (k_{-1} + k_2)(ES) - k_1(E)(S) - k_{-2}(P)(E) \tag{6-3}$$

$$= 0$$

dividing by (E), we have

$$\frac{(k_{-1} + k_2)(ES)}{(E)} = k_1(S) - k_{-2}(P) \tag{6-4}$$

Assuming $k_1(S) \gg k_{-2}(P)$ which is true in reality.

$$\frac{(E)}{(ES)} = \frac{(k_{-1} + k_2)}{k_1(S) - k_{-2}(P)}$$

$$\approx \frac{(k_{-1} + k_2)}{k_1(S)} \tag{6-5}$$

Let

$$K_S \approx \frac{(k_{-1} + k_2)}{k_1} \tag{6-6}$$

K_s here is the **Michaeles-Menten** constant or the **saturation constant**. Remember the unit is similar to the unit in concentration for the reactant.

Now E_0 is the starting enzyme concentration, which is

$$(E_0) = (E) + (ES) \tag{6-7}$$

Equation 6-5 becomes

$$[(E_0) = (E)](S) = K_s(ES) \tag{6-8}$$

$$(ES) = \frac{(E_0)(S)}{K_S + (S)} \tag{6-9}$$

Let

$$V_S = K(E_0)$$

here V_S is the maximum velocity.

$$K(ES) = \frac{V_S(S)}{K_S + (S)} = \frac{V_S}{1 + \dfrac{K_S}{(S)}}$$ [6-10]

Also let $V = K(ES)$

$$V = \frac{V_S}{1 + \dfrac{K_S}{(S)}} = \frac{V_S(S)}{K_S + (S)}$$ [6-11]

From the above equation

$$S = \infty \ or \ V = V_S$$ [6-12]
$$S = K_S \ or \ V = V_S/2$$

The condition set in Eq. [6-12] can be illustrated in Fig. 6-21.

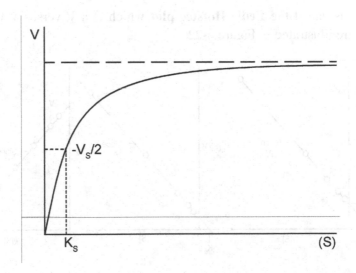

Figure 6-21. Schematic plot of initial velocity, V, versus substrate concentration, S, for an enzyme reaction involving one substrate.

Eq. [6-11] is often referred to as the **Monad form** and is extremely useful in biochemical kinetics. Arrangement of the Monad form can be

$$V^{-1} = V_S^{-1} + \frac{K_S}{V_S} S^{-1}$$

[6-13]

A plot of V^{-1} versus S^{-1} gives a straight line whereas the slope and intercept can evaluate K_S and V_S. This method is called the **Lineweaver-Barke double reciprocal plot**. This gives much weight at lower substance concentration.

Also the equation can be written as

$$\frac{S}{V} = \frac{K_S}{V_S} + \frac{S}{V_S}$$

[6-14]

A plot of S/V versus S also yields a straight line, which can evaluate K_S and V_S. Another method involves multiplying by V and V_S on both sides of Eq. [6-13].

$$V = V_S - V\frac{K_S}{S}$$

[6-15]

This is called the **Eedie-Hofstee plot** which is a V versus V/Vs. All three methods are illustrated in Figure 6-22.

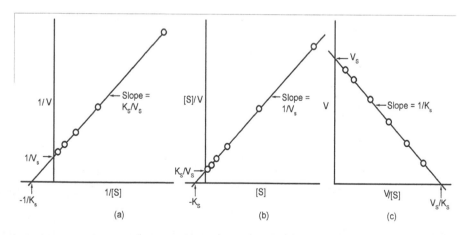

Figure 6-22. Plots used to determine V_s and K_s from initial velocity data for a single substrate reaction. (a) 1/V versus 1/[S]; (b) [S]/V versus [S]; (c) V versus V/[S]. Here V is the initial velocity and [S] is the substrate concentration.

[Example 6.1] When the monod form is applied to the substrate concentration of microbial growth related to waste water treatment process, K_s of *Zooglea ramigera*, is determined to be 0.3 mg/L. The maximum growth rate is 5.5 d^{-1}. For *Sphaalrotilus nateus*, K_s is 10 mg/L, and maximum growth ratio is 6.5 d^{-1}. Predict which bacteria will be predominant in a mixed culture at low substrate concentration (typically 1 mg/L) and at high substrate concentration (typically 10 mg/L).

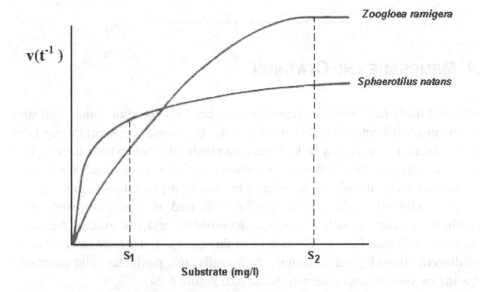

Figure 6-23. The effect of the concentration of growth-limiting substrate on the outcome of the competition between two organisms with different values of μ_m and K_s. At a low substrate concentrations (S_1), *Sphaerotilus natans* will be the predominant organism due to its higher substrate affinity (lower K_s). At a higher substrate concentration (S_2), *Zooglea ramigera* will predominate as it has a higher maximum specific growth rate (μ_m).

From Equation [6-11],

$$V = V_m \cdot \frac{S}{K_S + S}$$

Assuming *Zooglea ramigera* is 1 and *Sphaalrotilus nateus* is 2, then we have:

$$Ks_1 = 0.3 \text{ mg / L}$$
$$Vm_1 = 5.5 \text{ d}^{-1}$$
$$Ks_2 = 10 \text{ mg / L}$$
$$Vm_2 = 6.5 \text{ d}^{-1}$$

When substrate concentration is low (e.g., $S = 1$ mg/L), then $V_1 = 4.23$ d^{-1}, and $V_2 = 0.59$ d^{-1}. Therefore, *Zooglea ramigera* is predominant (about 7 times more) when substrate concentration is high, e.g., $S = 10$ mg/L. Consequently, $V_1 = 5.34$ d^{-1} and *Sphaalrotilus nateus* will be dominant (about 6 times more).

6.4 MEMBRANE AND CLATHRIN

Lipids and membrane-mimetic chemistry have been reviewed in colloid and surface chemistry (Chapter 5) earlier in this book. In general, transport through the cell membrane is not an easy task. Both **exocytosis** and **endocytosis** have to follow the natural principal. For example, when protein has been made from a cell, it is secreted across the plasma membrane by way of the Golgi apparatus through byways of **clathrin**. The clathrin can be visualized as a protein vesicle as a **triskelion**. Clathrin usually behaves as icosahedran and has exactly the same structure of fullerene C_{60} (see Chapter 11 of this book). It has been referred to as **biofullerene** (three legged structure). Apparently, this particular solid geometry helps the movement across the membrane (see Figure 6-24).

The same principle also applies to **phagocytosis** or **pinocytosis**, and **endocytosis** in which a coated protein similar to clathrin is prepared (Fig. 6-25).

6.4.1 Tubulin, Microtubule and Centriole

Proteins and protomers all have unique properties in symmetry. The lenticular shape, the triangle, the square and the pentagon respectively indicate the twofold, threefold, fourfold and fivefold rotational axes as shown in Fig. 6-26.

Clathrins are major components of **coated vesicles**. By electron micrographs, each vertex of the polyhedron is the center of the triskelion and its edge of ca. 150 Å in length. Such frameworks, which consists of 12 pentagons and a variable number of hexagons, are the most economical way of enclosing spheroidal objects in polyhedral cages (Fig. 6-27).

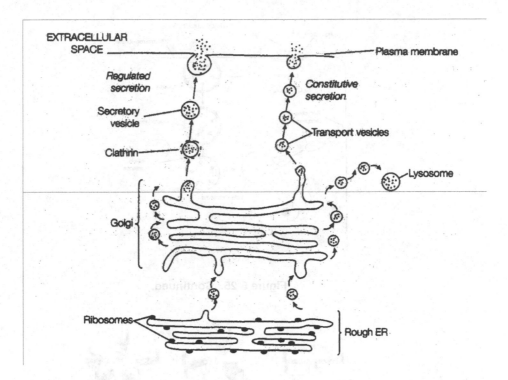

Fig. 6-24. Exocytosis of proteins by the constitutive and regulated seretory pathways.

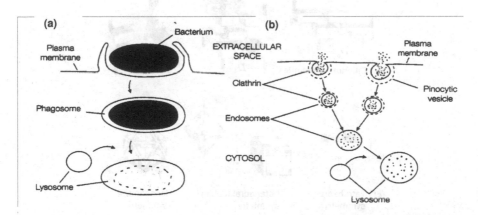

Figure 6-25. (a) Phagocytosis; (b) pinocytosis; (c) endocytosis.

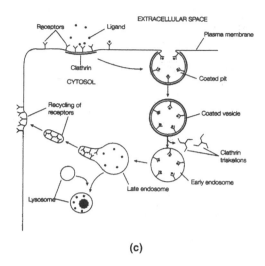

(c)

Figure 6-25. Continued.

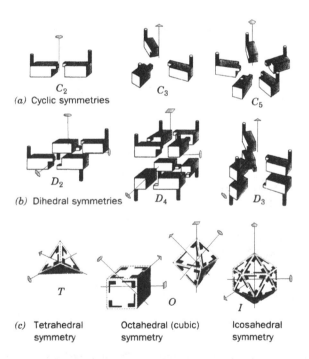

Figure 6-26. (a) Assemblies with the cyclic symmetries C_2, C_3 and C_5. (b) Assemblies with the dihedral symmetries D_2, D_4 and D_3. In these objects, a twofold axis is perpendicular to the vertical two-, four- and threefold axes. (c) Assemblies with T, O, and I symmetry. Note that the tetrahedron has some but not all of the symmetry elements of the cube, and that the cube and the octahedron have the same symmetry.

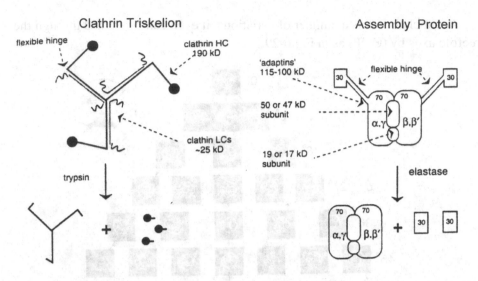

Figure 6-27. Summary of the structure of (a) clathrin and (b) assembly protein, the major cost constituents of clathrin-coated vesicles.

It has been found that the brain, liver and fibroblasts also contain different forms of polynomial shells. Figure 6-28 attempts to express these forms.

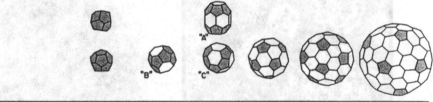

	Dodecahedron	Type "B"	Types "A" and "C"	Truncated icosahedron (brain)	(Liver)	Full icosahedron (fibroblasts)
No. of hexagons	0	4	8	20	30	60
No. of vertices	20	28	36	60	82	140
Predicted diameter (nm)	40	50	60	76	90	120
Predicted weight (MD)	13	18	23	38	53	77
No. of "facets" in deep etch	3–4	4–5	7–8	10	14	24
No. of "spikes" in neg. stain	4–5	5–6	6–8	8	12	16–17

Figure 6-28. Schematic drawings of several symmetrical basket designs of progressively larger sizes. The table catalogs their expected complement of polygons, expressed as a number of hexagons in addition to the 12 pentagons always needed to form a closed polygonal shell (pentagons are shaded in the drawings).

Again clathrin has a number of variations; it even can be rotated through the threefold axes by (Θ, Ψ) as in Fig 6-29.

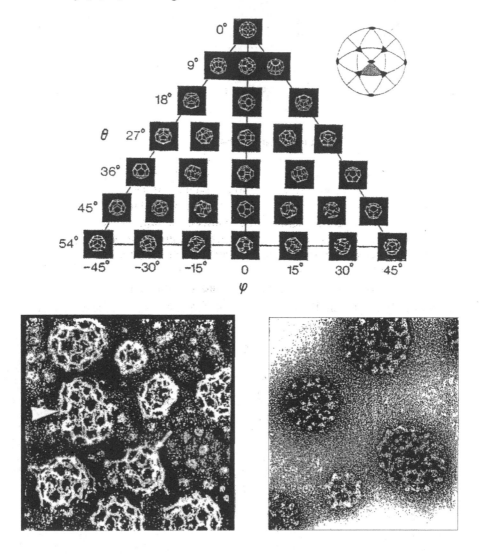

Figure 6-29. Gallery views covering the range indicated by the hatched area of the stereogram. The symmetry axes marked correspond to the 23 subgroup of 43m point group of the polyhedron. The full range of views can be filled by using the diagonal mirror lines passing through the threefold axes and then the rotation axes marked. It should be noted that the line φ=0° is an anti-mirror for the gallery of views. That is to say the view at (θ,-φ) is the same as that at (θ,φ) but reflected in a vertical mirror and rotated 180°.

The **microtubules** are cylindrical proteins present in all eukaryotic cells. They are composed of subunits of tubulin. These subunits are arranged on the cylinder by the same law of packing as shown in Fig. 6-30. Fig. 6-31 illustrates the formation of **centriole** from microtubule.

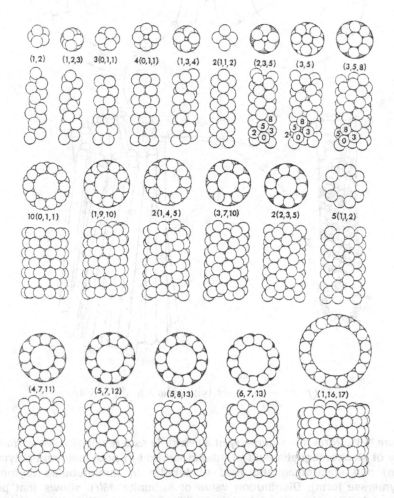

Figure 6-30. Tubular arrangement of spheres. Parallel projection onto the plane of tubular arrangements of spheres. Parameters are concerned with screw displacements along m- and n- parastichies. When the distance from 0 to *m+n* equals *2r*, the spheres are in contact alond the m-, n- and (m+n)- parastichies, the pattern is a triple contact or hexagonal packing pattern, and it may be designated (m, n, m+n), like (5, 8, 13).

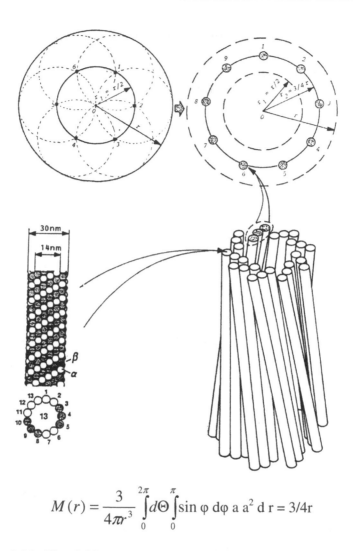

$$M\left(r\right)=\frac{3}{4\pi r^{3}}\int\limits_{0}^{2\pi}d\Theta\int\limits_{0}^{\pi}\sin\varphi\;d\varphi\;a\;a^{2}\;d\;r=3/4r$$

Figure 6-31. Nine-fold symmetry of a centriole as a result of *wave particle* synergy of electromagnetic field of tubulin subunit in pool (cell–non-polymerase form) and electromagnetic field of tubulin in microtubules (cytoplasm–polymerase form). Distribution value of subunits, M(r), shows that particle (microtubule) arrangement is 3/4 of r(bigger circle). The distance between microtubules on 3/4r is the same as the distance between the basic circle r_1=r/2. Oscillation (pulsing) of microtubules between r1 and r is the result of twisting. Only tubulin subunit on middle of length of centriole always has value 3/4r.

6.4.2 Superhelicity

Superhelicity is equivalent to **supercoiling** or **supertwisting** which arises from a biologically important topological property of covalently closed circular duplex DNA. The helical arrangement of the coat protein subunits of **tobacco mosaic virus (TMV)** is illustrated (Fig. 6-32). The circular duplex molecules as shown in Fig. 6-33 cannot alter without first clearing at least one of its polynucleotide strands. This type of topology can be expressed as

$$L = T + W \qquad\qquad [6\text{-}16]$$

in which L is the **Linking number**. This number indicates the number of times that one DNA strand winds about the other.

Figure 6-32. A model of TMV illustrating the helical arrangement of its coat protein subunits and RNA molecule. The RNA is represented by the chain exposed at the top of the viral helix. Only 18 turns (415 Å) of the TMV are shown which represent ~14% of the TMV rod. Based on Voet and Voet, *Biochemistry*, 1990.

T is the **twist** which is the number of complete revolutions that one polynucleotide strand makes about the duplex axis in the particular conformation. T is positive for a right handed duplex turn, so that the twist is normally the number of base pair divided by 10.5. Usually α is base pair and h° is 10.5 ± 0.1 bp for B-DNA in solution. Lastly, W is the **writhing number** which is the number of turns that duplex axis makes about the superhelix axis in the conformation. This is explained by Fig. 6-34. Packing of bacteriophage λ in the head with DNA is similarly done with the double strand material (Fig. 6-34A).

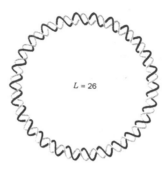

Figure 6-33. A schematic diagram of covalently closed circular duplex DNA that has 26 double helical turns. Its two polynucleotide strands are said to be topologically bonded to each other because, although they are not covalently linked, they cannot be separated without breaking covalent bonds.

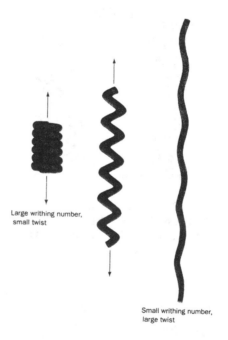

Large writhing number, small twist

Small writhing number, large twist

Figure 6-34. The difference between writhing and twist as demonstrated by a coiled telephone cord. In its relaxed state (*left*), the cord is in a helical form that has a large writhing number and a small twist. As the coil is pulled out (*middle*) until it is nearly straight (*right*), its writhing number becomes small as its twist becomes large.

(a)   (b)

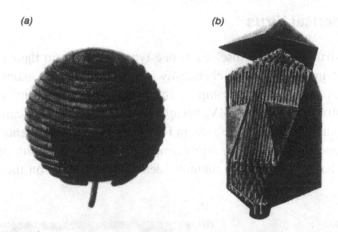

Figure 6-34A. Models for the packing of a double stranded DNA inside a phage head: (a) The concentric shell model in which the DNA is wound inward like a spool of twine about the phage's long axis. [After Harrison, S.C., *J. Mol. Biol.* 171, 579 (1982).] (b) The spiral-fold model in which the DNA strands run parallel to the phage's long axis with sharp 180° bends at the ends of the capsid. The folds themselves are radially arranged about the phage's long axis in spirally organized shells. [After Black, L.W., Newcomb, W.W., Boring, J.W., and Brown, J.C., *Proc. Natl. Acad. Sci.* 82, 7963(1985).]

[Example 6.2] For a DNA supercoil of $L=10$, T-10, $W=0$, what is the consequence of: a) unwind one turn   b) twist once
The consequence is that, they are interconvertible, as shown below.

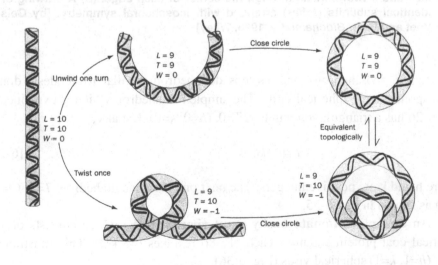

6.4.3 Spherical Virus

Spherical viruses limit themselves to one type of protein in their capsid. Since the coat protein substances are chemically identical, they must maintain the same configuration having relationship with their neighbors. Similar to quasi-equivalent strategy used by TMV, encapsulation of nucleic acid can be done using a polyhedral shell. We have seen in Fig. 6-26 that these elements in a tetrahedron, cube and icosahedron are equivalent, since the capsids with these symmetries would have 12, 24 or 60 subunits identically arranged on the surface of a sphere.

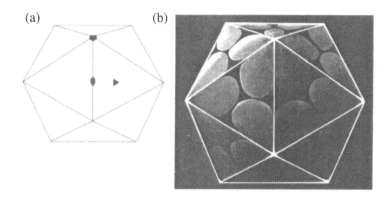

Figure 6-35. An icosahedron. (a) This regular polyhedron has 12 vertices, 20 equilateral triangular faces of identical size and 30 edges. It has a fivefold axis of symmetry through each vertex, a threefold axis through the center of each face and a twofold axis through the center of each edge. (b) A drawing of 60 identical subunits (*lobes*) arranged with icosahedral symmetry. [by Geis in Voet and Voet, *Biochemistry*, 1990, Wiley]

Large icosahedrons such as icosadeltahedron (similar to geodesic dome) are responsible for spherical virus. The simple icosahedron which is explained in Fig. 6-26 has a triangulation number, $T=0$, ($h=0$, $k=0$). Usually,

$$T=h^2+kh+k^2 \qquad\qquad [6\text{-}17]$$

where h and k are positive integers. The next one is icosadeltahedron, $T=1$, ($h=1$, $k=0$) as shown in Fig. 6-35.

An example is **tomato bushy stunt virus (TBSV)** which consists of 180 identical coat protein subunits, each of 386 residues (43 kD). This illustrates a $T=3$, ($h=1$, $k=1$) spherical virus (Fig. 6-36).

(a)   (b)

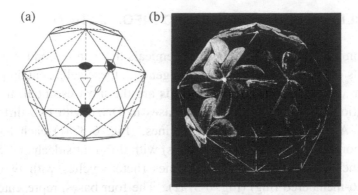

Figure 6-36. A T=3 icosadeltahedron. (a) This polyhedron has the exact rotational symmetry of an icosahedron (*solid symbols*) together with local sixfold, threefold and twofold rotational axes (*hollow symbols*). (b) A drawing of a T=3 icosadeltahedron showing its arrangement of 3 quasi-equivalent sets of 60 icosahedrally related subunits. [by Geis in Voet and Voet, *Biochemistry*, 1990, Wiley]

(c)

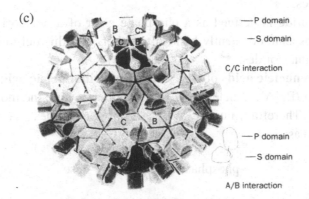

— P domain
— S domain

C/C interaction

— P domain
— S domain

A/B interaction

Figure 6-36. (c) The T=3 icosadeltahedral arrangement of TBSV's coat protein subunits. The subunits occur in three quasi-equivalent packing environments, A, B and C. The A subunits pack around exact fivefold axes, whereas the B subunits alternate with the C subunits about the exact threefold axes (local sixfold axes). The C subunits are also disposed about the local strict twofold axes, whereas the A and B subunits are related by local twofold axes.

6.5 NUCLEIC ACIDS AND GENETIC INFORMATION

Cells are complex localized capsules of chemicals that can reproduce or regenerate new cells of nearly-identical nature. In general, these cells contain permanent information. Nucleotides and **nucleic acids** are the most essential in structures and replications. There are four major **bases** in DNA with two different ring structures. Adenine and guanine are purines, (Fig. 6-37a1) each having two jointed carbon-nitrogen rings (heterocyclics) with different sidechains. Similarly, both thymine and cystosine are pyrimidines (heterocyclics with two nitrogen atoms in a 6-membered ring) (Fig. 6-37a2). The four bases, represented by one letter abbreviations, A, G, C, and T, are important since they carry genetic information. These four bases can bond covalently to a sugar. DNA is given the name deoxynucleoside because it contains the sugar deoxyribose (Fig. 6-37b1). There are four types of **deoxynucleoside** in DNA. They are: deoxyadenosine, deoxyguanosine, deoxythymidine, and deoxycytidine (Fig. 6-37b2). As a note, some unusual nucleosides are listed in Fig. 6-38. These are modified by transfer RNA molecules.

A **nucleotide** is defined as a phosphate ester of a nucleoside or a base, a sugar and phosphate covalently bonded together. By this definition, both DNA and RNA are nucleotides.

The term **nucleic acids** often refer to **deoxyribonucleic acid (DNA)** and ribonucleic acid (RNA). Nucleic acids are polymers, while the monomeric unit is a nucleotide. Therefore, nucleic acids are polynucleotides. A **nucleotide** can also be written as:

phosphate-5'-sugar-1'-N-base

In DNA, as indicated by Fig. 6-39, the sugar is deoxyribose. The sugar in RNA is a ribose (the 2'-OH present in RNA is susceptible to alkaline hydrolysis). In such a fashion, a **nucleoside** is:

-sugar-1'-N-base

Adenosine triphosphate (ATP) is:

P-P-P-s'-ribose-1-adenine

See Table 6-4 for further definition of nucleotide and nucleoside.

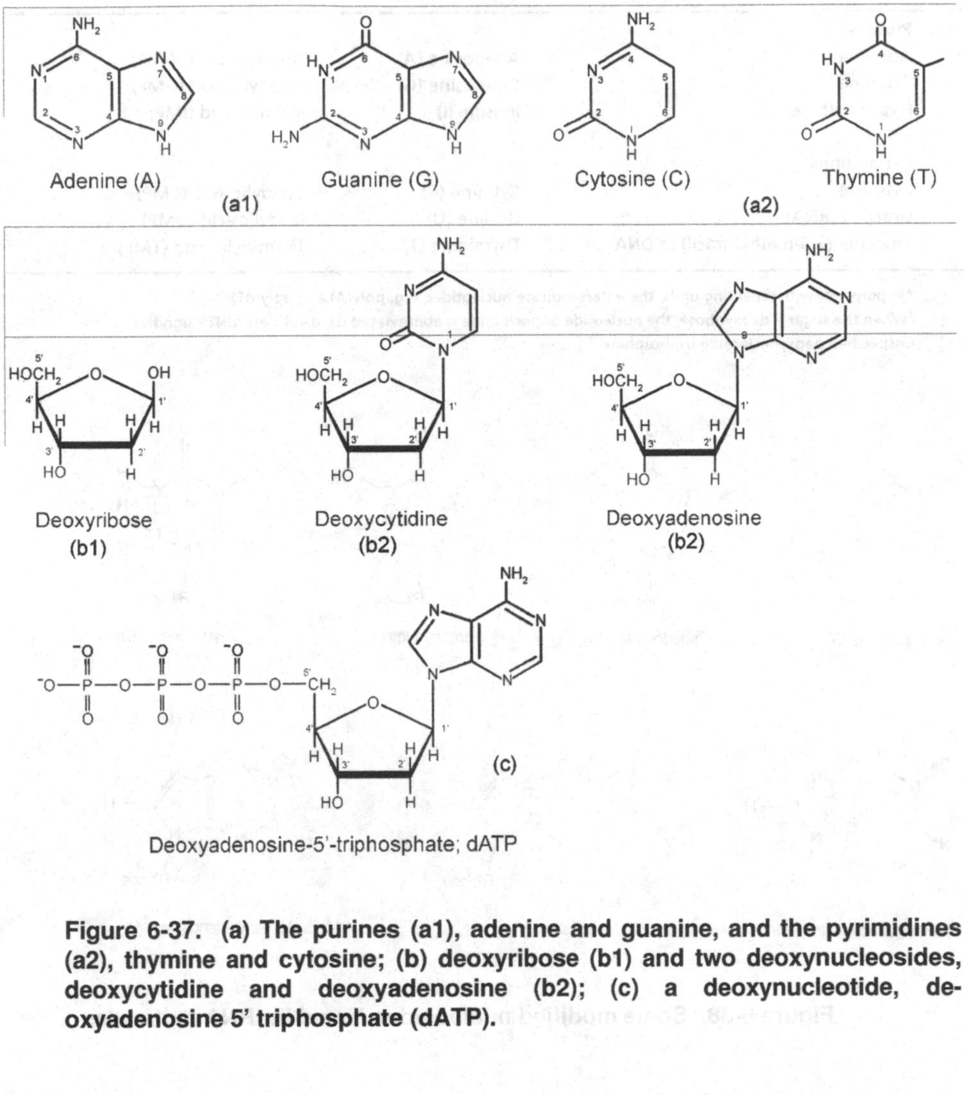

Adenine (A)   Guanine (G)   Cytosine (C)   Thymine (T)
(a1)   (a2)

Deoxyribose   Deoxycytidine   Deoxyadenosine
(b1)   (b2)   (b2)

(c)

Deoxyadenosine-5'-triphosphate; dATP

Figure 6-37. (a) The purines (a1), adenine and guanine, and the pyrimidines (a2), thymine and cytosine; (b) deoxyribose (b1) and two deoxynucleosides, deoxycytidine and deoxyadenosine (b2); (c) a deoxynucleotide, deoxyadenosine 5' triphosphate (dATP).

Table 6-4. Nomenclature of the Bases and Nucleosides and Nucleotides Derived from Them

Base	Nucleoside*	Nucleotide†
Purines		
Adenine	Adenosine (A)	Adenylic acid (AMP)
Guanine	Guanosine (G)	Guanylic acid (GMP)
Hypoxanthine	Inosine (I)	Inosinic acid (IMP)
Pyrimidines		
Cytosine	Cytidine (C)	Cytidylic acid (CMP)
Uracil (in RNA)	Uridine (U)	Uridylic acid (UMP)
Thymine (= 5-methyluracil) in DNA	Thymidine (T)	Thymidylic acid (TMP)

*In polymers with repeating units, the letters indicate nucleotides, e.g. poly(A) and poly(dT).
†When the sugar is deoxyribose, the nucleoside or nucleotide is abbreviated dT, dAMP, etc. dNTP signifies unspecified deoxynucleoside triphosphate.

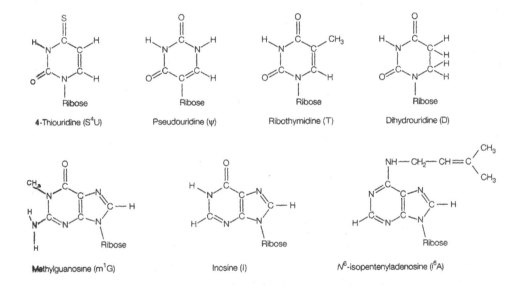

Figure 6-38. Some modified nucleosides found in tRNA molecules

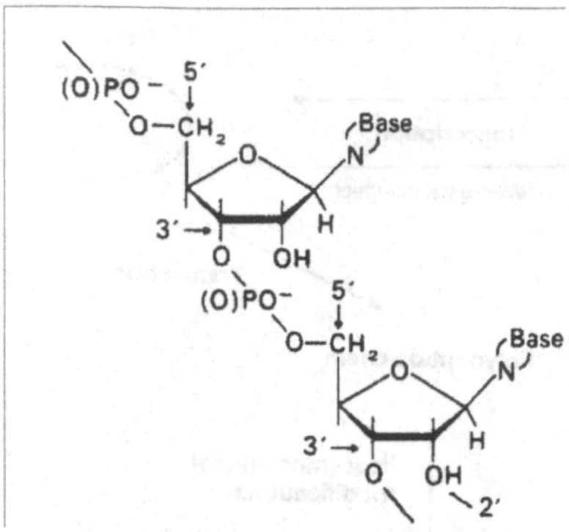

Figure 6-39. The linkage of the nucleotide in RNA. In DNA the 2'-OH is replaced by H.

Similarly, adenosine diphosphate (ADP) and adenosine monophosphate (AMP) are defined by their names. The primary structure of nucleic acids and conventional sequences are represented in the direction from 5' → 3' as indicated by Fig. 6-40.

The so-called **central dogma of Crick** states that information of the polypeptides has to pass from nucleic acid to polypeptides and can not flow from polypeptides to polypeptides. This indicates the importance of nucleic acids. The replication, transorption, and translation, as indicated by Crick, are illustrated in Fig. 6-41.

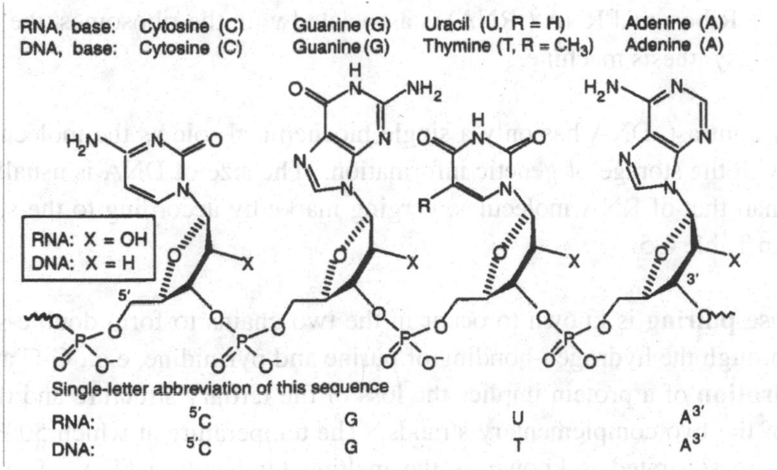

Figure 6-40. Representations of a specimen nucleic acid sequence.

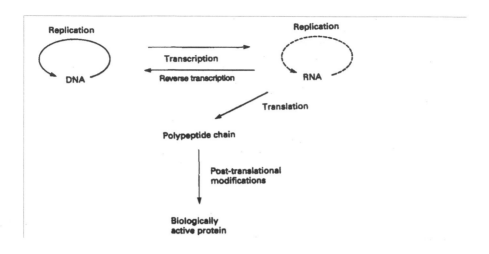

Figure 6-41. Representations of a specimen nucleic acid sequence.

There are three major classes of RNA molecules in cells, categorized according to different biochemical functions.

- Messenger RNA (mRNA) carries the information for the sequence of a protein from DNA.

- Transfer RNA (tRNA) decodes the information during biosynthesis of the protein.

- Ribosomal RNA (rRNA) is associated with the ribosomes, the protein-synthesis machine.

By contrast, DNA has only a single biochemical role as the molecule associated with the storage of genetic information. The size of DNA is usually much larger than that of RNA molecules, verging markedly according to the source as shown in Table 6-5.

Base pairing is known to occur in the two chains to form double-stranded DNA through the hydrogen-bonding of purine and pyrinidine, e.g., G-C and A-T. **Denaturation** of a protein implies the loss of the tertiary structure and the reparation of the two complementary strands. The temperature at which 50% of two strands are separated is known as the melting temperature (T_m). The reverse process, whereby the two strands reassociate, is called annealing. Denaturation can be followed by photometry for a C_0t plot as shown in Fig. 6-42.

Table 6-5. The Sizes of DNA Molecules from Various Sources

Source	Number of base pairs	Length (µm)
Viruses		
Polyoma SV40	5.1×10^3	1.7
Bacteriophage T2	1.66×10^5	55
Bacteria		
Escherichia coli	4.7×10^6	1360
Human	2.9×10^9	990 000

DNA from a wide range source of DNA is illustrated in Fig. 6-43.

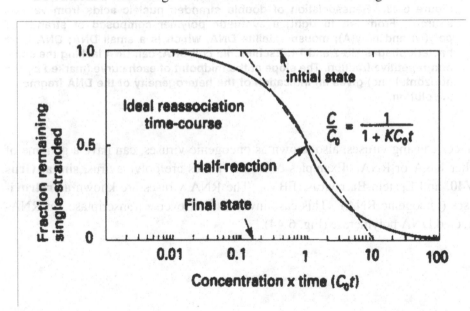

Figure 6-42. A C_0T curve.

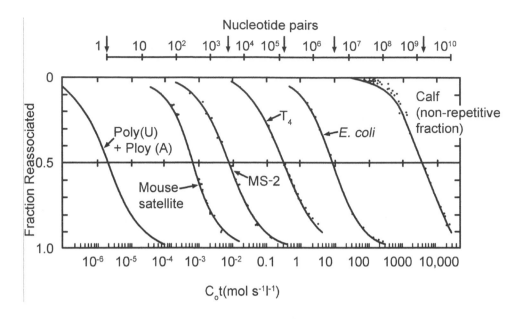

Figure 6-43. Reassociation of double stranded nucleic acids from various sources. From left to right; a synthetic polymer composed of strands of poly(U) and poly(A); mouse satellite DNA, which is a small DNA; DNA from bacteriophages MS-2 and T4; Escherichia coli DNA; calf DNA lacking the small non-repetitive fraction. The slope at the midpoint of each curve (marked by the horizontal line) gives an indication of the heterogeneity of the DNA fragments in solution.

Cancer-causing viruses, also known as oncogenic viruses, can have a genome of either DNA or RNA. Examples of DNA viruses are polymavirus, **simianvirus SV40**, and Epstein-Barr virus (EBV). The RNA viruses are known as **retroviruses** (oncogenic RNA). This case involves the reverse transcriptase, an RNA-directed DNA polymerase (Fig. 6-44).

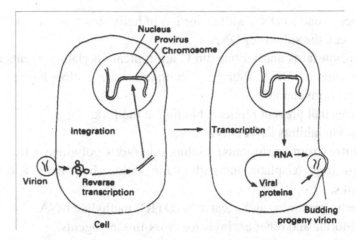

Figure 6-44. Incorporation of the information in the RNA of a retrovirus into the DNA of the host.

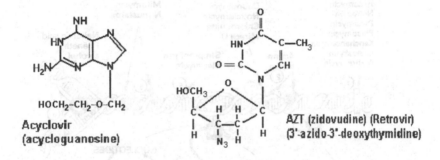

Figure 6-45. The structure of two antiviral drugs: acyclovir and AZT.

Human immunodeficiency virus (HIV-1) is the vitrovirus which is responsible for the **acquired immunodeficiency syndrome (AIDS)**. It is difficult to devise non-toxic antiviral drugs. **Acyclovir** and **AZT** (structures are shown in Fig. 6-45) are reactor non-specific in that they are designed to inhibit the activity of all rapidly-dividing cells, including the mucusal cells of the intestine and the hair follicles. Drugs that interact with DNA to inhibit the replication or **transcription** include the following mechanisms, also illustrated in Fig. 6-46:

- Drugs bind non-covalently to DNA and prevent replication.
- Actimycin D binds DNA without distortion of the helix structure to prevent transcription. This occurs with dexorubicus (adriamycin), chronomycin A, and distamycin as well.

- Substances bond to DNA with distortion of helix structure.
- Drugs break the strands of DNA.
- Nitrogen mustards and mitomycin C are typical alkylating agents and cause cross-linking of DNA strands: chlorambuncil, cycloophosphamide, and bushalphan (myeleran).
- Substances that prevent replicate binding at the fork.
- Rifampician inhibits RNA polymerase.
- α-Amanitin (from mushrooms) inhibits eukaryotic polymerase II.
- Platinum drugs, cisplatin and carboplatin, induce both inter- and intrastand cross-links.
- Hexamethylmelamine and dacabazin (DTIC) methylate DNA.
- Vinyl chloride and other α-vinyls for cross-linking agents.

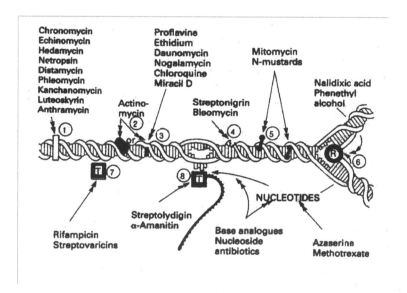

Figure 6-46. The site of action of chemotherapeutic drugs.

     Individual DNA can provide information for the construction of various RNA molecules and proteins. The RNA molecules of interest are produced by transcription through the mRNA from DNA template. During biosynthesis the relationship of the linear sequence of the mRNA and that of the corresponding rRNA is known as the genetic code. There are four types of nucleotide and twenty amino acid building blocks of proteins. It is not possible to have a one to one correspondence between the nucleotide order and the amino acid order. Instead, the RNA is read in triplets; three contiguous nucleotides will specify a single amino acid order. There are 64 (4^3) genetic codons as listed in Table 6-6.

Table 6-6. The Genetic Code

Amino acid/codons	Amino acid/codons	Amino acid/codons
Arg CGU, CGC, CGA, CGG, AGA, AGG	Ile AUU, AUC, AUA	Gln CAA, CAC
Leu UUA, UUG, CUU, CUC, CUA, CUG	Asn AAU, AAC	Glu GAA, GAG
Ser UCU, UCC, UCA, UCG, AGU, AGC	Asp GAU, GAC	Lys AAA, AAG
Ala GCU, GCC, GCA, GCG	Cys UGU, UGC	Phe UUU, UUC
Gly GGU, GGC, GGA, GGG	His CAU, CAC	Tyr UAU, UAC
Pro CCU, CCC, CCA, CCG		Trp UGG
Thr ACU, ACC, ACA, ACG		
Val GUU, GUC, GUA, GUG	*Stop* UAA, UGA, UAG	Met (*Start*) AUG

A schematic view of the transcription and translation between DNA, RNA, and proteins are located in Fig. 6-47.

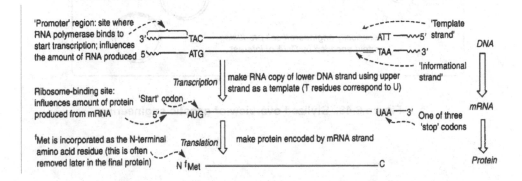

Figure 6-47. Schematic overview of transcription and translation.

Cells have the ability to replicate their DNA and produce protein via the process of transcription and translation. Occasionally, foreign DNA can hijack the replicative ability. For example, viruses infect cells and use cellular nucleic acid polymerases and ribosomes to generate more viral nucleic acids and then to self-assemble to produce more viruses. This replicative aility has been exploited by genetic engineering for the alteration of DNA content of the cells and subsequent modification of the organism. Introduction of DNA can encode a protein which can lead to a new organism that can produce extra protein. This can lead to the correction of any genetic defect of the organism as gene therapy. It is also

possible to hydrolyze DNA at specific sites, join pieces of DNA from different
sources, and recreate an organism to form a new organism—e.g., the genetic engineered microorganism (Fig 6-48).

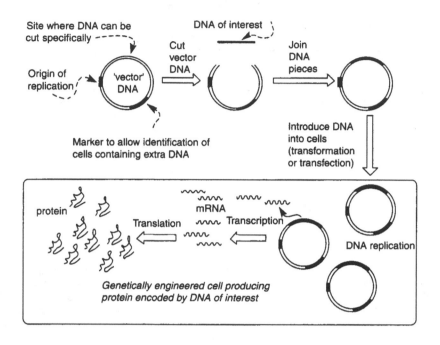

Figure 6-48. Stylized overview of genetic engineering.

REFERENCES

6-1. W. H. Elliott and D. C. Elliott, *Biochemistry and Molecular Biology*, Oxford University Press, 1997.

6-2. M. K. Campbell, *Biochemistry*, 3rd ed. 1995 Sonder College Publishers, Philadelphia, PA.

6-3. D. G. Knorre and S. D. Mysina, *Biochemistry: A Manual for Universities*, Nova Science Publishers, 1998, Comack, N. Y.

6-4. R. Werner, *Essential Biochemistry and Molecular Biology: A Comprehensive Review*, Appleton and Lange, Newark, CN, 1992.

6-5. M. Cerdorio and R. W. Noble, *Introductory Biophysics*, World Scientific, 1986, Singapore.

6-6. C. M. Dobson, J. A. Gerrard, and A. J. Pratt, *Fundamentals of Chemical Biology*, Oxford University Press, Oxford, UK, 2001.

6-7. D.Voet and T.G.Voet, *Biochemistry*, 2nd ed., John Wiley, New York, 1995.

6-8. B.D.Hames and N.M.Hooper, *Instant Notes of Biochemistry*, 2nd ed., Springer-Verlag, New York, 2000.

PROBLEM SET

1. Find out Michaelis constant from the following data:

Time (sec)	8	7	6	5	4	3	2
Disappearance of substrate (ppm)	1.7	1.6	1.5	1.4	1.3	1.2	1.1

2. Sketch a T=9 icosadeltahedron.

3. A closed cylinder duplex DNA has a 100 bp segment of alternating C and G residues. Upon transfer to a solution containing higher salt concentration, the segment undergoes a transition from the B conformation to the Z conformation. What is the accompanying change in its linking number, writhing number and twist?

CHAPTER 7

GEOCHEMISTRY

Geochemistry is perhaps one of the few branches of chemistry that was explored and studied earlier than other types of chemistry. Science of the Earth has a direct impact on the well being as well as future susceptibility of many human endeavors.

This chapter begins with the Earth's crust, mantle and the inner core. The essential components of plate tectonics including mid-ocean ridges, subduction zones and their relationship to mineral and petroleum resources will be outlined. The formation and development of the present Earth through the past geological time scale will be discussed. Following this, stable isotopes and their applications to radioactive dating and environmental forensics will be covered. Geochemical biomarkers and their progress are summarized for characterization of organic fossil fuels investigated during different stages of maturation and transformation.

7.1 PLATE TECTONICS AND NATURAL RESOURCES

Continental movement since the formation of the Earth has been closely tied to the events of geological times. Over time, the continents have been fused into one giant super continent landmass and later fragmented into many isolated floating islands. The units of solid earth can be summarized as follows (see Fig. 7-1):

- The base of the crust (also known as **Mohorovich discontinuity** or **Moho**) is 40 km below the continents but only 5-7 km below the oceans.

- The base of the lithosphere is on average 80 km below the oceans and still deeper beneath the continents. This is the lower boundary of the rigid tectonic plates. The softer part of the upper mantle underneath the lithosphere is called the **astherosphere**.

- The **440 km discontinuity** corresponds to a change of olivine structure (spinel or ringwoodite phase).

- The **660 km discontinuity** corresponds to the transformation of all minerals into perovskite and minor Fe-Mg oxide (magnesio-wustite). This is the base of the upper mantle.

- The mantle-core boundary is at about 2900 km. This layer is known as the **D" layer** (200 km thick).

- The **inner core** is made of solid Fe-Ni alloy that ends at 6370 km (1220 km thick).

The Earth's surface is covered with rigid lithospheric plates that may or may not carry continents. These plates move apart along **mid-ocean ridges** and cover **subduction zones** (marked by trenches), where one slides under the other (Fig. 7-2). **Plate tectonics** are surface expressions of mantle convection. Convection is a generalized movement of the mantle maintained by density inversion (heavy above light) brought about by thermal contrasts within the earth. Density inversions are maintained by heat from radioactive elements such as U, Th, and K contained in the mantle as well as by the heat released from the core.

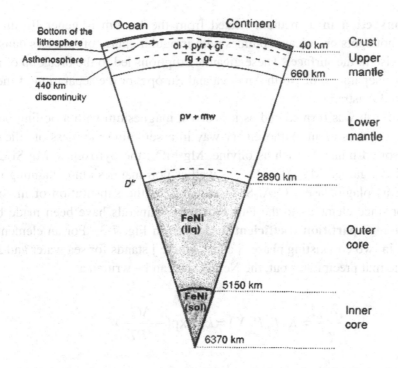

Figure 7-1. The internal structure of the earth determined by the propagation of seismic waves from earthquakes. Essential minerals are garnet (gr), magnesio-wustite (mw), olivine (ol), perovskite (pv), pyroxene (pyr), and ringwoodite (rg).

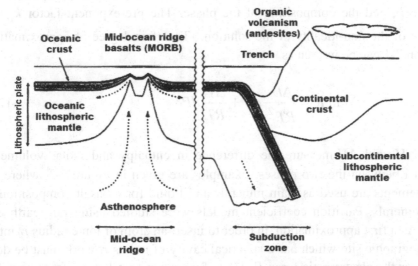

Figure 7-2. The essential components of plate tectonics: mid-ocean ridges, subduction zones, and rigid lithospheric plates. Oceanic crust is indicated by thick lines.

Convection in a medium heated from the bottom is generally unstable, causing hot spots or volcanic plumes with irregular spurts of hot, less dense material to rise to the surface. These gigantic eruptions led to the break up of a continent, the separation of North America and Europe, or the separation of the Antarctica and Australia.

As Earth was formed and as it evolves, magnesium with a boiling point of 1105 °C behaves in an extraordinary way in assembling the mass of silicates to form assorted minerals such as olivine, Mg_2SiO_4, the pyroxenes, $Mg_2Si_2O_2$ and $CaMgSi_2O_6$, garnet, $Mg_2 Al_2Si_3O_{12}$, as well as non-magnesium-containing minerals such as plagioclase - $CaAl_2Si_2O_8$. Studies of the substitution of minor elements or trace elements to the lattice of these minerals have been made by the conception of **partition coefficients** as shown in Fig. 7-3. For an element is in solution in two co-existing phases j and J, where j stands for sea water and J for a carbonate that precipitates out, the Nernst law can be written as:

$$\frac{X_j^i}{X_J^i} = K(T,P,X) = k_o \exp(-\frac{\Delta G_o}{RT}) \qquad [7\text{-}1]$$

where X_j^i is the molar proportion of element i in phase j, R is the gas-law constant, and ΔG is a measure of the energy of exchange of this element between the two phases j and J. The partition coefficient K depends on temperature T, pressure P, and the composition of the phase. The pre-exponent factor k_o is a measure of the non-ideality of the solution. To a fair degree of approximation, Equation 7-1 can be written as

$$d \ln K = \frac{\Delta H}{RT^2} dT + \frac{\Delta V}{RT} dP \qquad [7\text{-}2]$$

where ΔH and ΔV measure the difference in enthalpy and molar volume of element i between the two phases. Examples are given in Figure 7-3, where the trace elements are used as main minerals and liquid for a basalt composition of these minerals. Partition coefficient models were studied using rare earth elements. As a first approximation, in order to insert an atom of ionic radius r_0 into a crystallographic site which has a spherical cavity of radius r, work must be done to counter the electrostatic force F. These forces can be taken as proportional to the expansion or contraction of ionic radius for r and r_0.

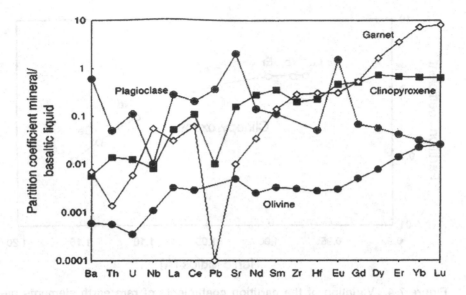

Figure 7-3. Typical partition coefficients for some important trace elements between the main minerals and the liquid for a basalt composition.

$$F \approx k(r - r_o) \qquad [7\text{-}3]$$

Equation 7-3 is Hooke's law, where k is a constant related to certain elastic properties of the medium known as Young's modulus and the Poisson's ratio. A change in the elastic energy U upon compression or expansion can be expressed as,

$$dU = -PdV$$
$$= -\frac{F}{4\pi r^2} 4\pi r^2 dr$$
$$= -Fdr \qquad [7\text{-}4]$$

Upon integration of Equation 7-4 between r and r_0,

$$\Delta U = -\frac{k}{2}(r - r_o)^2 \qquad [7\text{-}5]$$

Motions of identical charges fit a parabolic relationship of binding energy. Thus, a plot of lnK to the squared radius, as shown in Figure 7-4, reflects a control of partition coefficients to elastic properties of the crystalline lattice.

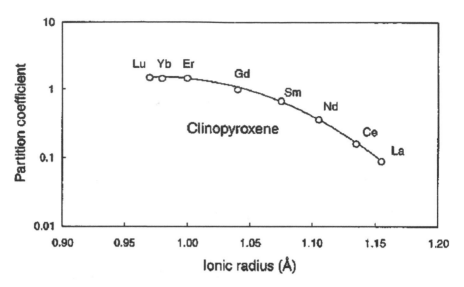

Figure 7-4. Variation of the partition coefficients of rare earth elements be-tween clinopyroxene and basaltic liquid by ion radius. The parabolic shape of the curve reflects a physical control of the partition coefficient by the elastic properties of the crystalline lattice (After J.D Blundy et al., 1998, Earth Planet. Sci. Lett. 160, 493-504).

7.1.1 Hydrothermal Reaction to Mineral Deposits

Typically, all the median temperature reactions from 100 to $500\,^{\circ}$C between the aqueous solution and the rock are hydrothermal. Transformations of minerals are often controlled by water pressure and temperature. For example, museovite (white mica), which disappears from gneiss. They also characterize the entry of metamorphic rocks into granulite facies,

$$KAl_2Si_3O_{10}(OH)_2 + SiO_2 \leftrightarrow KAlSi_3O_8 + Al_2SiO_5 + H_2O$$

(museovite)   (quartz)   (k-feldspar)   (silimanite)

$$\ln P_{H_2O} = \frac{\Delta H}{RT} + c \qquad\qquad [7\text{-}6]$$

i.e., the mass action law applies where ΔH is the enthalpy of the reaction. Fur-thermore, in the reaction between feldspar and the solution with K/Na fractiona-tion, c is constant

$$K^+ \quad + \quad Na^+ \quad \leftrightarrow \quad K^+ \quad + \quad Na^+$$
$$\text{(solution)} \quad \text{(feldspar)} \quad\quad \text{(feldspar)} \quad \text{(solution)}$$

Exchange of H^+ and cationic exchange between the lithosphere and hydrosphere is as follows:

$$2NaAlSi_3O_8 + 2H^+ + H_2O \leftrightarrow Al_2SiO_2(OH)_4 + 4SiO_2 + 2Na^+$$
$$\text{(albite)} \quad\quad\quad\quad \text{(solution)} \quad\quad\quad \text{(kaolinite)} \quad\quad \text{(silica)} \quad \text{(solution)}$$

For pure phase, we can write:

$$\ln\left(\frac{(Na^+)}{(H^+)}\right)_{solution} = \frac{\Delta H}{RT} + const. \qquad [7\text{-}7]$$

It is also possible to write:

$$\log_{10}\left(\frac{(Na^+)}{(K^+)}\right)_{solution} = \frac{0.908}{T} - 0.70 \qquad [7\text{-}8]$$

In the case of silica, the spa water is used as a thermometer:

$$d\ln(SiO_2)_{solution} / d(1/T) = \Delta H / R \qquad [7\text{-}9]$$

Equation 7-9 can be expressed as:

$$\log_{10}(SiO_2)_{solution} = -\frac{1306}{T} + 0.38 \qquad [7\text{-}10]$$

as shown in Figure 7-5.

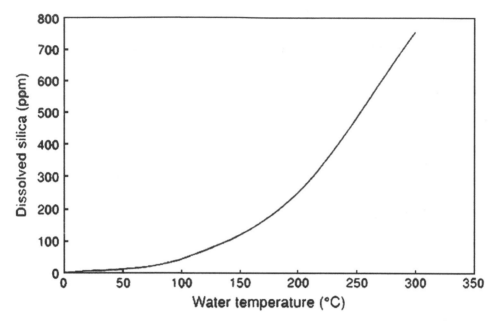

Figure 7-5. Solubility of silica in water. Silica contents of thermal waters can be used as a thermometer to a temperature of about 220ºC and thus indicate the depth of equilibration. Above this temperature, amorphous silica precipitates as the fluid rises.

Water is essential during diagenesic, metamorphic, and hydrothermal processes. Usually the water ratio can be estimated. It falls in the range 1-5 (very common) while much higher values (up to several hundreds) are not unusual. An example is illustrated below.

[Example 7-1] A hydrothermally resulted basalt sample has a $\delta^{18}O_{HR}$ value of –2 per mil. Examination of thin sections suggests that this rock was hydrothermally altered at 350°C in the greenschist facies. Paleography also suggests that when this basalt was erupted, the latitude was such that the meteoritic water had a $\delta^{18}O_{MW}$ of –10 per mil. No data are available for the $\delta^{18}O$ of hydrothermal water. Assuming the function factor of $^{18}O / ^{16}O$ between feldspar and water at 350°C is $\delta^{18}O_{feldspar} = \delta^{18}O_{HW} + 4$, and the $\delta^{18}O$ value of basalt is similar to that of feldspar, +5.5 per mil, the $\delta^{18}O$ of hydrothermal solution can be estimated. How much water interacted with the basaltic sample before it turned into meta basalt? (Here HR is hydrothermally meteoric rock, FR is fresh rock,; MW is meteoric water; HW is hydrothermal water.)

Let R and W denote the mass and water of rock. It is possible to achieve a mass balance,

$$R\delta^{18}O_{FR} + W\delta^{18}O_{MW} = R\delta^{18}O_{HR} + W\delta^{18}O_{HW}$$

or

$$\frac{W}{R} = \frac{\delta^{18}O_{FR} - \delta^{18}O_{HR}}{\delta^{18}O_{HW} - \delta^{18}O_{MW}}$$

Here for fresh rock,

$$\delta^{18}O_{FR} = 5.5/mil$$

For hydrothermally altered rock,

$$\delta^{18}O_{HR} = -2/mil$$

For meteoric water,

$$\delta^{18}O_{MW} = -10/mil$$

For hydrothermal water,

$$\delta^{18}O_{HW} = \delta^{18}O_{HR} - 4/mil = -2/mil - 4/mil = -6/mil$$

$$\therefore W/R = \frac{5.5 - (-2)}{-6 - (-10)} = 1.875$$

Today, most of the world's mineral deposits are located along former **convergent plate** boundaries where an oceanic lithosphere plate plunges under the margin of a continent (including the **continental shelf**) or under a chain of volcanic islands. Examples of metallic sulfides along convergent plate boundaries include the Kunoko deposit of Japan, the sulfide ores of the Philippines, and the deposits along the mountain belt of western North America and South America (the Coast Ranges, the Rockies, the Andes and from the eastern Mediterranean region to Pakistan). The role of **plate boundaries** in the accumulation of mineral resources can be found in Figure 7-6.

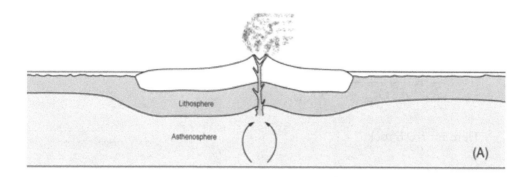

Figure 7-6. Role of plate boundaries in the accumulation of mineral deposits is exemplified in this sequence of cross-sectional views of the development of the South Atlantic Ocean. The position of Africa is assumed to be stationary throughout the sequence of cross sections. (A) In stage 1, a single ancestral continent called Pangaea is rifted into two continents (South America and Africa) about a divergent plate boundary.

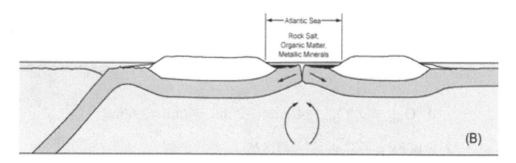

Figure 7-6. Continued. (B) In stage 2, the oceanic crust created by the process of sea-floor spreading from the divergent plate boundary (a precursor of the Mid-Atlantic Ridge) rafts South America westward and is compensated for by consumption of oceanic crust at a trench (a convergent plate boundary) that develops to the west of South America. Thick layers of rock salt, organic matter, and metallic minerals accumulate in the Atlantic Sea during this early stage of continental drift.

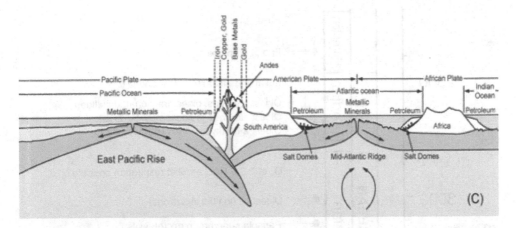

Figure 7-6. Continued. (C) In stage 3, continued sea-floor spreading from the Mid-Atlantic Ridge widens the Atlantic into an ocean, and rafts South America westward over the trench, reversing the inclination of the trench and producing the Andes mountain chain as a consequence of the deformation that develops at the convergent plate boundary along the western margin of South America. Metallic minerals that are melted from the Pacific plate as it plunges under South America ascend through the overlying crustal layers and are deposited in them to form the metal-bearing provinces of the Andes. Meanwhile in the Atlantic Ocean metallic minerals continue to accumulate about the Mid-Atlantic Ridge. Salt originating in the thick layers of rock salt that have been buried under the sediments of the continental margins rises in large, dome-shaped masses that act to trap the oil and gas that are generated from the organic matter that was preserved in the former Atlantic sea. **(Modified after A.Rona, Scientific Am. July, 1973).**

7.1.2 World Phosphate Deposits

A glimpse of the **reconstructed paleogeographic maps** of the world would reveal significant events, including the biotic history that has occurred since the beginning of the Earth. The events depicted in Figure 7-7(A) represent some important milestones encountered worldwide. The continuation of Fig 7-7(B) to 7-7(D) have been included in the appendix. This is a useful reference for the geological timescale.

First, the world-wide phosphate deposits, the phosphorites, are formed usually within 45° of their paleo equator, like those of today. During **the Permian period**, on the west side of Paleozoic Panguea in what is today western USA, there existed the largest concentration of phosphoria formation. Modern day patterns of phosphorite deposits are similar to those of the past (see Fig. 7-8).

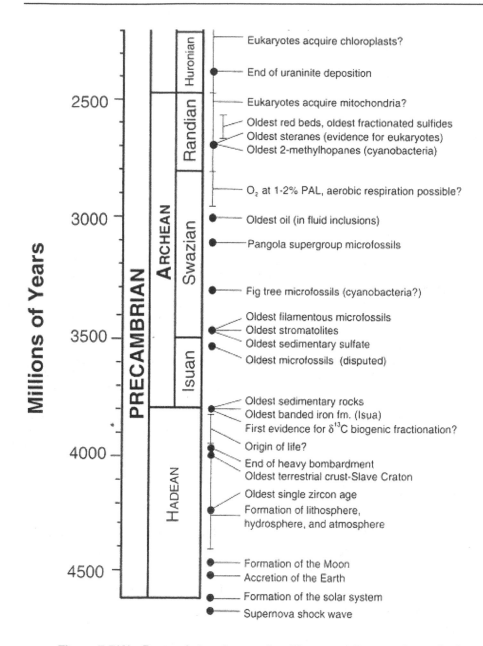

Figure 7-7(A). Precambrian time scale with events. Some paleontologists take 4500 Ma to 500 Ma.(After Schopf _et al._,1983). For Figures 7-7(B) to 7-7(D), please refer to appendix.

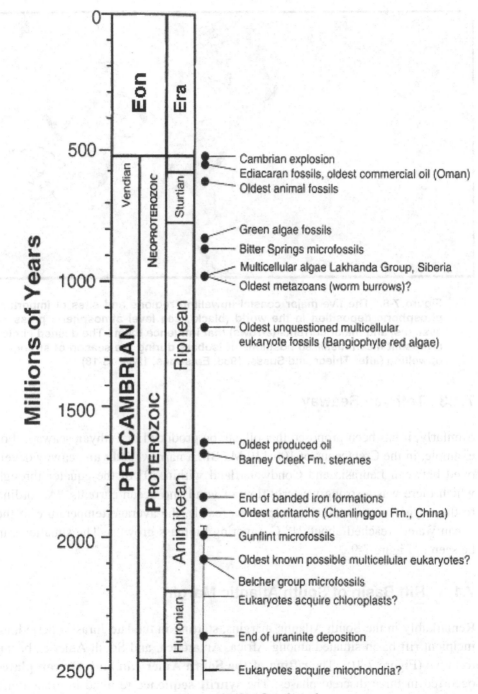

Figure 7-7(A). Continued.

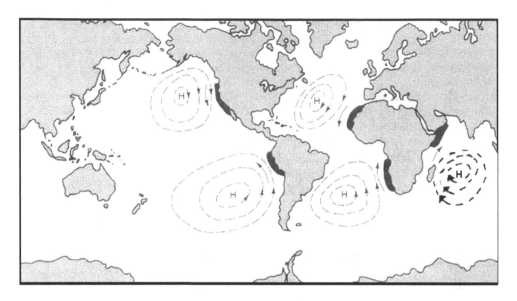

Figure 7-8. The five major coastal upwelling regions and sites of important phosphoric deposition in the world (black), sea level atmospheric pressure systems, and major currents (arrows) that influence them. The dashed circles represent mean idealized positions of isobars during the season of strongest upwelling (after Thiede and Suess, 1983, Episodes, 1983, 15-18).

7.1.3 Tethyan Seaway

Similarly, it has been proposed that oil can be produced in Tethyan seaway. For example, in the Cretaceous of 100-110 Ma BP, a narrow **Tethyan seaway** developed between Laurasia and Gondwana land within 30° of the equator through which there was a strong westerly flow of wind and ocean currents. According to the oxygen isotope ratio in calcareous fossils, the average temperature of the ocean waters reached about 21° C, favoring biomass growth. The situation can be seen in Figure 7-9.

7.1.4 Rift Basin of South Atlantic Margin

Remarkably in the South Atlantic margins, starting in the late Jurassic period, an incipient rift basin situated among Africa, Antarctica, and South America began to move (Figure 7-10). The **rifting of the South American and African plates** occurred in three discrete phases. The **synrift sequence** resulted in graben and half-graben troughs filled with lacustrine sequences. It is possible to correlate the oil produced in Nigeria with that produced in Brazil (see Figure 7-11).

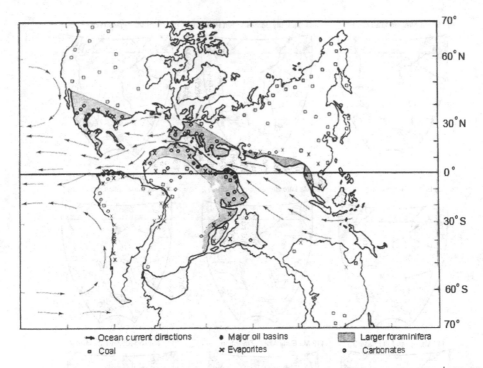

Ocean current directions • Major oil basins ▨ Larger foraminifera
▫ Coal ✗ Evaporites ○ Carbonates

Figure 7-9. The distribution of the continents in the Cretaceous, 100 ± 10 Ma BP (after Smith et al., 1973) showing the major Tethyan seaway. Oil basins after Irving et al., 1974a, foraminifera after Dilley (1973), ocean currents from Luyendyk et al. (1972) and sediments after Seyfert and Sirkin (1979) and Habicht (1979) (After B.F. Windley).

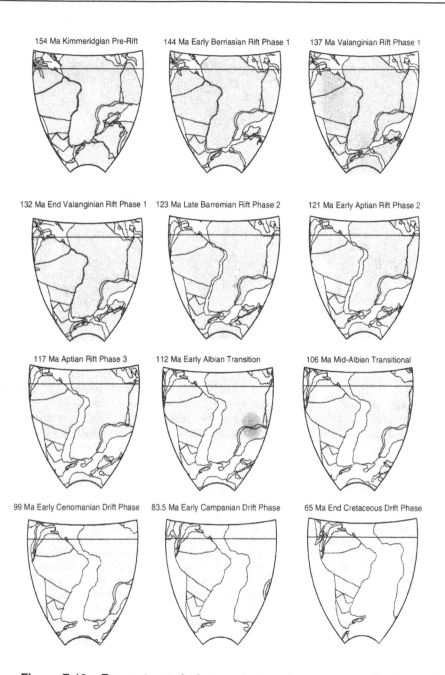

154 Ma Kimmeridgian Pre-Rift   144 Ma Early Berriasian Rift Phase 1   137 Ma Valanginian Rift Phase 1

132 Ma End Valanginian Rift Phase 1   123 Ma Late Barremian Rift Phase 2   121 Ma Early Aptian Rift Phase 2

117 Ma Aptian Rift Phase 3   112 Ma Early Albian Transition   106 Ma Mid-Albian Transitional

99 Ma Early Cenomanian Drift Phase   83.5 Ma Early Campanian Drift Phase   65 Ma End Cretaceous Drift Phase

Figure 7-10. Reconstructed plate tectonics views showing the formation of the South Atlantic. Drawings generated from the GEOMAR/ODSN (Ocean Drilling Stratigraphic Network) website, based on data in Hay et al. (1999). www.odsn.de/odsn/services/paleomap/paleomap.html from (a) to (l) stretching from 154 Ma before the drift to 65 Ma to the drift phase of Cretaceous.

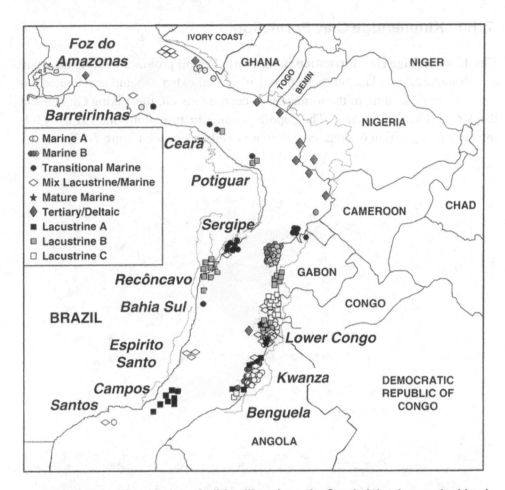

Figure 7-11. Correlation of oil families along the South Atlantic marginal basin. Lacustrine A oils originated from source rocks deposited in balanced-fill to underfilled conditions. Lacustrine B oil originated from source rocks deposited in balanced-fill freshwater conditions. Lacustrine C oils originated from the Bucomazi Formation. Marine A oils were derived from late rift transgressional marine shales and marls. Marine B oils, primarily located offshore Gabon and Kwanza basins, originated from early rift calcareous shales. Modified from Schiefelbein et al. (2000).

7.1.5 Estonian Kukersite Oil Shale

Estonian kukersite oil shale is the largest energy source in the former Soviet Union with a structure of poly-*n*-alkyl resorcinol or 5-*n*-alkylbenzene-1,3-diol unit (Figure 7-12). It comes from marine algae, *Glucocapsomorpha*.

7.1.6 Kimmeridge Clay Formation

The Kimmeridge clay formation in the North Sea oil province is very important. It is dominated by a late Jurassic episode of crustal extension and accelerated basin subsidence leading to the formation of central Graben, the Viking Graben and the Moray Firth rift system. The deposit became mature along the rift axes following the deposition of cretaceous-tertiary overburden (see Figure 7-13).

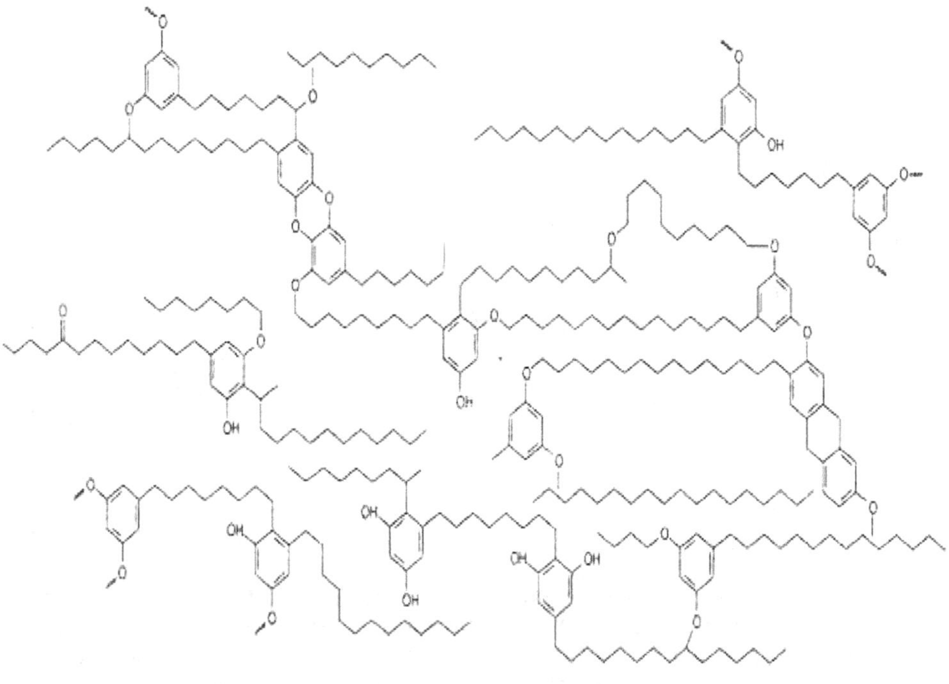

Figure 7-12. Hypothesized kerogen structure of kukersite showing predominance of the suggested poly-n-alkyl resorcinol structure, i.e., 5-n-alkyl-benzene-1,3-diol subunit. Reprinted from Blokker et al. (2001).

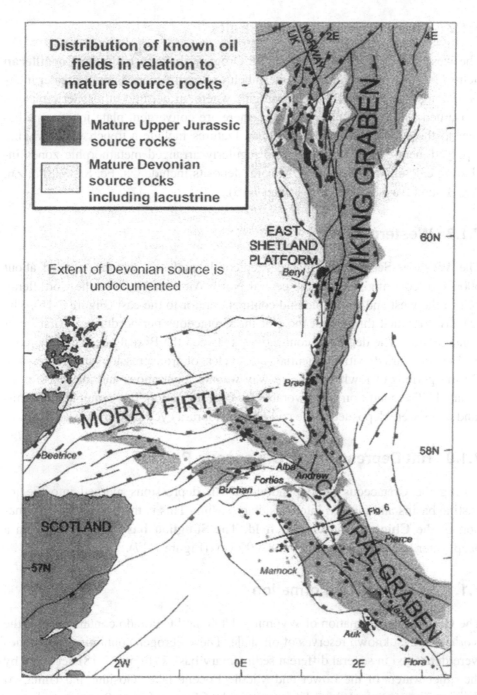

Figure 7-13. Principal reservoirs for oil and gas fields in North Sea oil province (Brooks et al., 2001). The Kimmeridge Clay Formation was deposited over much of the North Sea. It is mature only along the axis of the rift system.

7.1.7 Cordilleran and Andean Belts

The west American continental margin Orogenic belts, usually the **Cordilleran belt** of North America and the Andean belt of South America are of greater interest. Along the seismically active margins where an oceanic lithospheric plate is consumed beneath a continental plate, there are convergent plate junctions adjacent to the subduction zones. The main features include a trench with turbidite type sediments, high heat flow, and regularly arranged metamorphic zones including calc-alkaline volcanics. Mineral deposits include Cu, Fe, Sn, W, Pb, Zn, Ag, Bi, and Au (see Figures 7-14 and 7-15).

7.1.8 Western Seaway

The **Western Seaway** was a broad epicontinental sea, which extended about 6000 km, covering most of the central North America between the Cordilleran belt to the west and the stable mid continent crator to the east (Figure 7-16). The seaway stretched throughout most of the Cretaceous period during a first order transgression. The degree of connectivity between the Boreal ocean and the Gulf of Mexico varied with sequential sub cycles of transgression and regression. During periods of lowlands, the seaway was not continuous and may have been isolated. The condition was favorable for coal formation. For example, the Fruitland coals were deposited at the end of the Niobrara cyclotherm.

7.1.9 Rift Depressions in Cretaceous Period

During the Cretaceous period, individual rift depressions merged into the locustine basins and merged into subsidiary basins. This is the basis for the formation of the **Chinese Daquing oil field**. The **Songliao basin** originated from a deep water lake that covered about 87000 km^2 (Figure 7-17).

7.1.10 Green River Formation

The Green River formation of Wyoming, Utah, and Colorado contains one of the world's largest known reserves of oil shale. These kerogen containing marlstones were deposited in several different sedimentary basins (Figure 7-18) occupied by the quiet waters of the Lower and Middle Eocene lakes Gosiute and Uinta. At their maximum extent these lakes occupied 16,000 and 20,000 square miles respectively. Oil shale bearing rocks of Green River Formation presently cover an area of about 16,500 sq miles.

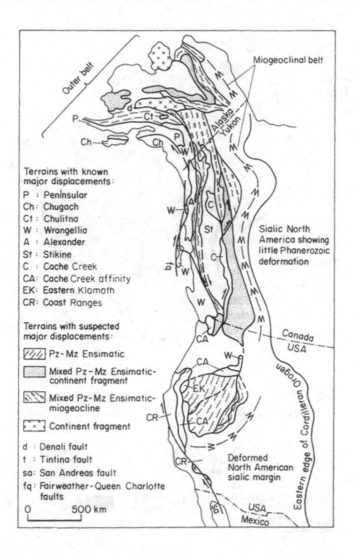

Figure 7-14. Map showing generalized distribution of the main allochthonous terrains and the miogeocline of the North American Cordillera. Pz: Palaeozoic, Mz: Mesozoic (from Saleeby, 1983, Ann. Rev. Earth Planet Sci., 11, Fig. 1, p.47; modified after Coney et al., 1980).

Figure 7-15. Mesozoic-Cenozoic Andean metallogenic-palaeogeographic scheme: 1: Andean foreland, 2: Andean geosynclinal area, 3: iron-apatite belt, 4: polymetallic deposits, 5: porphyry copper belt, 6: tin belt, 7: Peru-Chile trench, 8: plate movement direction, 9: Patagonian polymetallic belt, 10: Precambrian ranges (Pampean Arc and Patagonian Arc) and shield areas in the Andean foreland, 11: Altiplano polymetallic belt (Cu, Ag, Pb, Zn and also sedimentary Cu), 12: Fe-P sedimentary deposits, mainly Cretaceous, 13: Auplacer deposits, 14: Mo-(U)-belt (mainly hydrothermal veins) (After Frutos, 1982, in G.S Amstutz et al. (Eds), Ore Genesis: the state of the Art, Springer Verlag, Berlin, Fig. 1, p. 497).

The Green River Formation was primarily deposited as two lenticular bodies separated by Uinta uplift consisting of fine-grained organic rich sediments that appear as layers. These layers, like leaves of a book, can be seen from retrieved core samples and they can be used as indicators of paleo-temperature and weather patterns.

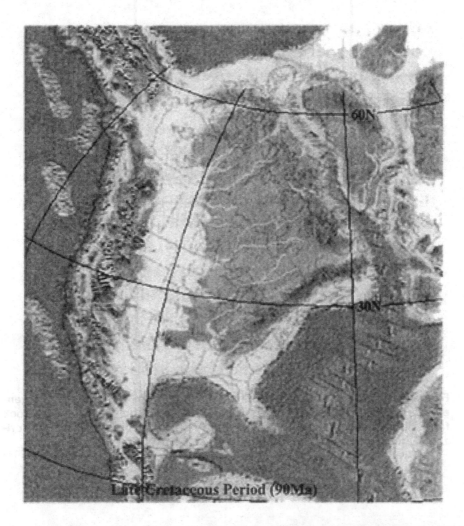

Figure 7-16. Paleo-reconstruction of the Western Interior Seaway during the late Cetaceous period (~90 Ma) (After Peters, Waters and Moedowan, 2005).

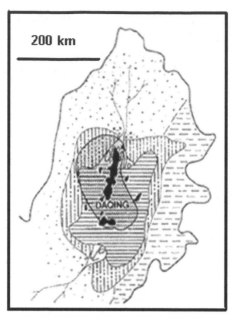

SONGLIAO BASIN
(LOWER CRETACEOUS)

Diluvial to Flood Plain Facies

Alluvial Facies

Shallow Lake Facies

Deep Lake Source Facies

Mature Zone

Immature Zone

Figure 7-17. The giant Daqing oil field lies updip from the center of the generative Songliao Basin, where the lower Cretaceous deep lake source rock facies is thermally mature (Demaison, 1984). The maximum distance of horizontal migration is < 40 km (After Peters et al. 2005).

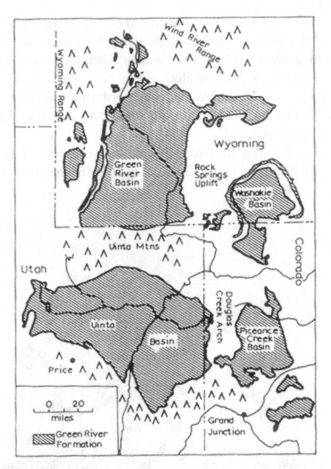

Figure 7-18. Map of the Colorado-Wyoming-Utah area. Shows present outcrop extent (shaded) and depositional basins of the Green River Formation (After Yen and Chilingar, Oil Shale, 1976).

7.1.11 Collision of Floating Islands of Panthalassa Ocean in Eastern Asia

One type of reconstruction of modern Asia occurred during the fragmentation of the large supercontinent of Gondwana. There are many smaller isolated fragments floating in the **Panthalassa Ocean**. The borderline of these floating islands may trap the potential to become oil bearing, especially for Tarim, north China, Tibet, Kazakhstan, Tuva Mongolia, Siberia, etc. (see Fig. 7-19).

Yin and Nie have reconstructed the eastern Asia by vertical sections with 15 time slices from the late Proterozoic (638 Ma) to late Miocene 10Ma. The northern Tarim block and north China collided with the combined Pamir- South

Tarim-Quidam block during earlier Silurian and earlier Devonian. Afterwards the south China block collided with the combined north China-north Tarim and Pamir-south Tarim-Quidam blocks at late Permian. Then the Qiangfang- Indochina block collided with the south China block at late Triassic to create the Red River fault. As the Lhasa –Sibumasn block moved northward, the Mongolo-Okhotsk plate subducted southward to close the Mongolo-Okhotsk Ocean and formed the Mongolo-Okhotsk suture.

At last India moved northward to collide with Eurasian continent in Miocene and caused the uprising of the Himalaya **opening of the South China Sea**.

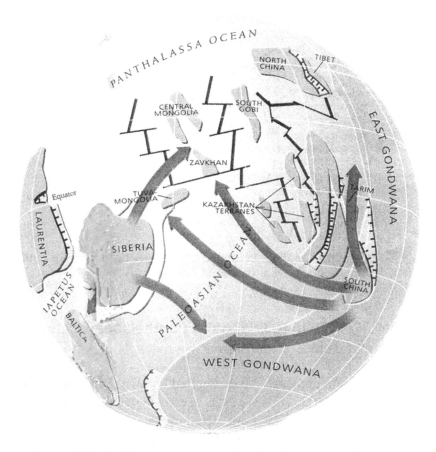

Figure 7-19. Continental fragmentation is a recurring theme of the Earth's evolution. The assembly and break-up of a single supercontinent occurring about every 250 million years since the Late Proterozoic is one of the major cycles in a permanent, ongoing transfiguration of the Earth's surface. The supercontinent may split into several large continents or many smaller fragments.

7.2 STABLE ISOTOPES

As determined in the chapter of Nuclear Energy in the Environmental Chemistry book, Z is the number of protons in a chemical element called a **nuclide**. Usually, the mass number A of a given nuclide consists of the number of protons, Z and neutrons, N.

$$A = Z + N \qquad\qquad [7\text{-}11]$$

Isotope is referred to, in Greek, as the element that has the same (iso) place (topes), same protons Z, but different neutrons, N. Nuclides of the same A but different Z are called **isobars**. Nuclides with same N but different mass number, A are called **isotones**. Finally, nuclides with same N-Z or A-2Z are termed **isotopic number** or **isodiasphere**.

$$N - Z = A - 2Z \qquad\qquad [7\text{-}12]$$

A plot of N vs. Z is termed the **Segre chart**, as shown in Figure 7-20. All stable nuclides fall on the N/Z = 1 line. This symmetry rule holds true for low atomic numbers. In stable nuclides with more than 20 protons or neutrons, the N/Z value could reach 1.5. Another rule, the **Oddo-Harkins rule**, states that nuclides of even atomic numbers are more abundant than those with odd numbers. The N-Z number of combinations is small if one atomic number is odd. It is even smaller if both are odd.

Isotopes affect the properties strongly. With water, for example, $H_2{}^{16}O$, $D_2{}^{16}O$, and $H_2{}^{18}O$ exhibit temperature of greatest density at 3.98, 11.24, and 4.30 °C respectively.

An isotope can undergo an exchange reaction. For example, if A and B contain the light or heavy isotope 1 or 2, then

$$aA_1 + bB_2 = aA_2 + bB_1 \qquad\qquad [7\text{-}13a]$$

For equilibrium constant K,

$$K = (A_2 / A_1)^a / (B_2 / B_1)^b \qquad\qquad [7\text{-}13]$$

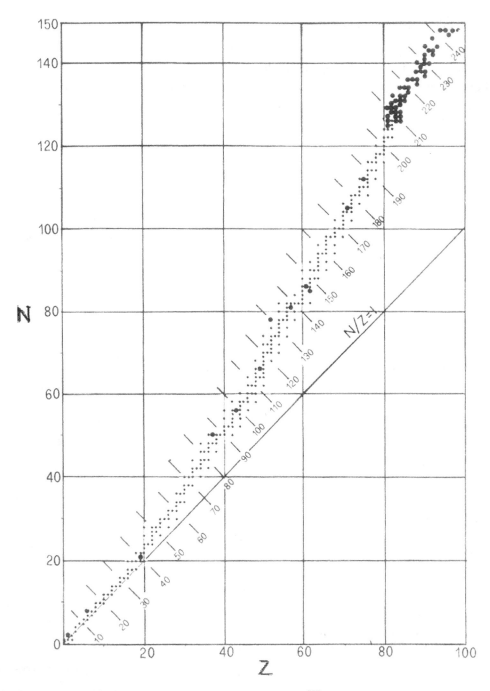

Figure 7-20. Segre chart of nuclides. ∘: stable, ●: unstable. Reproduced with modifications from K. Rankama, Isotope Geology, Pergamon press, London, 1954.

In terms of statistical mechanics, the equilibrium constant can be expressed by partition function Q, or

$$K = (\frac{Q_{A2}}{Q_{A1}}) / (\frac{Q_{B2}}{Q_{B1}})$$ [7-14]

The partition function can be defined as:

$$Q = \sum_i [g_i \exp(-E_i / kT)]$$ [7-15]

where E_i is the energy levels of the molecules, g_i the statistical weight of the ith level and k the Boltzmann's constant. A common expression is the fractionation factor, α, which is defined as

$$\alpha_{A-B} = R_A / R_B$$ [7-16]

the ratio of the numbers of any two isotopes in one compound A divided by the corresponding ratio for another compound B. In relation to the equilibrium constant,

$$\alpha = K^{1/n}$$ [7-17]

where n is the number of atoms being exchanged. If n = 1, then $\alpha = K$.

For example, the exchange of ^{18}O and ^{16}O between water and carbonate is expressed as:

$$H_2^{18}O + \frac{1}{3}CaC^{16}O_3 = H_2^{16}O + \frac{1}{3}CaC^{16}O_3$$

The $\alpha_{CaCO_3H_2O}$ can be written as:

$$\alpha_{CaCO_3H_2O} = \left(\frac{^{18}O}{^{16}O}\right)_{CaCO_3} / \left(\frac{^{18}O}{^{16}O}\right)_{H_2O} = 1.031 \text{ at } 25\ °C$$ [7-18]

In isotope geochemistry, it is common to express in terms of **delta (δ) value**

$$\delta_A = (R_A / R_{St} - 1)10^3 \ 0/00$$

and

$$\delta_B = (R_B / R_{St} - 1)10^3 0/00 \qquad\qquad [7\text{-}19]$$

where R_A and R_B are respective isotope ratios for compounds A and B and R_{St} is the default isotope ratio in a standard sample.

For the two compounds A and B, the δ values and fractionation factor α are related by

$$\delta_A - \delta_B = \Delta_{A-B} \approx 10^3 \ln \alpha_{A-B} \qquad\qquad [7\text{-}20]$$

The accepted unit of isotope ratio measurements is the **delta value** (δ) or **d value** given in **per mil** (0/00). The δ-value is defined as:

$$\delta \text{ in } 0/00 = \frac{R_{sample} - R_{standard}}{R_{standard}} \cdot 10^3 \qquad\qquad [7\text{-}21]$$

where R represents the measures isotope ratio. Some absolute isotope ratios of instructional standards are listed in Table 7-1.

Table 7-1. Absolute Isotope Ratios of International Standards (After J.M Hayes 1983)

Standard	Ratio	Accepted value($\times 10^6$) (with 95% confident interval)
SMOW	D/H	155.76 ± 0.10
	$^{18}O/^{16}O$	2005.20 ± 0.43
	$^{17}O/^{16}O$	373 ± 15
PDB	$^{13}C/^{12}C$	11237.2 ± 2.9
	$^{18}O/^{16}O$	2067.1 ± 2.1
	$^{17}O/^{16}O$	379 ± 15
Air nitrogen	$^{15}N/^{14}N$	3676.5 ± 8.1
Canyon Diablo Troilite (CDT)	$^{34}S/^{32}S$	45004.5 ± 9.3

7.2.1 Geochronology — Radiometric Dating

Some commonly occurring isotopes with their naturally occurring abundances are listed in Table 7-2. For instance, the $\delta\ ^{18}O$-values for a number of geological waters and rocks and their interactions between fluid-rock are very important. They can come about from solution-precipitation, chemical reactions, or diffusion. In Figure 7-21, the δ ranges of geological materials are compared. The ocean water has a very narrow range. As indicated in Table 7-2, carbon has two stable isotopes: ^{12}C and ^{13}C. Some carbon occurs in a wide variety of compounds on Earth, from highly reduced organic compounds in the biosphere to the highly oxidized inorganic compounds such as CO_2 or carbonates, various carbon derived compounds that are the raw materials for a number of products. This is ideal for finding applications in environmental forensics. For example, as illustrated in Figure 7-22, geochemical materials have distinctive ranges of delta values.

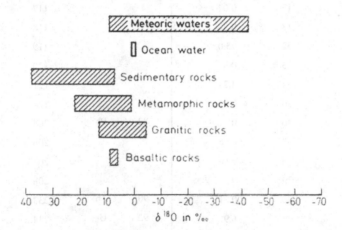

Figure 7-21. $\delta\ ^{18}O$-values of important geological reservoirs.

It is possible to determine if a natural gas is from the biogenic (fermentation from landfill) or from the thermogenic source (abundant used oilfield) (see Table 7-3).

Table 7-2. Selected Isotopic Abundances

Atomic No.	Symbol	Mass No.	Abundance (%)	Atomic No.	Symbol	Mass No.	Abundance (%)
1	H	1	99.9	37	Rb	85	72.2
		2	0.01			87	27.8
2	He	3	10^{-4}-10^{-5}	38	Sr	84	0.5
		4	100			86	9.9
6	C	12	98.9			87	7.0
		13	1.1			88	82.6
		14	10^{-10}	50	Sn	112	1.0
7	N	14	99.6			114	0.6
		15	0.4			115	0.3
8	O	16	99.8			116	14.2
		17	0.04			117	7.6
		18	0.2			118	24.0
16	S	32	95.0			119	8.6
		33	0.8			120	33.0
		34	4.2			122	4.7
		36	0.02			124	6.0
18	Ar	36	0.3	82	Pb	204	1.4
		38	0.06			206	25.2
		40	99.6			207	21.5
19	K	39	93.1			208	52.0
		40	0.01	90	Th	232	100
		41	6.9	92	U	234	0.006
20	Ca	40	97.0			235	0.72
		42	0.6			238	99.28
		43	0.1				
		44	2.1				
		46	0.003				
		48	0.2				

Source: Clark. S. P., Jr., ed., 1966, *Handbook of Physical Constants*, Geological Society of America Memoir 97, pp, 12-17.

Note: Values have been rounded to the nearest 0.1 per cent, except in cases of very rare isotopes.

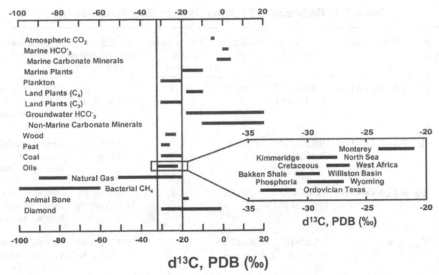

PDB = standard derived from fossil data obtained from the Crustaceous age Pee Dee Formation in South Carolina (see Table 7-1)

Figure 7-22. d^{13}C values of geochemical materials.

Table 7-3. Range of d ^{13}C and dD for Methane Gas Derived from Different Sources

	Origin of Methane	d^{13}C(‰)	dD(‰)
Biogenic	CO_2 reduction	-100 to -60	-150 to -250
	Organic acid decomposition	-60 to -50	-250 to -350
Thermogenic	Early maturity	-50 to -40	-300 to -220
	Optimum maturity	-40 to -30	-220 to -160
	Late maturity	-30 to -15	-160 to -90

7.2.2 Rb–Sr System

Many systems have been used in dating materials (Table 7-4). Usually, Rb-Sr geochronological systems are used for dating rocks. The element rubidium has 17 isotopes, two of which are found in nature in the following isotopic proportions:

$$^{85}Rb = 0.72165$$
$$^{87}Rb = 0.27835$$

Table 7-4. Radioactive Systems Used in Geochronolology

Parent/daughter	Type of decay	λ (yr^{-1}) *	Half-life (yr)	Effective range (yr) (T_0=age of earth)	Crystal abundance of parent and daughter	Typical materials dated
^{238}U/^{206}Pb	8 Alpha + 6 beta	1.5369×10^{-10} (1.55125×10^{-10})	4.50×10^9	10^7-T_0	0.9928 g/g U 0.252 g/g Pb	Zircon, uraninite, monazite lead-bearing minerals
^{235}U/^{207}Pb	7 Alpha + 4 beta	9.7216×10^{-10} (9.8485×10^{-10})	0.71×10^{10}	10^7-T_0	0.0072 g/g U 0.215 g/g Pb	Zircon, uraninite, monazite lead-bearing minerals
^{232}Th/^{208}Pb	6 Alpha + 4 beta	4.987×10^{-11} (4.9475×10^{-11})	1.39×10^{10}	10^7-T_0	1.00 g/g Th 0.520 g/g Pb	Zircon, uraninite, monazite lead-bearing minerals
^{87}Rb/^{87}Sr	Beta	1.39×10^{-11} (1.42×10^{-11})	50×10^{10}		0.278 g/g Rb 0.07 g/g Sr	Biotite, musco-vite, microcline, whole rocks
^{40}K/^{40}Ar	Electron Capture	0.584×10^{-10} (0.581×10^{-10})	1.30×10^9 (Total)	5000-T_0	0.0001 g/g K 0.996 g/g Ar	Biotite, musco-vite, hornblende, whole rocks
^{40}K/^{40}Ca	Beta	4.72×10^{-10} (4.962×10^{-10})				
^{14}C/^{14}N	Beta	1.21×10^{-4}	5730	0-70,000	10-12 g/g C 0.996 g/g N	Charcoal, wood, peat

*The numbers given for the decay constants are the traditional decay constants used for many years. Because of uncertainties in these values, attempts have been made in recent years to obtain widespread agreement on more accurate values for the constants. It appears that the values given in the parentheses will be formally accepted for use in the future. Use of the new constants will cause some confusion, since all ages obtained with the new constants cannot be directly compared to previously published ages. However, the changes involved do not exceed 2.6 percent of all the age values.

Many systems have been used in dating material (Table 7-4). Usually the Rb-Sr geochronological system is used for dating rocks. The element Rubidium has 17 isotopes, two of which has found in nature in the following proportions:

$$^{85}\text{Rb} = 0.72165$$
$$^{87}\text{Rb} = 0.27835$$

Strontium has 18 isotopes, four of which occur in nature with the following average weights:

$$^{84}\text{Sr} = 0.0056$$
$$^{86}\text{Sr} = 0.0986$$
$$^{87}\text{Sr} = 0.0700$$
$$^{88}\text{Sr} = 0.8258$$

In general, ^{87}Rb decays to ^{87}Sr by β^- emission according to

$$^{87}_{37}\text{Rb} \rightarrow {}^{87}_{38}\text{Sr} + \beta^- + \gamma^-$$

with a decay constant $\lambda = 1.42 \times 10^{-11}\,a$ (a is the Latin word for anna – year) and a half life of

$$t_{1/2} = 4.9 \times 10^{10}\,a \qquad [7\text{-}22]$$

Because of the quantity of unstable parent nuclides P remaining at time t, in relation to the decay constant, we can write:

$$P = N_{o,p}\exp(-\lambda t) \qquad [7\text{-}23]$$

and the quantity of the daughter nuclides produced at time t is:

$$D = P[\exp(\lambda t) - 1] \qquad [7\text{-}24]$$

or the amount of radiogenic strontium ^{87}Sr$_{\text{rad}}$ produced at time t is:

$$^{87}\text{Sr}_{\text{rad}} = {}^{87}\text{Rb}[\exp(\lambda t) - 1] \qquad [7\text{-}25]$$

The bulk amount of ^{87}Sr at time t is:

$$^{87}\text{Sr}_t = {}^{87}\text{Sr}_0 + {}^{87}\text{Rb}[\exp(\lambda t) - 1] \qquad [7\text{-}26]$$

where ^{87}Sr$_0$ is the amount of ^{87}Sr present at time $t = 0$.

In practice, the use of mass spectrometry is convenient to measure mass ratios, rather than the absolute values. Thus, Eq. 7-26 is written as:

$$\frac{^{87}\text{Sr}}{^{86}\text{Sr}} = \left(\frac{^{87}\text{Sr}}{^{86}\text{Sr}}\right)_0 + \frac{^{87}\text{Rb}}{^{86}\text{Sr}}[\exp(\lambda t) - 1] \qquad [7\text{-}27]$$

Whenever it is possible to determine in a given rock at least two minerals that crystallize in the same time t = 0, then Equation 7-27 can be solved in t.

Alternatively, if one plots $\dfrac{^{87}\text{Sr}}{^{86}\text{Sr}}$ vs. $\dfrac{^{87}\text{Rb}}{^{86}\text{Sr}}$, then Equation 7-27 appears as a

straight line with slope $(\exp(\lambda t)-1)$ and intercept $\left(\dfrac{^{87}\text{Sr}}{^{86}\text{Sr}}\right)_0$. As indicated in

Figure 7-23(A), the straight line is known as an **isochron**. All minerals were

crystallized at the same time from initial composition $\left(\dfrac{^{87}\text{Sr}}{^{86}\text{Sr}}\right)_0$ resting on the

same isochron, whose slope $(\exp(\lambda t)-1)$ increases progressively with t. In this

manner, the slope of the isochrones may define the actual age of the crystallization (Figure 7-23B). The K-Ar system is based on the natural decay of ^{40}K into
^{40}Ar (electron capture, $\beta^+ = 10.5$ %). However, ^{40}K also decays into ^{40}Ca
$(\beta^- = 89.5\%)$ (see Figure 7-24). The decay constant of the first order transforming

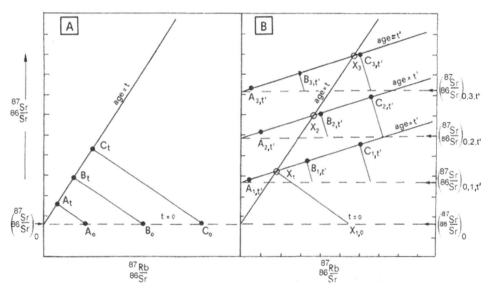

Figure 7-23. (A) Internal Rb-Sr isochron for a system composed of three crystalline phases of initial compositions A_0, B_0, and C_0 formed after time $t = 0$ and
thereafter closed to isotopic exchanges up to the time of measurement t,
when they acquired compositions A_t, B_t and C_t. **(B)** Effects of geochronological resetting resulting from metamorphism of interaction with fluids X_1, X_2, and
X_3; bulk isotopic compositions of the three rock assemblages. In cases of
short-range isotopic re-equilibration, the three assemblages define crystalliza-
tion age and original $(^{87}\text{Sr} / {}^{86}\text{Sr})_0$ of the system; the three internal isochrones
(concordant in this example) define resetting age.

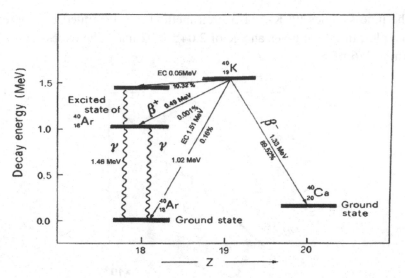

Figure 7-24. Decay scheme for ^{40}K, showing double decay to ^{40}Ca and ^{40}Ar.

^{40}K into ^{40}Ar is $\lambda_{EC} = 0.581 \times 10^{-10} a^{-1}$, and the β^- decay constant producing ^{40}Ca is $\lambda_\beta = 4.962 \times 10^{-10} a^{-1}$. The resulting branching ratio $\lambda_{EC}/\lambda_\beta = 0.117$, and the half-life for the dual decay $= 1.250 \times 10^9 a^{-1}$. Due to the double decay involved in the parent P disintegration, the decay equation can be written as:

$$_{18}^{40}\text{Ar} = {}_{18}^{40}\text{Ar}_0 + \left(\frac{\lambda_{EC}}{\lambda_{EC} + \lambda_\beta}\right) {}_{19}^{40}\text{K}\{\exp[(\lambda_{EC} + \lambda_\beta)t - 1]\} \qquad [7\text{-}28]$$

Equation 7-28 can be written as:

$$\frac{^{40}Ar}{^{36}Ar} = \left(\frac{^{40}Ar}{^{36}Ar}\right)_0 + \left(\frac{\lambda_{EC}}{\lambda_{EC} + \lambda_\beta}\right)\left(\frac{^{40}K}{^{36}Ar}\right)\{\exp(\lambda_{EC} + \lambda_\beta)t - 1]\} \qquad [7\text{-}29]$$

[Example 7-2] The volcanic rocks at Olduvai Gorge, Tanzania, are of paramount importance because the remains of early man have been found there. Nine samples of Tuff 1B by the whole rock using K-Ar isochron are shown by Fitch et al. in 1976, as shown in Figure 7-25. Obtain the date as well as the initial ^{40}Ar/^{26}Ar.

The nine samples by K-Ar isochron method are illustrated in Figure 7-25. From the plot, the slope gives an age of 2.04 ± 0.02 ma. The excess argon intercept term is 276 ± 29.

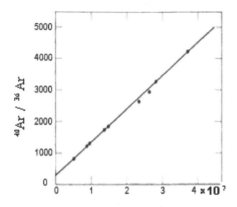

Figure 7-25. Whole rock K-Ar isochron of Tuff 1B from Olduvai Gorge, Tanzania. These rocks were originally dated by Curtis and Hay (1972) using the conventional K-Ar method. Nine samples yielded an average date of 1.976 $\pm$ 0.034 Ma. Fitch et al. (1976) subsequently used the measurements to construct a K-Ar isochron for these rocks.

7.2.3 Re–Os System

A system that is often used is Re-Os. Actually, most iron meteorites and the metallic phases of chondrites fit for the Re-Os isochron.

$$\frac{^{187}Os}{^{186}Os} = \left(\frac{^{187}Os}{^{186}Os}\right)_0 + \frac{^{187}Re}{^{186}Os}[\exp(\lambda t) - 1] \qquad [7\text{-}30]$$

7.2.4 ^{14}C Dating

The half life of ^{187}Os ^{187}Os by β^- decay is $(4.23 \pm 0.13) \times 10^{10} \alpha$. An example of the isochron application is seen in Figure 7-26. Even the Earth's mantle falls on the same isochron. This suggests that the solar nebula must be homogenous (e.g. $^{187}Os / ^{186}Os = 0.805 \pm 0.006$) and the sources of all meteorites and the Earth must be formed within a short period.

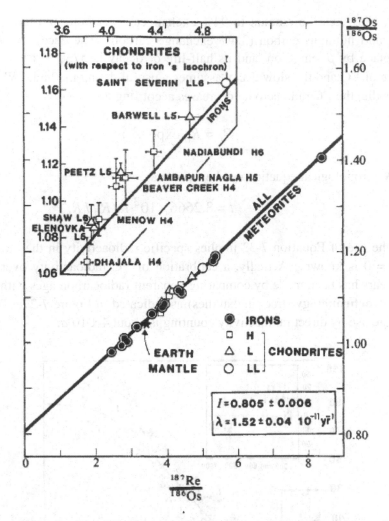

Figure 7-26. Re-Os isochron for iron meteorites and metallic phase of chondrites and earth's mantle (After J.M Luck and C.J. Allegre, Nature, 302, 130-132, 1983 Macmillan Magazines Limited).

Another often used isotope for dating is ^{14}C. The radiogenic ^{14}C is continuously formed from (n,p) reaction of thermal neutron (slow) with ^{14}N.

$$^{14}_{7}N + n \rightarrow ^{14}_{6}C + p$$

The ^{14}C isotope directly reacts with atmospheric O_2 to produce $^{14}CO_2$ and undergo isotope exchange with $^{12}CO_2$ in the atmosphere. The atmospheric carbon (CO_2, 1.4% exchangeable carbon) equilibrates with oceanic carbon (H_2CO_3,

HCO_3^-, CO_3^{2-}, etc. to amount to 94% exchangeable carbon) and terrestrial biosphere and hummus carbon (3.4% exchange carbon). The decay of ^{14}C to ^{14}N takes place by β^- emission, and its half-life of decay is $5730 \pm 40a$. The rapid mixture of ^{14}C and the slow decay assumes a useful dating method. When living matters die, the ^{14}C radioactivity decreases according to

$$R_t = R_0 \exp(-\lambda t) \qquad\qquad [7\text{-}31]$$

A chronological equation can be written as:

$$t = 8.2666 \times 10^3 \ln(R_0 / R_t) \qquad\qquad [7\text{-}32]$$

The use of Equation 7-32 implies specific radioactivity in the biosphere at time t = 0 is known. Actually, a calibration of ^{14}C radioactivity over the past 7500 years has been made by comparing apparent radiocarbon ages with the ages of **dendrochronology** (tree ring studies) as indicated in Figure 7-27. The maximum age use by direct radioactivity counting is about $4 \times 10^4 a$.

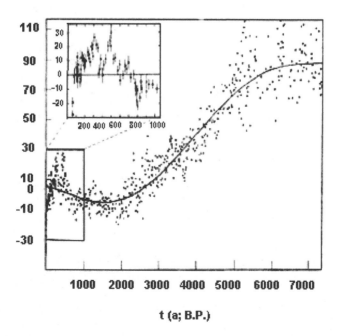

Figure 7-27. Specific radioactivity of ^{14}C expressed as per-mil deviation (Δ o/oo) from present-day radioactivity over the last 7500 years, derived from comparisons with dendrochronological studies (After G. Ottonello, Principles of Geochemistry, 1997, Columbia Univ. press).

7.3 GEOCHEMICAL BIOMARKERS

Geochemical biomarkers can be thought of as organic compounds derived from biochemical precursors in a geological environment. This term is synonymous with a number of other terms used in the literature such as **chemical fossils**, molecular indicators, or biological markers. They are often stable organic molecules related to living organisms. This detailed characterization of biomarkers permits the assessment of contributing species of extinct or extant life. Known biomarkers include alkanes, terpanes, steroids, and porphyrins.

Porphyrins can be described as a tretrapyrole (four-pyrole system) pigment. For other biomarkers, the isoprene rule is essential. In this section, we will introduce the concept of these 5-C inputs. Actually, all types of terpane molecules can be derived from it. We can divide terpanes into the following types:

> Hemiterpane (C_5)
> Monoterpane (C_{10})
> Sequiterpane (C_{15})
> Diterpane (C_{20})
> Sesterterpane (C_{25})
> Triterpane (C_{30})
> Tetraterpane (C_{40})
> Polyterpane (C_{40+})

7.3.1 Monoterpanes

For each type, the components can be classed as either acyclic or cyclic. Isoprene is a diene molecule, and isoprene unit is the skeleton of this 5-C compound. Namely,

Isoprene Unit   Isoprene

For the monoterpanes (C_{10}), the acyclic hydrocarbons are:

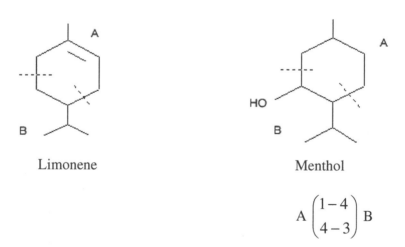

2,6-Dimethyloctane   Mercene
(1,4) or head to tail connection
A (1-4) B

Both hydrocarbons can be linked by C1 of an A unit to C4 of a B unit. Usually, the function group can be added, such as in the case of mercene, A (1—4) B—(3=)$_2$. For the cyclic form,

Limonene   Menthol

$$A \begin{pmatrix} 1-4 \\ 4-3 \end{pmatrix} B$$

The skeleton can be represented by $A \begin{pmatrix} 1-4 \\ 4-3 \end{pmatrix} B$, but specifically for lio-mone $A \begin{pmatrix} 1-4 \\ 4-3 \end{pmatrix} B\text{-}2_A=$ and for menthol $A \begin{pmatrix} 1-4 \\ 4-3 \end{pmatrix} B\text{-}4_B OH$. Functional graphs are indicated by the position (2 or 4) and the unit (A or B). For two isoprene units, AB, bonds can be broken for forming the acyclic series. For example,

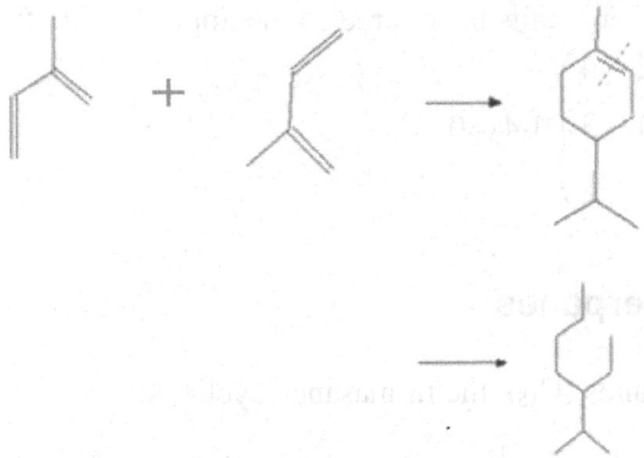

2-methyl-3-ethyl octane

Both 2,6-dimethyl octane and 2-methyl-3-ethyl heptane can be accounted for C_{10}-monoterpanes. They may be thermal breakdown products from the A $\begin{pmatrix} 1-4 \\ 4-3 \end{pmatrix}$ B skeleton. Furthermore, there are tricyclic structures for the monoterpanes, such as:

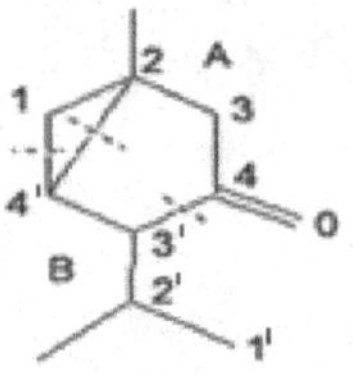

Thujone

The skeleton is $\qquad$ A $\begin{pmatrix} 2-4' \\ 4-3' \\ 1-4' \end{pmatrix}$ B

Or in this case, the apostrophe is omitted

$$A \begin{pmatrix} 2-4 \\ 4-3 \\ 1-4 \end{pmatrix} B$$

The two isoprene units have three connectors. Hence, the specific structure

for thujone is $A \begin{pmatrix} 2-4 \\ 4-3 \\ 1-4 \end{pmatrix} B\text{-}4_A = 0$.

7.3.2 Sesquiterpanes

For the sesquiterpanes (C_{15}), the farmasane acyclic is:

Farmasane
A (1,4) B (1,4) C(1,4)

The following two structures are examples of bicyclics:

Bulgarene

$A \begin{pmatrix} 3-4 \\ 4-1 \end{pmatrix} B \begin{pmatrix} 3-4 \\ 1-4 \end{pmatrix} C$

Eudesmane

$A \begin{pmatrix} 1,4 \\ 2,5 \end{pmatrix} B(3-4)C(4-5)A$

7.3.3 Diterpanes

The next series consists of diterpanes (C_{20}). Usually the acyclic is phytane or the related hydrocarbons of phytane (such as pristine, C_{19} or norpristane, C_{18}):

Phytane
A(4-1)B(4-1)C(4-1)D

Another drawing of phytane may be ring-like

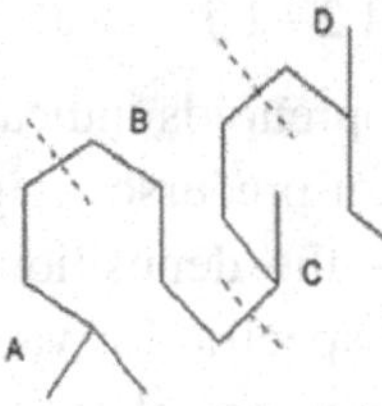

For the tricyclics,

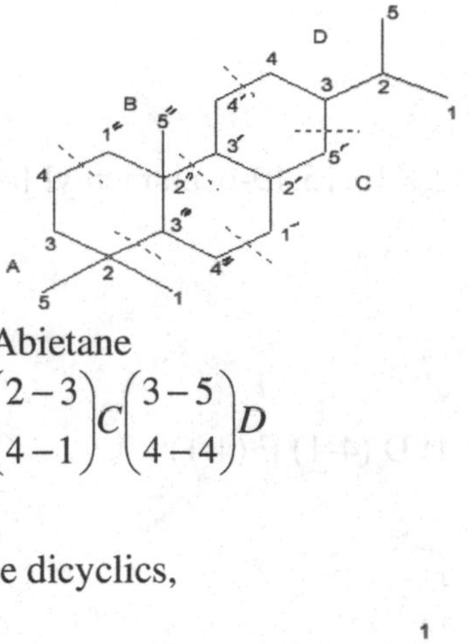

Abietane

$$A\begin{pmatrix}2-3\\4-1\end{pmatrix}B\begin{pmatrix}2-3\\4-1\end{pmatrix}C\begin{pmatrix}3-5\\4-4\end{pmatrix}D$$

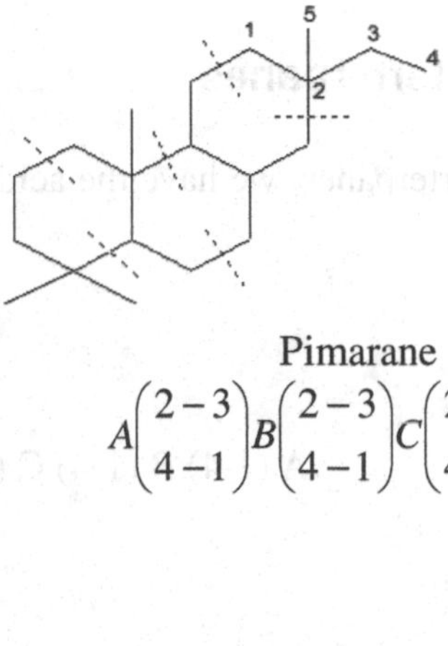

Pimarane

$$A\begin{pmatrix}2-3\\4-1\end{pmatrix}B\begin{pmatrix}2-3\\4-1\end{pmatrix}C\begin{pmatrix}2-5\\4-4\end{pmatrix}D$$

For the dicyclics,

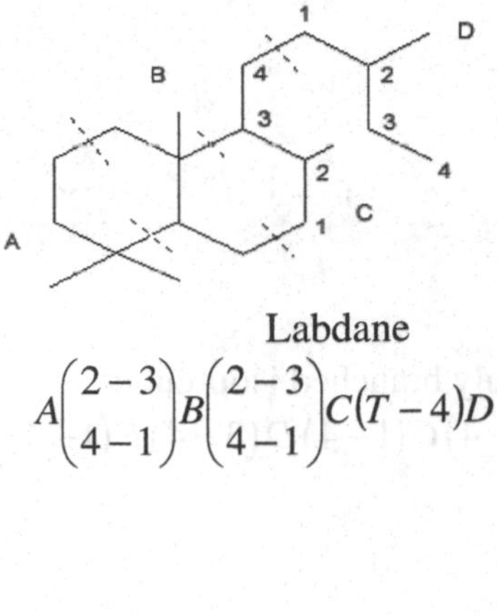

Labdane

$$A\begin{pmatrix}2-3\\4-1\end{pmatrix}B\begin{pmatrix}2-3\\4-1\end{pmatrix}C(T-4)D$$

For the tetracyclics,

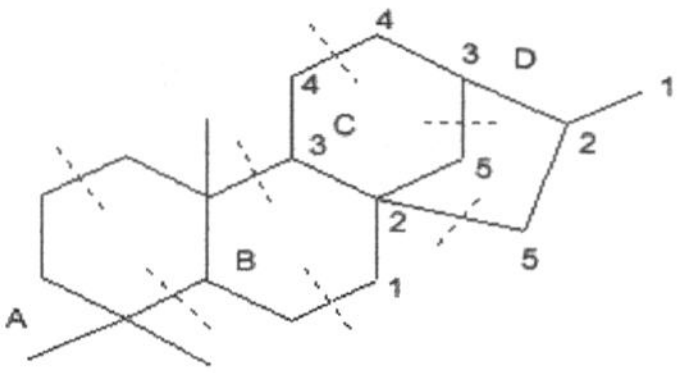

Kaurane

$$A\begin{pmatrix} 2-3 \\ 4-1 \end{pmatrix} B\begin{pmatrix} 2-3 \\ 4-1 \end{pmatrix} C\begin{pmatrix} 3-5 \\ 4-4 \\ 2-1 \end{pmatrix} D$$

For diterpenoids, the macrocyclic isoprenoids indicate immature lacusterine kerogen samples. For example, cemprene of precursor originated from melvonate, and geranylgeranyl diphosphate indicates the deposition history of the sample (Figure 7-28). The bicyclic diterpanes are specific for various microbial, gymnosperm and angiosperm imparts. They have to correlate the source with one from the well. Other tricyclic diterpenoids are also important (Figure 7-28A).

7.3.4 Sesterterpanes

For the sesterterpanes, we have the acidic 2,6,10,15,19-pentamethyl licosane, C_{25}.

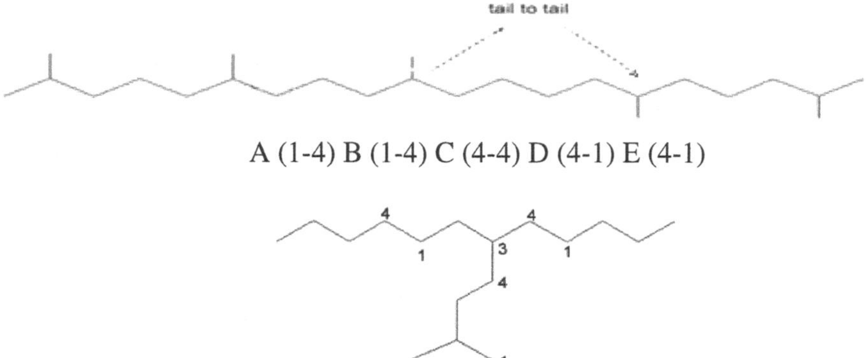

A (1-4) B (1-4) C (4-4) D (4-1) E (4-1)

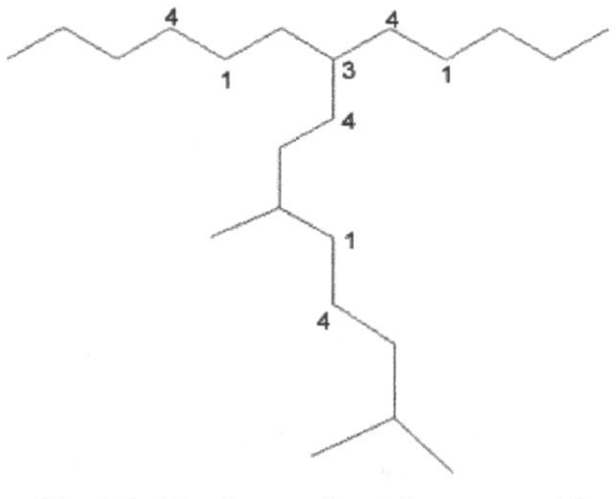

C_{25} highly branched isoprenoid
$$A(1-4)B(1-4)C(1-4)D(3-4)E(1-4)$$

The following is an example of a cyclic form:

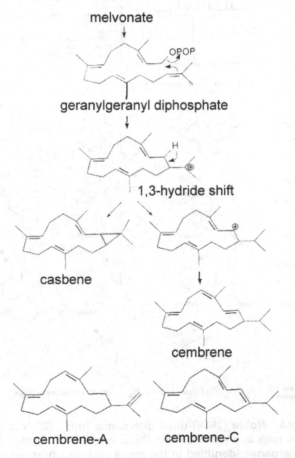

9 asordic acid

$$A(2=,4-1)B\begin{pmatrix}2-3\\4-4\end{pmatrix}C\begin{pmatrix}1-1\\3-3\end{pmatrix}D(4-1)E(2=,1-COOH,4-3)C$$

melvonate

↓

geranylgeranyl diphosphate

↓

1,3-hydride shift

casbene

cembrene

cembrene-A   cembrene-C

Figure 7-28. Synthetic pathway for cembranoids and the related casbenoids.

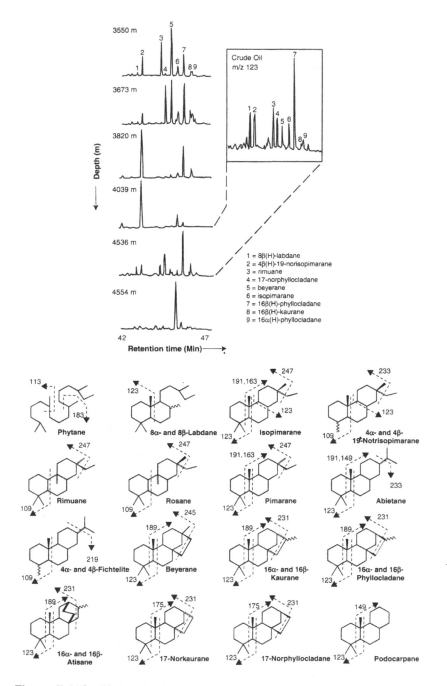

Figure 7-28A. Noble (1986) used diterpanes (m/z 123) to correlate crude oil with source rock in a well from the Gippsland Basin, Australia. Structures for tricyclic diterpanes identified in the mass chromatogram show their principal fragment ions.

7.3.5 Triterpanes

For the triterpanes, a vast amount of molecules of terpanes are constituents.

Squalane
A (4-1) B (4-1) C (4-4) D (1-4) E (1-4) F

The 34-carbon botryococcane is a tetramethyl substituent of an isopro-panoic of 5 units. Due to the tetrasubstitution, this is not an isoprenoid.

C_{34}-Botryococcane

A (4-1) B (4-1) C (4-2) D (5-4) E (1-4) (($^{+}$3Me)$_4$ $_{-ABDE}$)

(The o sign indicates the substitute at 3-positions)

The C_{30}-botryococcane can be expressed as a perfect isoprenoid.

C_{30}-Botryococcane
A (4-1) B (4-5) C (2-3) D (1-4) E (1-4) F

The tetracyclic chiolestane consists of major classes of stearanes. For cholestane C_{27}, always counting as derivatives of C_{30} isoprenoid.

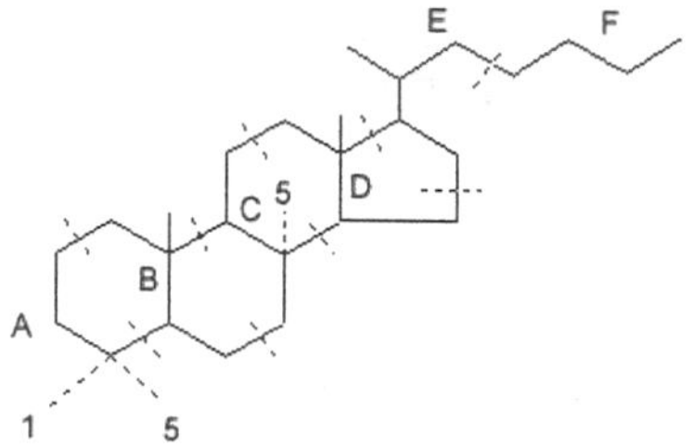

C_{27} Cholestane

$$A \begin{pmatrix} 4-1 \\ 2-3 \end{pmatrix} \quad B \begin{pmatrix} 4-1 \\ 2-3 \end{pmatrix} \quad C \begin{pmatrix} 4-1 \\ 2-3 \end{pmatrix} \quad D \begin{pmatrix} 4-4 \\ 2-3 \end{pmatrix} \quad E\,(1\text{-}4) \quad F\text{-}1_A \text{ -}5_A\text{-}5_C$$

In this case, three carbons, A(1), A(5), and C(5) have been eliminated. The pentacyclic series will be no problem. For example,

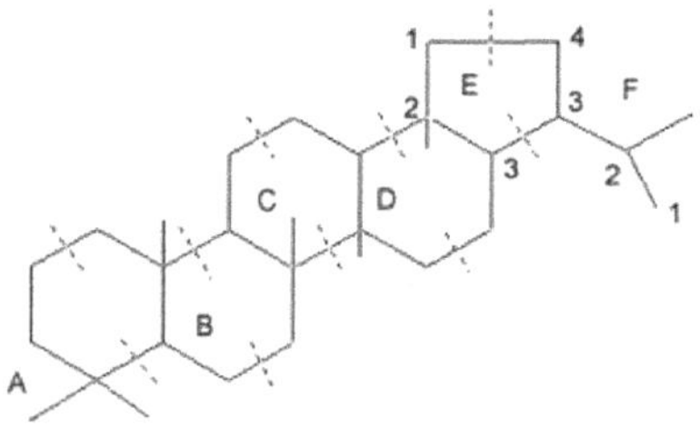

Hopane

$$A \begin{pmatrix} 4-1 \\ 2-3 \end{pmatrix} \quad B \begin{pmatrix} 4-1 \\ 2-3 \end{pmatrix} \quad C \begin{pmatrix} 4-4 \\ 2-2 \end{pmatrix} \quad D \begin{pmatrix} 3-2 \\ 1-4 \end{pmatrix} \quad E \begin{pmatrix} 3-3 \\ 1-4 \end{pmatrix} \quad F$$

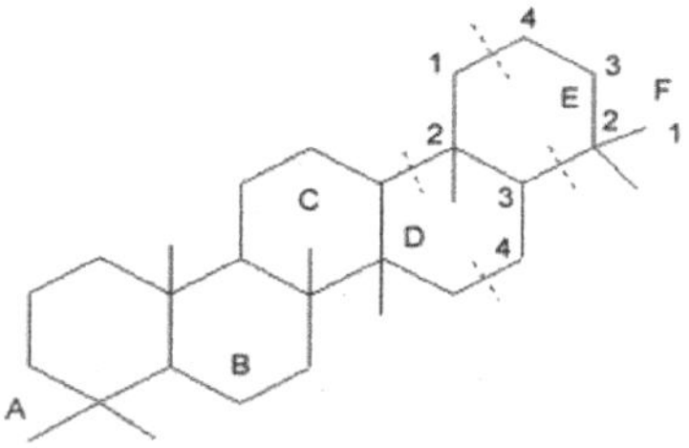

Gammacerane

$$A \begin{pmatrix} 4-1 \\ 2-3 \end{pmatrix} \quad B \begin{pmatrix} 4-1 \\ 2-3 \end{pmatrix} \quad C \begin{pmatrix} 4-4 \\ 2-2 \end{pmatrix} \quad D \begin{pmatrix} 3-2 \\ 1-4 \end{pmatrix} \quad E \begin{pmatrix} 3-2 \\ 1-4 \end{pmatrix} \quad F$$

Oleanane

$$A \begin{pmatrix} 4-1 \\ 2-3 \end{pmatrix} B \begin{pmatrix} 4-1 \\ 2-3 \end{pmatrix} C \begin{pmatrix} 4-4 \\ 2-2 \end{pmatrix} D \begin{pmatrix} 3-2 \\ 1-4 \end{pmatrix} E (2\text{-}4) \ F \ (1\text{-}1) \ D$$

7.3.6 Tetraterpanes

The last class is tetraterpanes, C_{40}. These are represented by lycopane below, followed by unsaturated examples.

Lycopene

A (4-1 B (4-1) C (4-1) D (4-4) E (1-4) F (1-4) G (1-4) H

$$A \begin{pmatrix} 4-1 \\ 2-3 \end{pmatrix} B (4\text{-}1) \ C (4\text{-}1) \ D (4\text{-}4) \ E (1\text{-}4) \ F (1\text{-}4) \ G \begin{pmatrix} 3-2 \\ 1-4 \end{pmatrix} H$$

β-Carotene

7.3.7 Polyterpanes

Finally for polyterpanes, $C_{40>}$, examples include all-trans rubber or polyisoprene.

$$A\ (1\text{-}4)_n\ X,\ \ X = B\ldots N$$
Trans-rubber

$$(OMe)_2\text{-Quinone-A}\ (1\text{-}4)_{10}\ X$$
Co-enzyme Q

We like to summarize the isoprene rule as:
The terpenoid hydrocarbons as $I_i(j) = A,\ B\ldots X$

$$i = 1,\ I_1(j) = A$$
$$i = 1,\ I_2(j) = AB$$
$$\ldots\ldots$$
$$i = 1,\ I_n(j) = AB..X$$
$$j = 1,2\ldots5\ (1\ \text{and}\ 5\ \text{are equivalent})$$

$i = 1,$ Hemi
$i = 2,$ Mono
$i = 3,$ Sesqui
    … and so on.

In general the biomarkers can be used as a source related parameter. For example, it can relate to the depositioned environment. Sometimes it can be used as a maturation parameter and also to study the migrating petroleum and biodegration as well as basin modeling.

7.3.8 Porphyrins

Beyond the isoprenoid compounds, the pigment porphyrin has also been studied. Porphyrin in petroleum or sediment is in vanadyl chelates (See Figure 7-29 VII or VIII). The precursor originates from chlorophyll (I) as follows:

The structure of chlorophyll, which is a Mg chelate of the porphyrin skeleton, also carries an isoprenoid phytal group. Both chlorophyll a and heme has the basic etioporphyrical structure of Fischer convention III. The latter is an Fe chelate. The gradual chemical transformation from chlorophyll a (I) to the deoxophylloerythrin porphyrin DPEP (VII) can be seen from the scheme expressed in Figure 7-29. Most petroporphyrin contain both DPEP (VII) or etio (VIII). Occasionally they also contain an aromatic, either substituted or fused on the etio structure of VIII. Actually, this was observed in a sample of Nonesuch shale (1.2 Ba). Thus, it is possible to use $\Sigma E \Pi D / \Sigma DPEP$ as a maturity index. An approximate plot of this sort from a number of asphaltene extracts can be found in Figure 7-30.

The detection of porphyrins in trace quantity can be done photometrically by fluorescence. The absence of lunar soil has been used to explain the absence of life on the moon.

7.3.9 Others

Figure 7-31 is a GC-MS of an Australian crude oil at various depths; the biomarkers can be related to the source rock. Many times, a fossil extract of a particular plant will give some specific biomarkers such as sugiol and chamaecydin.

There is a gammacearane index which is used for water-column stratification during some rock deposition. Oils from lacustrine source rocks in Angola indicate that increased water salinity during deposition results in higher gammacerane to pristine/plytane ratio (See Figure 7-32). For the application in the characterization of coal deposit or the formation water please see Yen's Environmental Chemistry of this series, Imperial College Press, 2005.

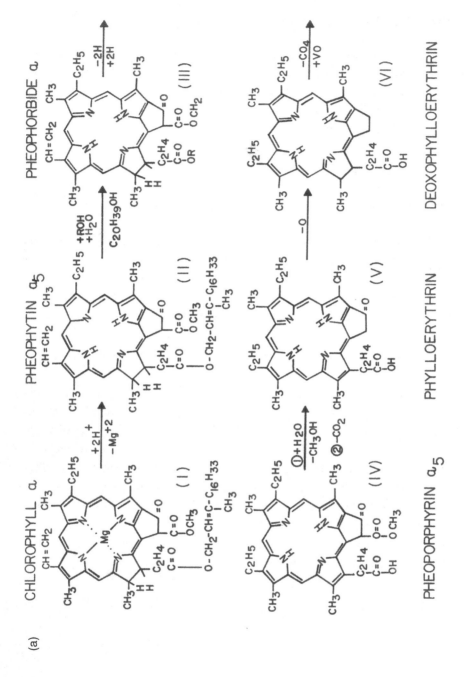

Figure 7-29. Proposed geochemical transformation of chlorophyll to vanadyl DPEP and other stable vanadyl chelates. (a) Chlorophyll to dexophylloerythin.

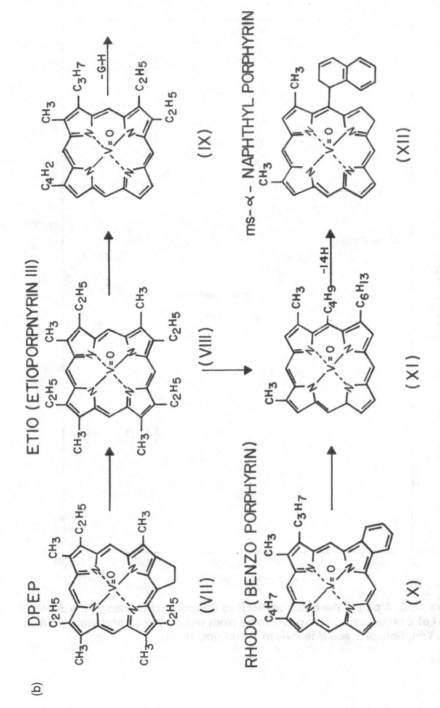

Figure 7-29. Proposed geochemical transformation of chlorophyll to vanadyl DPEP and other stable vanadyl chelates. (b) DPEP to ms--naphthyl porphyrin. (Source: T. F. Yen and G. V. Chilingarian, 1994.)

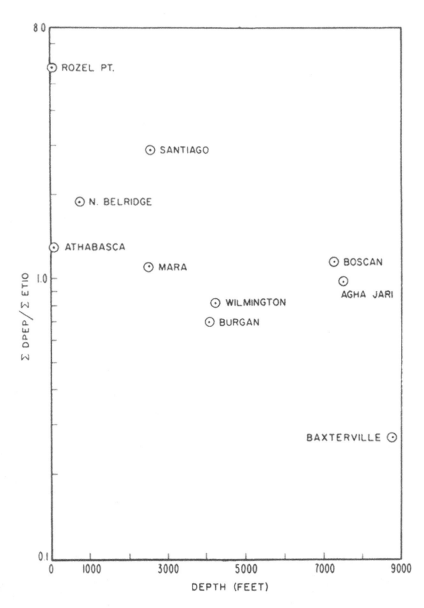

Figure 7-30. A plot of the ratios of DPEP to Etioporphyrins versus the depth of burial of a number of native petroleums from which these porphyrins derived. (T. F. Yen, Role of Trace Materials in Petrolium, 1975).

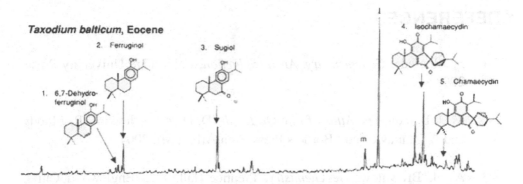

Figure 7-31. Gas chromatography/mass spectrometry (GCMS) traces of total ion current (TIC) for the aromatic compound fraction from the extract of the seed cone of Eocene Taxodium balticum. I, an isomer of taxoquinone, and m, an isomer of inuroyleanone. After Otto et al. Science, 297, 1543, 2002.

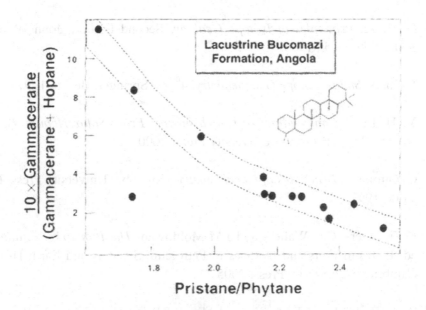

Figure 7-32. Variations in pristine/phytane (Pr/Ph) (redox) and gammacerane index (salinity stratification) for oils derived from lacustrine source rocks in Angola. Inset shows gammacerane. Increased water salinity in the source-rock depositional environment results in higher gammacerane indices. Higher salinity is typically accompanied by density stratification and reduced oxygen content in bottom waters (i.e. lower Eh), which results in lower Pr/Ph. Biomarker evidence suggests a marine co-source for the point that lies off the trend. Figure courtesy of B. J. Huizinga.

REFERENCES

7-1. F. Alberede, *Geochemistry, An Introduction*, Cambridge University Press, 2003.

7-2. M. J. Benton, *The Atlas of Life On Earth* (D. Dixon, J. Jenkins, R. Moody and A. Zhuravlev ed.) Borders Press, Ann Arbor, MI, 2004.

7-3. A. H. Brownlow, *Geochemistry*, Prentice-Hall, Inc., Englewood Cliffs, N.J., 1979.

7-4. W.G. (Editor) Ernst, *Frontiers in Goechemistry: Organic, Solution, and Ore Deposit Geochemistry*, Konrad Krauskopf Volume 2, International Book Series, Volume 6, Bellwether Publishing, Ltd. 2002.

7-5. G. Faure, *Principles of Isotope Geology,* Second Edition, John Wiley & Sons, New York, 1986.

7-6. J. Hoefs, *Stable Isotope Geochemistry*, 4th ed. Springer, Berlin, 1997.

7-7. Y. H. Li, *A Compendium of Geochemistry: From Solar Nebula To The Human Brain*, Princeton University Press, 2000.

7-8. G. Ottonello, *Principles of Geochemistry*, Columbia University Press, New York, 1997.

7-9. K. E. Peters, C.C Walters and J.M Moldowan, *The Biomarker Guide*, 2nd ed. II. Biomarkers and Isotopes in Petroleum Systems and Earth History, Cambridge University Press, 2005.

7-10. P. A. Rona, *Plate Tectonics and Mineral Resources*, Scientific America, July 1973.

7-11. H. D. Schulz, and A. Hadeler, (Editors), Deutsche Forschungsgemeinschaft, *Geochemical Processes in Soil and Groundwater: Measurement-Modelling-Upscaling.* GeoProc 2002, WILEY-VCH GmbH & Co. KGaA, 2003.

7-12. http://www.scotese.com/images (PALEOMAP Project).

7-13. J. Thiede, and E. Suess, *Sediment Record of Ancient Coastal Upswelling*, *Episodes*, 1983, 15-18.

7-14. J. T. Wilson, *Continents Adrift and Continents Aground*, W.H. Freeman San Francisco, 1976.

7-15. J. T. Wilson, *Readings from Scientific American: Continents Adrift and Continents Aground*, W. H. Freeman and Company, San Francisco, 1976.

7-16. B. F. Windley, *The Evolving Continents*, 2nd ed. John Wiley, Chichester, 1984.

7-17. A. Yin, and M. Harrison, *The Tectonic Evolution of Asia*, Cambridge University Press, 1996.

7-18. Stryer, Lubert, *Biochemistry,* W. H. Freeman and Company, San Francisco, 1975.

7-19. Sybesma, C., *An Introduction to Biophysics*, Academic Press, New York, 1977.

PROBLEM SET

1. Use Rb-Sr isochron to date a suite of whole-rock samples. The initial $^{87}Sr/^{86}Sr$ ratio is 0.7071 ± 0.0005 and λ of ^{87}Rb is 1.42×10^{-11} y^{-1}. What is the age of these samples?

$$\text{Assuming } \frac{^{87}Sr}{^{86}Sr} = \left(\frac{^{87}Sr}{^{86}Sr}\right)_i + \frac{^{87}Rb}{^{86}Sr}[e^{\lambda t} - 1]$$

$$e^{\lambda t} = 1 + \lambda t + \frac{(\lambda t)^2}{2!} + \frac{(\lambda t)^3}{3!} + \dots$$

Since λ is small,

$$1 + \lambda t \gg \frac{(\lambda t)^2}{2!} + \frac{(\lambda t)^3}{3!} + \dots$$

and,

$$\frac{^{87}Sr}{^{86}Sr} \approx \left(\frac{^{87}Sr}{^{86}Sr}\right)_i + \frac{^{87}Rb}{^{86}Sr}\lambda t$$

(Ans: 18.43 Ga)

2. With reference to paleothermometry, gastropods do have the dependence of mineral water surrounding them. $\delta^{18}O$ of calcite and water based on *Patinopecten yessoensis* in a bay has the correlation,

$$t°C = 17.04 - 4.34(\delta_c - \delta_w) + 0.161(\delta_c - \delta_w)^2$$

If the corresponding difference in $\delta^{18}O$ between calcite and water is around one $°/_{00}$, what is the temperature of water at the bay?

(Ans: 12.86 °C)

3. Gallegos cited three possible routes of C_{20} stachane, kaurane, and tachylobane to give tricyclic diterpanes at m/e of 197 in Green River shale as shown in the figure. Test the isoprene rule for it.

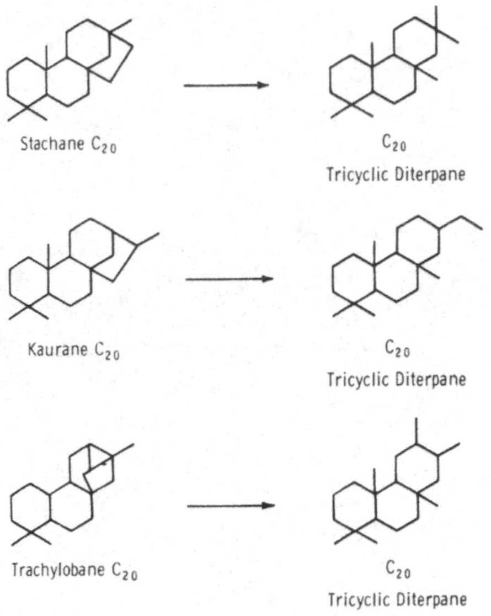

Stachane C_{20}

C_{20}
Tricyclic Diterpane

Kaurane C_{20}

C_{20}
Tricyclic Diterpane

Trachylobane C_{20}

C_{20}
Tricyclic Diterpane

(from *Oil Shale*, by Yen and Chilingarian, Elsevier, Amsterdam, 1976.)

CHAPTER 8

FUEL CHEMISTRY

*T*he value of fuel is directly related to the sources of energy which is critical to the responsibility of engineers. In short, fuel science has a direct impact on the well being of human society. Chemistry can modify and refine fuel through chemical reactions. In some instances, synthetic fuel can be manufactured using simple material such as carbon and hydrogen. One of the significant accomplishments is that raw fuels can be cracked (fragmented) and synthesized to chemical precursors, the petrochemical industry.

Catalyst chemistry will be reviewed with emphasis on surface characterization methods. For example, the BET equation will be used for surface area measurement. The catalyst carrier supports molecular size, zeolites and other shapes and size selectors will be covered. For major petrochemical processes, reactor designs and catalytic cracking will be reviewed in comparison to physical cracking (thermal and ultrasound).

A cleaner fuel, i.e. a highly "pure" fuel that does not contain heterocyclic elements (such as metals, N, S and O) can be obtained using not only synthetic methods like Fischer-Tropsch synthesis, but can also be accomplished by desulfurization, demetalization, denitrogenation and deoxygenation.

8.1 CATALYST CHEMISTRY

A catalyst is usually defined as a material that enhances the rate and selectivity of a chemical reaction and in the process is cyclically regenerated. A **homogenous catalysis** is a process in which all reactants and catalysts are present in the same place. On the other hand, **heterogeneous catalysis** is the process in which reactants and products adsorb into the surface of the catalyst during the corrosion process so that the catalyst is instantaneously charged. Adsorbed reactants are activated by interaction with the catalyst surface and selectively transformed to adsorbed products. After the product is desorbed from the surface of the catalyst, the catalyst can repeat the catalytic cycle.

A typical heterogeneous catalyst is made up of three components: an **active catalytic phase**, a **promoter** which increases activity or stability, and a high-surface area **carrier (support)** which serves to facilitate the dispersion and stability of the active catalyst. Examples of the three components of heterogeneous catalysts are given in Table 8-1.

Table 8.1. Components of a typical heterogeneous catalyst: material types and examples (After Farrauto and Bartholomew, 1997).

Component	Material types	Examples
Active phase	metals	noble metals (Pt, Pd), base metals[a] (Ni, Fe)
	metal oxides	transition metal oxides (MoO_2, CuO)
	metal sulfides	transition metal sulfides (MoS_2. Ni_3S_2)
Promoter Textural	metal oxides	transition metal and Group IIIA (Al_2O_3, SiO_2, MgO, BaO, TiO_2, ZrO_2)
Chemical	metal oxides	alkali or alkaline earth (K_2O, PbO)
Carrier (or support[b])	stable, high surface area metal oxides, carbons	Group IIIA, alkaline earth transition metal oxides (Al_2O_3, SiO_2, MgO) zeolites and activated carbon

[a] The term base metal derives from the jewelry industry where Fe and Ni serve as the base metal for coating with noble metals such as Au, Pt, and Rh.
[b] Refers to a high surface area carrier or matrix which is an integral part of the catalyst; this carrier is distinct from low surface area metal or ceramic monolithic supports or substrates upon which catalysts are sometimes coated.

Transition metals and their oxides, sulfides, nitrides, and carbides can be the active catalytic phases as seen in Table 8-2.

A number of commonly used supports or carriers are tabulated in Table 8-3 along with their physical properties. In Table 8-3, the surface area, pore size, and pore volume are the most fundamentally important properties in catalysts.

Table 8-2. Active catalytic phases and reactions they typically catalyze (Ref. 8-1).

Active phase	Elements / compounds	Reactions catalyzed
Metals	Fe, Co, Ni, Cu, Ru, Rh, Pd, Ir, Pt, Au	hydrogenation, steam reforming, hydrocarbon reforming, dehydrogenation, synthesis (ammonia, Fischer-Tropsch), oxidations
Oxides	oxides of V, Mn, Fe, Cu, Mo, W, rare earth, Al, Si, Sn, Pb, Bi	complete and partial oxidation of hydrocarbons and CO, acid-catalyzed reactions (e.g. cracking, isomerization, alkylation), methanol synthesis
Sulfides	sulfides of Co, Mo, W, Ni, V	hydrotreating (hydrodesulfurization, hydrodenitrogenation, hydrodeoxygenation), hydrogenation
Carbides	carbides of Fe, Mo, W	hydrogenation, Fischer-Tropsch synthesis

Table 8-3. Physical properties of common carriers (supports).

Support / catalyst	BET area ($m^2\ g^{-1}$)	Volume ($cm^3\ g^{-1}$)	Pore diameter (nm)
Activated carbon	500-1500	0.6-0.8	0.6-2.0
Zeolites (molecular sieves)	500-1000	0.5-0.8	0.4-1.8
Silica gels	200-600	0.40	3-20
Activated clays	150-225	0.40-0.52	20
Activated Al_2O_3	100-300	0.4-0.5	6-40
Kieselgurh (Celite 296)	4.2	1.14	2200

8.1.1 Surface Area and the Brunauer, Emmett, and Teller (BET) Equation

The quantity of a species adsorbed is characterized by **isotherm**, a plot of θ versus pressure at constant temperature where θ is the fraction of surface covered with a specific adsorbed atom or molecule. Isotherms relate surface concentration (unmeasurable) to gas plane concentration (measurable). Usually fractional coverage θ is defined as volume adsorbed per volume required for monolayer coverage (V/V_m). Generally,

$$\theta = f(T, P)$$

or

$$\theta_T = f(P) \qquad [8\text{-}1]$$

where T is a constant.

For a monolayer absorption of an elementary equilibrium reaction

$$A + S \leftrightarrow A - S \qquad [8\text{-}2]$$

where A is **adsorbate**, S is a **site** on a metal surface, and $A - S$ is adsorbed A. If P_A is the pressure of the gas A, θ_A is the fraction covered by A, the rate constant for adsorption is K_a, and rate constant for desorption is K_d then the overall rate is the sum of the forward and reverse rate.

$$R = K_a P_A (1 - \theta_A) - K_d \theta_A \qquad [8\text{-}3]$$

At equilibrium, $r = 0$.

$$(K_a / K_d) = P_A (1 - \theta_A) = \theta_A \qquad [8\text{-}4]$$

Now if K is the equilibrium constant, by Arrhenius law:

$$K = (K_a / K_d) = (A_a / A_d) e^{-\Delta H_a / RT} \qquad [8\text{-}5]$$

when

$$\Delta H_a = E_a - E_d \qquad [8\text{-}6]$$

assuming that $\Delta H_a = f(\theta_A)$.

From Eq. [8-4],

$$KP_A (1 - \theta_A) = \theta_A \qquad [8\text{-}7]$$

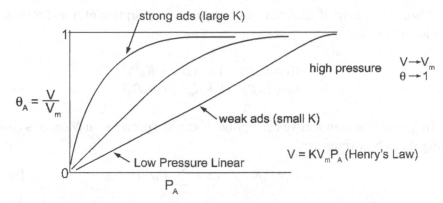

Figure 8-1. Isotherms or coverage versus pressure curves for weak, moderately strong and strong adsorption of a species *A* on a surface.

Solving for θ_A,

$$\theta_A = KP_A / (1 + KP_A) \qquad [8\text{-}8]$$

This is the form of **Langmuir isotherms**. Different strengths of adsorption are depicted in Fig. 8-1.

Since $\theta_A = V / V_m$, Eq. [8-8] can be rearranged to obtain

$$V / P_A = -KV + KV_m \qquad [8\text{-}9]$$

If V / P_A is plotted against V, a straight line is obtained with a slope of $-K$ and an intercept of KV_m.

Other forms of Langmuir isotherms, such as

$$A_2 + 2S \leftrightarrow 2A - S \qquad [8\text{-}10]$$

can also be obtained. Again by treating under equilibrium,

$$r = K_a P_{A_2} (1 - \theta_A)^2 - K_d \theta_A^2 \qquad [8\text{-}11]$$

$$(K P_{A_2})^{1/2} (1 - \theta_A) = \theta_A \qquad [8\text{-}12]$$

$$\theta_A = (K P_{A_2})^{1/2} / [1 + (K P_{A_2})^{1/2}] \qquad [8\text{-}13]$$

Also, a treatment of simultaneous competitive adsorption of A and B (non-dissociative) can be:

$$\theta_A = K_A P_A / (1 + K_A P_A + K_B P_B)$$
$$\theta_B = K_B P_B / (1 + K_A P_A + K_B P_B) \qquad\qquad [8\text{-}14]$$

In general for competitive adsorption, if the ith species with N species (including i) can be written as:

$$\theta_i = P_i K_i \Big/ \left(1 + \sum_{i=1}^{N} P_i K_i\right) \qquad\qquad [8\text{-}15]$$

Brunauer, Emmett, and Teller successfully attempted to **model** a multilayer adsorption with the following assumptions:

- Langmuir adsorption is valid
- Adsorbed species in first layer serve as sites for the second layer
- Rate of adsorption r_a on the ith layer is equal to the rate of desorption r_d of the (i +1)th layer
- ΔH_{ads} is the same for the second and succeeding layers and equal to the heat of condensation of the gas

$$\Delta H_{a_1} \neq \Delta H_{a_2}$$

$$\Delta H_{a_2} = \Delta H_{a_3} = \ldots = \Delta H_{a_c} \qquad\qquad [8\text{-}16]$$

The following isotherm can be derived:

$$\theta = V / V_m = \frac{cx}{(1-x)[1(c-1)x]} \qquad\qquad [8\text{-}17]$$

where $x = p / P_o$, P_o is the vapor pressure of the absolving gas at a given temperature and V_M is the volume of gas required to provide a complete monolayer. Also,

$$C = C_o \exp\left(\frac{\Delta H_{a_1} - \Delta H_c}{RT}\right) \qquad\qquad [8\text{-}18]$$

where ΔH_{a_1} is the heat of adsorption on the first layer and ΔH_c is the heat of condensation of the gas. After rearranging Equation 8-17, it can be written as:

$$\frac{x}{V(1-x)} = \frac{1}{cV_m} + \frac{(c-1)x}{cV_m}$$

[8-19]

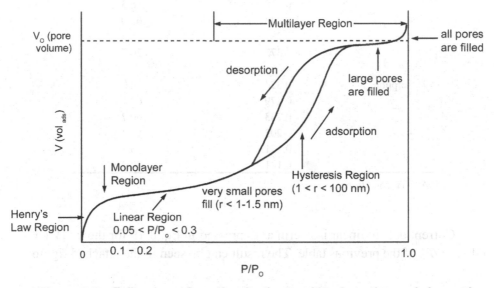

Figure 8-2. Full range adsorption isotherm with adsorption and desorption branches (After Farrarto and Bartholomew, 1997).

Experimentally in both adsorption and desorption, isotherm data versus P/P_o in the form of volume adsorbed or described are collected at the boiling point of the adsorbant, usually N_2 at -196°C. From the value of V_M, the volume of monolayer coverage, then the surface area of the adsorbant (usually the catalyst) can be computed. A full range adsorption isotherm with information of surface area, pore volume, and pore size can be seen in Fig. 8-2.

[Example 8-1] A sample of bayerite aluminum was investigated under N_2 at 77K for adsorption and desorption. Data in the P/P_o range of 0.05-0.30 were selected for the BET equation. Data are listed in the table:

Table for Example 8-1. Adsorption and Desorption Data

	P / P_o	$Vol^* (cm^3 g^{-1})$
Adsorption	0.049	39.7
	0.079	43.6
	0.109	47.1
	0.138	50.2
	0.168	53.2
	0.198	56.3
	0.243	61.0
	0.283	65.5
	0.372	70.7
Desorption	0.324	70.0
	0.279	64.7
	0.243	60.7
	0.202	56.4
	0.171	53.3
	0.141	50.2

*At STP conditions

Current data in linear isotherm as expressed in Eq. 8-19 for the $x / [V (1 - x)]$ vs. P/P_o of the previous table. The result can be seen in the attached figure.

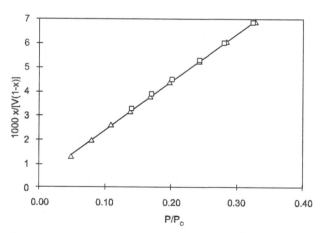

Figure for Example 8-1. Linearized BET plot of the N adsorption data for bayerite Al₂O₃. Triangles denote adsorption data; squares are for desorption; slope is $(c-1)/cV_m$ and intercept is $1/cV_m$.

In the figure, a linear-least squares fit yields a slope, (c-1) / cV_M, of 0.0199 and an intercept, 1 / cV_M of 4.25 x 10^{-4}. The slope-intercept ratio, $c - 1$, is 46.86 and here c is 47.86. The monolayer volume, V_M, is equal to 1 / (c x intercept). The calculated V_M is 49.22 cm^3 (STP) g^{-1}.

The BET surface is obtained using the widely accepted assumption that each N_2 molecule occupies an area of 0.162 nm^2 and Surface area $(m^2 g^{-1})$ = $(V_M cm^3 g^{-1} \times 6.02 \times 10^{23}$ molecules $mol^{-1} \times 16.2 \times 10^{-20}$ m^2 / molecules) / 22400 $cm^3 g^{-1}$ = 214 $m^2 g^{-1}$.

8.1.2 A Universal Carrier — Aluminas

One of the most important producers as well as carriers are the **aluminas**. The aluminas family consists of more than a dozen well-characterized amorphous or crystalline structures which depends upon its preparation and thermal history. Hydrated aluminas formed by precipitation at different conditions of pH and low temperatures such as gibbsite, boehmite, bayerite, and diaspore will yield different aluminas (see Figure 8-3). Some of the hydrates can be dried at 110°C to remove excess water and other volatile species, such as NH_3. Next calcinations, or heating in air at different temperatures, determines the final crystal structures. Table 8-4 illustrates the temperatures of **calcinations** from 250-1200°C.

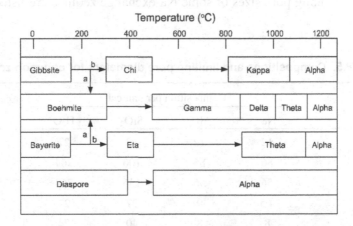

Figure 8-3. Alumina phases present at different temperatures: (a) the path favored for fine crystals; (b) the path favored for moist of coarse particles (Oberlander, 1984).

Table 8.4. Physical and structural characteristics of common aluminium oxides (data from Bartholomew et al., 1991).

$T_{calc}(^{\circ}C)$	Alumina phase	SA (m^2g^{-1})	$V_{pore}(cm^3g^{-1})$	d_{pore} (nm)	Crystal Structure
250	Pseudoboehmite	390	0.50	5.2	$Al_2O_3.H_2O$
450	γ-alumina	335	0.53	6.4	cubic defect spinel
650		226	0.55	9.80	$Al_{12}[Al_{12}H_4]O_{32}$
850		167	0.58	14.0	(ABC-ABC stacking)
950	δ-alumina	120	0.50	16.6	orthorhombic
1050	θ-alumina	50	0.50	28.0	
1200	α-alumina	1-5			hcp (ABAB stacking)

8.1.3 Molecular Sieves and Zeolites — Shape Selectivity

Molecular sieve refers to a class of crystalline materials having a range of composition that exhibits shape-selective adsorption and reaction properties. **Zeolites** are shape-selective materials composed only of aluminosilicates. Zeolites have the general formula $M_v(AlO_2)_x(SiO_2)_y \cdot 2H_2O$. The AlO_2 and SiO_2 species are the basic units that share oxygen ions to form tetrahedral AlO_4 and SiO_4 building blocks for zeolite unit cell. The framework of the zeolite is made up of aluminum and silicon tetrahedral while metal of hydrogen cations occupy exchangeable cation sites. Limiting pore sizes of some Na-exchange zeolites are listed in Table 8-5.

Table 8.5. Compositions and limiting pore diameters for common zeolites.

Type	Composition per unit cell				Aperature size (Au)
	Na^*	AlO_2	SiO_2	H_2O	
A	12	12	12	27	4.2
Faujasite X	86	86	106	264	8.0
Faujasite Y	56	56	136	264	8.0
Erionite	4.5	9	27	27	4.4
Mordenite	8	8	40	24	6.6
Pentasil (ZSM-5)	9	9	87	16	55
Pentasil (Silicate)	0	0	96	16	5.5

* the number of Na^+ ions required for charge equalization = to the number of Al^{3+} ions.

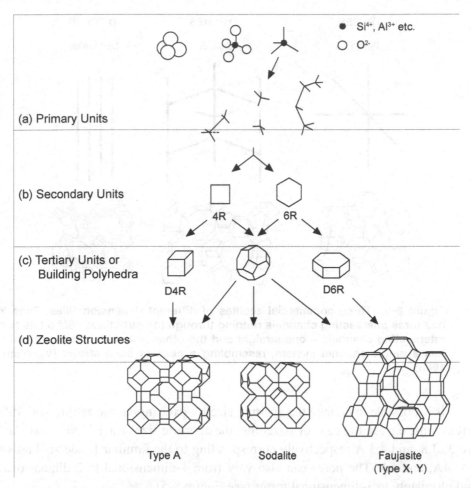

Figure 8-4. Formation of three common zeolites from primary SiO_4 and AlO_4 tetrahedral units through a combination of secondary ring units, and ultimately different mixes of tertiary polyhedra; note however that all three use the same structural polyhedron (cubo-octahedron) in the final construction (adopted from Vaughan, 1988).

The building up of three common zeolites is illustrated by Figure 8-4. The aluminum silicate is formed by polymerization of SiO_4 and AlO_4 tertrahedra to form sheet-like polyhedra (squares and hexagons) secondary units. Those in turn will form three-dimensional tertiary building-blocks (cubes, hexagonal prisms, and truncated octahedral of 14-sided. Finally, superstructures are formed with certain pores and supercages. These characteristic window size apertures will block entry of sufficiently large molecules which explain the sieve effect.

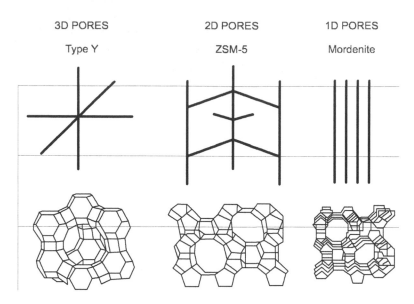

Figure 8-5. Three commercial zeolites of different dimensionalities. Type Y has three intersecting channels running through the structure; ZSM-5 has two intersecting channels – one straight and the other sinusoidal; and mordenite has a single channel system, resembling a pack of soda straws (Vaughan, 1988).

It is possible to change the aperture size by exchanging the zeolite with different ions; e.g., in the case of A-zeolite, the aperture sizes for K^+, Na^+, and Ca^{2+} are 3, 3.8, and 4.3 Å respectively corresponding to the familiar Linde zeolites of 3A, 4A, and 5A. The pores can also vary from 1-dimensional to 2-dimensional and ultimately to 3-dimensional forms (see Figure 8-5).

Often the molecule sieve structure can bel classified in terms of pore diameter and ring size By convention, ring size is specified by the number of T atoms or TO_4 units where T = Si, Al, P, or B. Newly developed molecular sieves include the **aluminophosphates** (ACPOs) and **titanium silicates** (TSs) having a large number rings and large pore diameters (see Table 8-6).

A novel gallophosphate molecular sieve has a 20-ring with 4 terminal hydroxy groups protruding into the opening in a clover frame geometry. Figure 8-6 demonstrates this molecule of **cloverite** , which is stable at 700°C and has a spacious 30Å super cage accommodating large molecules to be utilized.

Table 8.6. Pore diameters and ring sizes of typical molecular sieves.

Type	Pore diameter (Au)	No. of rings	Largest molecules adsorbed
Small pore			
3A	3.0	8	H_2O, NH_3
4A	3.8	8	N_2, C_2H_6
5A	4.3	8	n-alkanes, alkenes
Erionite	3.6 × 5.2	8	
Titanium silicate	3.7-4.0	12	
Medium pore			
Pentasil (e.g. ZSM-5)	5.1 × 5.6	10	CCl_4, m-xylene
Mordenite	2.9 × 5.7	8	
Large pore			
Faujasite (X,Y)	7.4-8.1	12	$(C_4H_9)_3N$, naphthalenes
Titanium silicate	8.0	12	
ALPO (e.g. ALPO$_4$-5)	3-10	12	
ALPO-VPI(e.g. VPI-5)	10-15	18	tris (isopropyl) benzene
Cloverite	17-18?	20	model enzymes

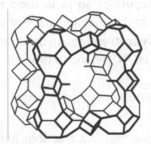

Figure 8-6. Cloverite, a stable gallophosphate molecular sieve of cloverleaf geometry with a ring opening of 17-18Au and a supercage diameter of 30Au (Estermann et al., 1991).

8.2 PETROCHEMICAL PROCESS CHEMISTRY

Crude oil refining and petrochemical processing continue to be essential in our chemical industry. Catalysis still plays a pivotal role in the development of efficient processes for upgrading petroleum crude to chemical feedstocks and premium transportation fuels. In fact, it was the development of catalytical cracking technology that contributed greatly to the Allied victory in the battle of Britain during World War II. New cracking catalysts developed by Hondry of Sun Oil and new alkylation catalysts developed by Pines, Ipatieff and Bloch of Universal Oil Products (UOP) made it possible to produce 100-octane fuel, providing British planes with 50% faster bursts of acceleration, thereby allowing them to out-maneuver the Luftwaffe fighters using 88-octane gasoline. This resulted in the loss of 915 British planes as opposed to 1733 for the Germans.

Conversion processes in the petroleum industry (or can hold true for the coal liquid or shale oil) have the following general principles:

- Upgrading lower value materials such as heavy residues or bottom fractions to more valuable products such as naphtha and liquified petroleum gas (LPG) cracking and **hydrocracking** belong to this task, for example, **fluid catalytic cracking** (FCC).

- Improving the characteristics of a fuel. A lower octane naphtha fraction is reformed to a higher-octane reformate product. **Reforming**, steam reforming, isomerization, and some forms of alkylation belong to this category.

- Reducing harmful impurities so as to meet pollution control regulatory standards and to avoid the contamination of other catalysts to be used in later process. All **hydrotreating**, hydrodesulfurization (HDS) and hydrodenitrogenation (HDN) belong to this class.

8.2.1 Reactor Design

Simulating a process involving multiple reactions and turbulent multiphase flow is a formidable task. To make our effort easier, we will adopt the concept of ideal reactors. Such as a well-mixed batch reactor, a **well-mixed constant flow stirred-tank reactor** or (**CSTR**) and an ideal tubular or **plug-flow reactor (PFR)** having orderly flow in the axial direction reactor design is based on the simultaneous solution of mass balance, rate law, and energy balance. Let us review the three types of reactors:

(a) Batch Reactor

$$\text{Accumulation of A} = \text{rate of production of species A} \qquad [8\text{-}20]$$

or

$$\frac{dn_A}{dt} = r_A V = -(-r_A V) \qquad [8\text{-}21]$$

Here n_A is the number of moles of species A, $-r_A$ is the disappearance of species A, V is the reactor volume which can be related to catalyst weight by $W = \rho_b V$, where ρ_b is bulk density of the catalyst. Note that moles of A at any time can be expressed in terms of the initial number of moles η_{A_0}, and the conversion of A, X_A.

$$n_A = n_{A_0}(1\text{-}X_A) \qquad [8\text{-}22]$$

Combining Equations 8-22 and 8-21, we have

$$\frac{-d\left[n_{A_0}\left(1\text{-}X_{A_0}\right)\right]}{dt} = n_{A_0}\frac{dX_A}{dt} = -r_A V \qquad [8\text{-}23]$$

Upon integration from time 0 to t and 0 conversion to X_A,

$$t = n_{A_0}\int_0^{X_A}\frac{dX_A}{(-r_A)V} \qquad [8\text{-}24]$$

(b) CSTR

$$\text{inport} - \text{output} = \text{disappearance by reaction} \qquad [8\text{-}25]$$

Or in terms of flow rates F_{Ao} and F_{At},

$$F_{Ao} - F_{At} = -r_A V \qquad [8\text{-}26]$$

Assuming well-mixed and

$$X_A = X_{At} \qquad [8\text{-}27]$$

$$F_{Ao} - F_{Ao}(1 - X_A) = -r_A V \qquad [8\text{-}28]$$

Collecting terms and solving,

$$\frac{V}{F_{A_0}} = \frac{X_A}{-r_A} \qquad [8\text{-}29]$$

This equation allows one to calculate the volume if conversion, feed rate, and reaction rate are known. This equation is termed the **reactor sizing equation.**

(c) Tubular Flow or Plug Flow Reactor

$$\text{input} - \text{output} - \text{disappearance by reaction} = 0 \qquad [8\text{-}30]$$

In terms of molar flow rate F_A and the change in molar flow rate across the differential volume element of V is the tube, i.e., F_A will result to $F_A + dF_A$, or Equation 30 becomes

$$0 = F_A - \left(F_A + dF_A\right) - (-r_A)\, dV \qquad [8\text{-}31]$$

Since the converted fraction and unconverted fraction are equal,

$$dF_A = d\left[F_{A_0}\left(1 - X_A\right)\right] = -F_{A_0}\, dX_A \qquad [8\text{-}32]$$

When Equation 8-32 is combined with Equation 8-31,

$$F_{A_0}\, dX_A = \left(-r_A\right) dV \qquad [8\text{-}33]$$

At any fixed axial distance of X_A with parallel flow of fluid elements. Integration over the volume and length with separating variables of Equation 8-33 yield

$$\frac{V_A}{F_{A_0}} = \int_0^{X_{A_f}} \frac{dX_A}{-r_A} \qquad [8\text{-}34]$$

This is the basic equation of plug flow.

(d) Time

Operation of a reactor can be defined by the following parameters.

Space time (τ) is defined as the time required to process one reactor volume at specified feed conditions.

$$\tau = \frac{C_{A_0}}{F_{A_0}} = \frac{V}{v_0} \qquad \text{[8-35]}$$

Here v_0 is the inlet volumetric flow rate.

Space velocity (SV) is the inverse of space time can be given as reciprocal of hours; for example, SV for gas hourly, $GHSV$, SV for liquid loading is $LHSV$.

$$SV = \frac{v_0}{V} \qquad \text{[8-36]}$$

Residence time (θ) is the actual time that fluid element spent at the reactor. For a batch reactor, referring to Eq. 7-24 the residue time can be

$$\theta = n_{A_0} \int_0^{X_A} \frac{dX_A}{-r_A V} \qquad \text{[8-37]}$$

For a mixed flow reactor, a mean residence time is

$$\theta_{mixed} = \frac{V}{v} = \frac{V}{v_f} \qquad \text{[8-38]}$$

The value of v_f can be obtained through the expansion coefficient, ε_A:

$$v_f = v_0 (1 + \varepsilon_A X_A)$$

and

$$\varepsilon_A = \frac{\sum_i v_i}{-\sum_r v_r} \qquad \text{[8-39]}$$

v_i and v_r are stoichiometric coefficients of reactants and products.

Finally, for the mixed reactor time,

$$\theta_{mixed} = \frac{V}{v_0 \left(1 + \varepsilon_A X_A\right)}$$ [8-40]

Similarly for a plug flow reactor,

$$\theta_{plug} = F_A \int_0^{X_A} \frac{dX_A}{-r_A v_0 \left(1 + \varepsilon_A X_A\right)}$$ [8-41]

(e) Derivation from Ideal Flow Behavior

In this instance, a pictorial description is better to illustrate the usually encountered non-ideal flow behavior in reactors. This is illustrated in Figure 8-7.

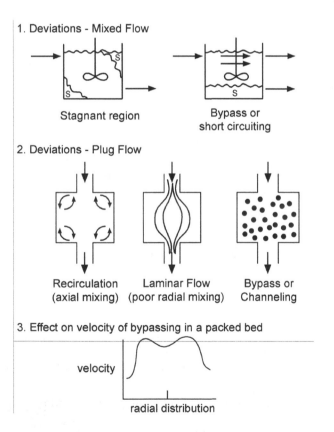

Figure 8-7. Non-ideal flow behaviors in flow reactors.

The key concept for reactor design will include both the energy balance and the kinetic rate laws. The latter include not only surface reaction control, but also pore diffusion and waste transfer. The following is a short summary of the key relationships in reactor design of a cylindrical geometry in z and r.

For material balance and conservation of mass,

Accumulation of A = (mass of A_{in})–(mass A_{out})+(rate of production of A)   [8-42]

$$\frac{\partial C_A}{\partial t} = -v_z \frac{\partial C_A}{\partial z} - v_r \frac{\partial C_A}{\partial r} + \frac{\partial}{\partial z}\left(D_{AB} \frac{\partial C_A}{\partial z} \right)$$
$$+ \frac{1}{r}\frac{\partial}{\partial r}\left(rD_{AB} \frac{\partial C_A}{\partial r} \right) + r_A$$

[8-43]

$$X_A = f(t, z, r, r_A) \rightarrow X_A(z, r_A) \text{ one-dimension}$$

$$\frac{dw}{F_{A_0}} = \frac{dX_A}{-r_A}$$

[8-44]

For energy balance and conservation of energy,

Accumulation = (energy$_{in}$) – (energy$_{out}$) + rate of heat production   [8-45]

$$\rho C_P \frac{\partial T}{\partial t} = -\rho C_P\left(v_z \frac{\partial T}{\partial z} + v_r \frac{\partial T}{\partial r} \right) + \frac{\partial}{\partial z}\left(k \frac{\partial T}{\partial z} \right)$$
$$+ \frac{1}{r}\frac{\partial}{\partial r}\left(rk \frac{\partial T}{\partial r} \right) + q_{gen} + q_{add}$$

[8-46]

$$T = f(t, z, r, r_A) \longrightarrow T = f(z, r_A)$$
$$F_{A_b} C_P dT = (-\Delta Hr)(-r_A dV) + q_{add}$$

[8-47]

For rate equation:

(a) of surface reactions control,

$$-r_A = f(T, C_i) = kTf(C_i) = A\,exp(-E/RT)f(C_i)$$

[8-48]

$$f(C_i) = \frac{K_A^\alpha K_B^\beta C_A^\alpha C_B^\beta}{\left(1 + K_A^\alpha C_A^\alpha + K_B^\beta C_B^\beta\right)^\gamma}$$ [8-49]

For Equation 8-48, α is activity. For Eq. 8-49, α, β, and γ are typically 0.5-2.0.

(b) of pore diffusion influence,

$$\eta = \frac{\tanh \phi_s}{\phi_s}$$

and $$\phi_s = L\left(\frac{k\rho_p}{D_{eff}}\right)^{1/2}$$ [8-50]

Strong pore resistance, first order

and $$L = \frac{V_P}{S_{ex}} \; ; L = \frac{r}{3} \text{ for a sphere}$$

(c) of film mass transfer control,

$$-r_{A_{abs}} = k_{abs} f(i)$$
$$= k_m a_m (C_{Ab} - C_{As})$$ [8-51]

[Example 8.2] For an isothermal tubular cracking reactor, the inlet volume flow rate is 20,000 Lh⁻₁, and the conversion is 50%. Assuming the first order irreversible cracking reaction of $A \rightarrow 4R$ with k' (constant pressure) = 0.11h⁻¹g_cat⁻¹(kV=k'W). Find catalyst requirement.

At first we shall find out the expansion coefficient.

$$-r_A' = k'C_A = k'C_{A_0}\frac{1-X_A}{1+\varepsilon_A X_A}$$

$$X_{A_f} = 0.5$$

From Eq. 8-39

$$\varepsilon_A = -\frac{\sum v_i}{\sum v_r} = -\left(\frac{-1+4}{-1}\right) = 3$$

Performance equation for integral reactor is

$$\frac{W}{F_{A_0}} = \int_0^{X_{A_f}} \frac{dX_A}{k'\frac{C_{A_0}(1-X_A)}{1+\varepsilon_A X_A}} = \frac{1}{k'C_{A_0}} \int_0^{X_{A_f}} \frac{1+\varepsilon_A X_A}{1-X_A} dX_A$$

Then using integration by parts, also using integral table,

$$\frac{W}{F_{A_0}} = \frac{F_{A_0}}{k'C_{A_0}}\left[-(1+\varepsilon_A)\ln(1-X_{A_f})-\varepsilon_A X_{A_f}\right]$$

$$= \frac{V_0}{k'}\left[-4\ln(0.5)-3(0.5)\right]$$

$$= \frac{20,000Lh^{-1}}{0.11h^{-1}g^{-1}_{cat}}(2.77-1.50)$$

$$= 255kg$$

8.2.2 Cracking

Catalytic cracking is the most important catalytic process in the last century, especially since the development of **fluidized catalytic cracking (FCC)**. FCC is a process in which the heavy feed stacks such as the heavy gas oil (C_{20}-C_{40}, 350-550°C) or vacuum resid (>C_{40}, > 550°C) are cracked to the gasoline range of hydrocarbons.

The discussion of molecular downsizing can be found in the series of books by Yen, (in Chapter *Pyrolysis and Catalysis*, Chemical Processes in Environmental Engineering, 2007).

Most catalytic cracking catalysts are solids including SiO_2-Al_2O_3 and Y-Zeolites. The origin of the activity with both Bronwsted sites and Lewis sites has

been discussed previously. In that chapter, paraffin cracking, naphthalene pyrolysis and pyrolysis of aromatics and others are discussed. **Autothermal Reforming
(ATR)** is also briefly discussed. A simple sketch is illustrated in Figure 8-8.

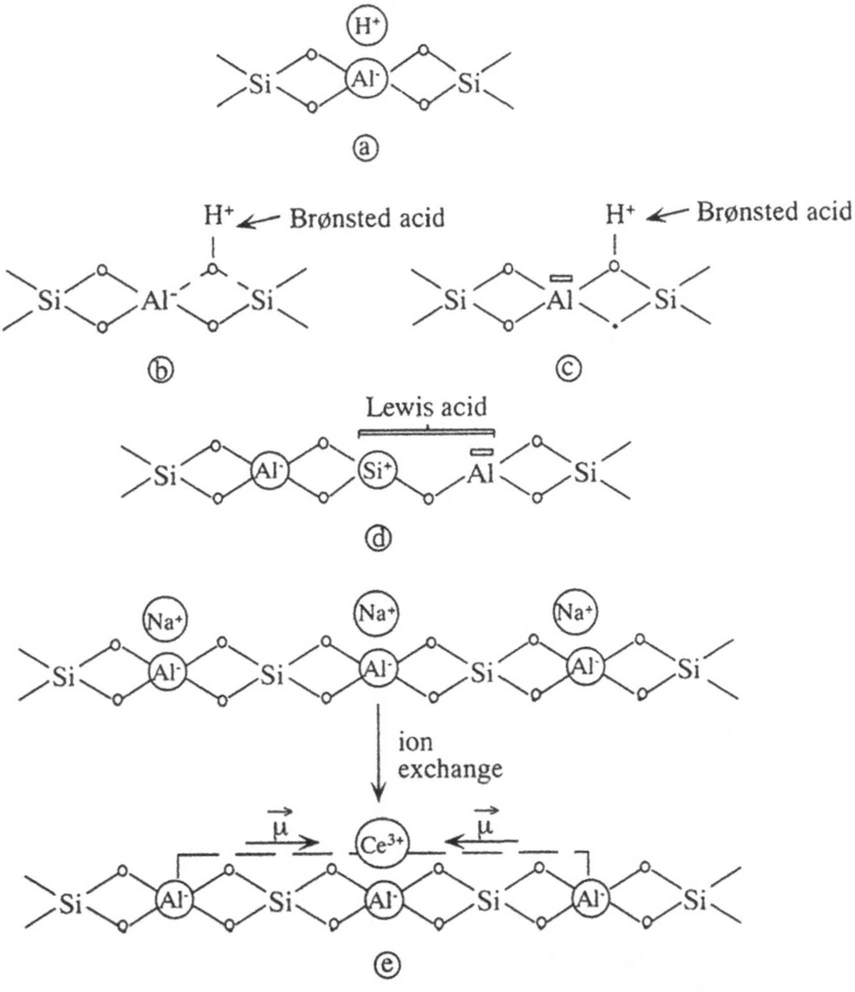

Figure 8-8. Origin of acidity in SiO_2-Al_2O_3 and stabilization by Ce^{3+}

8.3 SYNTHESIS GAS AND CHEMICAL STOCKS

8.3.1 Fisher–Tropsch Synthesis

The **Fisher-Tropsch synthesis** is the production of liquid hydrocarbons from the synthesis gas, most of which is CO and H_2. In view of the large coal and gas reserves, and the dwindling petroleum reserve, this process is an attractive and environmentally sound alternative process. Following the discovery that CO could be hydrogenated over copper, iron and nickel to methane, BASF reported the production of liquids over cobalt. Much of this work was conducted in Germany during the Second World War until Ruhrchemie, Lurgi and Kellog Co. commissioned the large scale Sasal plant in South Africa. Although the practice is quite flexible, the economics of the process involve both bond scission to single carbon compounds (e.g. coal) and the bond synthesis from single carbon compounds (e.g. usually hydrocarbons of C_5 to C_7) . The energy balance of the reactions has been calculated by Yen. Thermal efficiency suffers from an inefficient heat removal. Figure 8-9(a) and (b) describe the process flowchart for the production of liquid and gaseous fuels and chemicals from coal and natural gas by means of the Fisher-Tropsch synthesis.

The main reactions are as follows:

$$CO + 3H_2 \longrightarrow CH_4 + H_2O \qquad [8\text{-}52]$$

$$CO + 2H_2 \longrightarrow \tfrac{1}{n}(C_nH_{2n}) + H_2O \qquad [8\text{-}53]$$

They represent the formation of methane and hydrocarbons higher than methane. There is a water-shift reaction

$$CO + H_2O \longrightarrow CO_2 + H_2 \qquad [8\text{-}54]$$

Also there is a Boudouard reaction of carbon deposit

$$2CO \longrightarrow C + CO_2 \qquad [8\text{-}55]$$

Generally thermodynamics favors the formation of methane and other hydrocarbons. For example, at 200°C, ΔH^o values for methane, ethane and propane are about –50. –42 and –29 kcal mol^{-1} respectively. The distribution of hydrocarbon products are controlled by a chain of polymerization kinetic model involving stepwise addition of single carbon atoms to the growing chain. This is termed as the **Anderson-Schultz-Flory (ASF) model**.

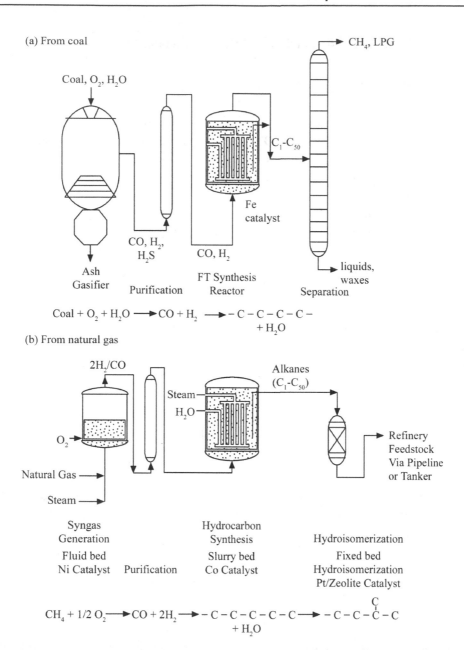

Figure 8-9. A simplified process overview for production of liquid and gaseous fuels and chemicals (a) from coal and (b) from natural gas, by means of Fischer-Tropsch synthesis.

$$\frac{W_n}{n} = (1 - \alpha)^2 \alpha^{n-1} \qquad\qquad [8\text{-}56]$$

where n is the number of carbon toms in the product, W_n is the weight fraction of the product containing n carbon atoms, α is the chain growth propagation probability. Traditionally the value of α is obtained by least squares linear regressional log form of the above equation.

$$\ln\!\left(\frac{W_n}{n}\right) = \ln(1 - \alpha)^2 + (n-1)\ln\alpha \qquad\qquad [8\text{-}57]$$

The value of α increases with decreasing H:CO ratio, decreasing reaction temperature and increasing pressure. Values of α are higher for unpromoted Ru and Co catalysts relative to unpromoted Fe catalysts as illustrated in Table 8-7.

Table 8.7. CO hydrogenation activities, selectivities and propagation probabilities of representative unpromoted Co, Fe and Ru synthesis catalysts at 480K ($H_2CO = 2$, 1atm, Bartholomew (1991)).

Catalyst	$N_{CO} \times 10^{3}$ [a] (s^{-1})	CO_2 in product (%)[b]	alkene[c] C_3-C_7(%)	α[d]
15% Co / Al_2O_3	17	1	54	0.90
Unsupported Fe	1.4	31	94	0.44
10% Ru / Al_2O_3	1.8	-	88	0.69
3% Ru / Al_2O_3	1.5	4	65	0.70

[a] Turnover frequency in molecules of CO converted per catalytic site per second.
[b] Mole percentage of CO2 in product (excluding unconverted reactants).
[c] Mole percentage of alkenes in C3-C7 product.
[d] Propagation probability determined from the slope of a mole percentage hydrocarbon versus carbon number plot.

Based on the distribution of the products obtained, the synthesized homologs can be illustrated by Figure 8-10 (a) and (b). Especially for the predominant portion of Fig. 8-10(a), the wax can be further cracked down to lower fractions.

Modern synthetic gas production can also originate from biomass, especially by steam reforming. A good example is illustrated in Fig. 8-11.

There are two types of steam reforming. They are called primary of secondary. The steam reforming of hydrocarbons over the catalysts used are summarized here from reference (Table 8-8).

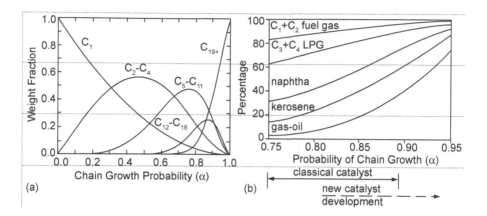

Figure 8-10. (a) Weight fraction of hydrocarbon products as a function of chain growth (propagation) probability during Fisher-Tropsch synthesis. (b) Percentage of different hydrocarbon product cuts as a function of chain growth (propagation) probability showing range of operation for classical and developing Fisher-Tropsch catalysts and synthesis.

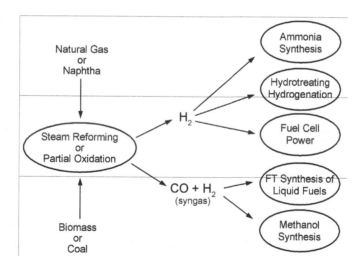

Figure 8-11. Processes based on production of H_2 and synthesis gas by steam reforming.

Table 8.8. Steam reforming of hydrocarbons: process steps, reactions and catalysts.

Process step	Reaction(s)	Catalysts
Desulfurization	$R-S+H_2 \longrightarrow H_2S+R-H$	$CoMo/Al_2O_3$
Primary steam reforming	$HC+H_2O \longrightarrow H_2+CO+CO_2+CH_4$	Ni/Mg (naphtha)
		$Ni/CaAl_2O_4$ (CH_4)
Secondary steam reforming	$2CH_4+3H_2O \longrightarrow 7H_2+CO+CO_2$	$Ni/CaAl_2O_4$, $Ni/\alpha\text{-}Al_2O_3$
Water gas shift (HT)	$CO+H_2O \longrightarrow H_2+CO_2$	Fe_3O_4
Water gas shift (LT)	$CO+H_2O \longrightarrow H_2+CO_2$	CuO
Methanation	$CO+3H_2 \longrightarrow CH_4+H_2O$	Ni/Al_2O_3

The removal of sulfur is necessary since subsequent catalysts can be poisoned, usually by a combined HDS and ZnO scrubbing process.

8.3.2 Organometallics

A most significant areas of homogeneous catalysis are generally characterized by a transition metal complex which involves metal-carbon bonds. These catalyzed reactions include carbonlyations, decarbonylations, hydroformylations and partial oxidations. organometallic complexes consist of transition metal ions or atoms bonded to organic ligands (as shown in Table 8-9).

These complexes posses an octahedral or tetrahedral geometry. Catalysis occurs by ligand dissociation followed by coordination of reactants with the metal ion. Reactions of transition metal ions or mononuclear complexes occur by exchange of d-electrons with ligands in accordance with the 18-electron rule (See chapter 2 of this book).

$$n+2(CN)_{max}=18 \qquad\qquad [8\text{-}58]$$

where n is the number of electrons and CN is the coordination number (number of ligands per metal ion or atom). The transition metals and their d-electrons in various oxidation states are listed in Table 8-9.

An example is illustrated by **Willkinson hydrogenation catalysis** of alkenes using PPh_3. As illustrated by Fig. 8-12, the active Wilkinson catalysts consist of a $Rh_2Cl_2(PPh_3)_4$ dimer and a $RhCl(PPh_3)_3$ monomer.

Table 8.9. Ligands and their typical charges and coordination numbers (Collman *et al.*, 1987).

Ligand	Charge[a]	Coordination number[b]
X (Cl, Br, I)	-1	1(2)
H	-1	1(2,3)
CH_3	-1	1(2)
CO	0	1(2,3)
$R_2C=CR_2$	0(-2)	1(2)
$RC\equiv CR$	0(-2)	1(2)
η^3-Allyl/[c]	-1	2
η^6-Benzene	0	3(2,1)
η^5-Cyclopentadienyl	-1	3
RCO	-1	1(2)
R_3N	0	1
R_3P	0	1
O	-2	2
O_2	-2(-1)	2(1)

[a] Less common charges are stated in parentheses.
[b] Less common coordination numbers are stated in parentheses.
[c] The subscript 3 implies that all three carbon atoms of allyl ligand interact with metal.

Table 8.10. Transition metals and their numbers of d electrons in various oxidation states (Collman *et al.*, 1987).

Row	Group number							
	4	5	6	7	8	9	10	11
First, 3d	Ti	V	Cr	Mn	Fe	Co	Ni	Cu
Second, 4d	Zr	Nb	Mo	Tc	Ru	Rh	Pd	Ag
Third, 5d	Hf	Ta	W	Re	Os	Ir	Pt	Au
Oxidation State	Number of d electrons							
Zero	4	5	6	7	8	9	10	-
+1	3	4	5	6	7	8	9	10
+2	2	3	4	5	6	7	8	9
+3	1	2	3	4	5	6	7	8
+4	0	1	2	3	4	5	6	7

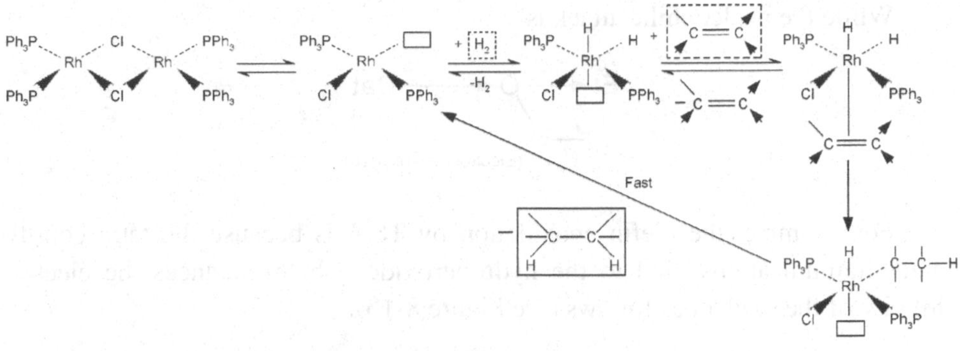

Figure 8-12. Catalytic cycle for the Wilkinson hydrogenation, as determined by Halpern *et al.* (Collman *et al.*, 1987).

[**Example 8-3**]   What is the charge of Cr in $Cr(CO)_6$ assuming 6 to be the maximum number of CO ligands?

According to Eq. 8-58

$$(CN)_{max} = 6$$

From Table 8-9, the coordination number of CO is 1. Each CO ligand supplies 2 bonding electrons giving $2 \times 6 = 12$. Substituting into Eq. 8-58, $N = 18 - 12 = 6$.

From Table 8-10, Cr is zero valent.

8.3.3 Selective Oxidation and Desulfurization

Either the combustion substrate or the oxidizing agent can form an acid-base adductor thus enhancing reactivity and acting as a catalyst for the oxidation. The electrophilic attack is

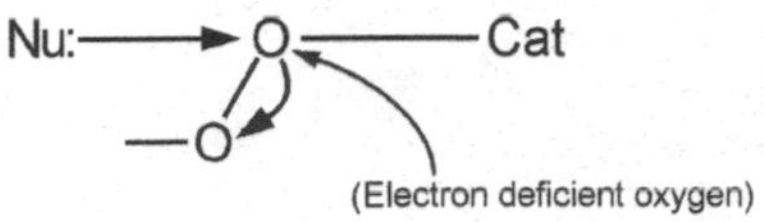

(Electron deficient oxygen)

While the nucleophilic attack is

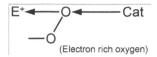

For example, the olefin epoxidation by TS-1 is because the tetra coordinated titanium atoms bind to the hydroperoxide, which enhances the electrophilicity of the oxidant as follows (see Figure 8-13):

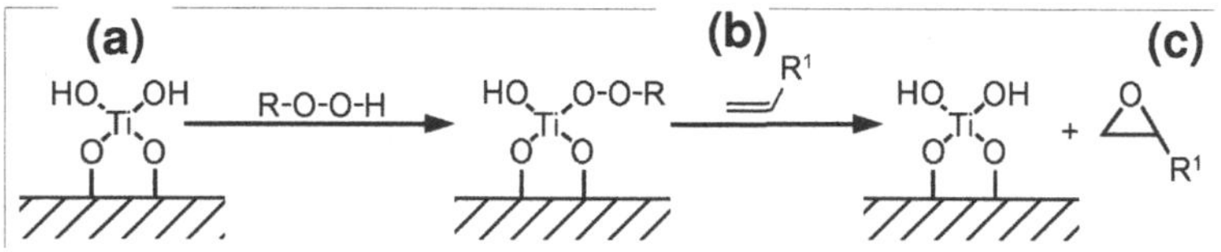

Figure 8-13. Oxidation of olefin by TS-1. (a) is TS-1. (b) is α-olefin and (c) is an epoxide product.

Polyoxymetalates and heteropolyacids are obtained usually by the condensation of two or more different types of polyanions, for example

$$12WO_4^{2-} + HPO_4^{2-} + 23H^+ \longrightarrow PW_{12}O_{40}^{3-} + 12H_2O$$

In these compounds, the anion contains a central atom, typically Si or P, tetrahederically coordinated to oxygen and surrounded by 2-18 oxygen linked hexavalent peripheral transition metals. These are usually Mo or W, but can also be Nb, Ta, U, either alone or in combination. The most common structure is the Kaggin type heteropolygonal $(XM_{12}O_{40})^{n-}$ as shown in Fig. 11-30. The structure of Keggin ions vs. Dawson ions will be further discussed in Chapter 11 under the topics of nanotechnology.

REFERENCES

8-1. Absi-Halabi, M., J. Beshara, et al., *Catalysts in Petroleum Refining and Petrochemical Industries*, Elsevier, Amsterdam, 1996.

8-2. Farranto, R. J. and C. H. Bartholomew, *Fundamentals of Industrial Catalytic Processes,* Blackie Academic and Professionals, London, UK, 1997.

8-3. Jahn, Frank, Mark Cook, and Mark Graham, *Developments in Petroleum Science, Volume 46, Hydrocarbon Exploration and Production.*

8-4. Speight, James G., Editor, *Petroleum Chemistry and Refining*, Taylor and Francis, Laramie, WY, 1998.

8-5. W.Wei, C.A.Bennett and R.Tanaka, G.Hou and M.T.Klein, *Preprints*, ACS Div Pet. Chem., 2005 **50(4)** 386-387.

PROBLEM SET

1. Butadiene can be oligomerized to cyclic diene and trienes using TiCl$_4$ and AlCl$_3$(Et)$_3$. For example, 70% of triene of

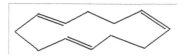

   called 1,5,9-cyclodecatrine can be obtained. This triene can be the precursor of nylon 12 or nylon 6/12. How can these be achieved?

2. Lawrence and Rawlings are pioneers in the isomerization process. From their studies, compare the C$_6$ vapor phase equilibrium at 200°C and 600°C.

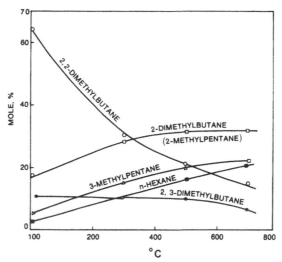

(PA Lawrence and AA Rawlings, Proc. The
World Petroleum Congress, P. 137, 1967)

3. Super acids can isomerize alkanes at lower temperature and yet can yield branched paraffins for better octane number. From data of Asinger's table examine the difference between -6°C and 38°C.

Hydrocarbon	Temperature in °C			
	-6	38	66	93
Butanes				
Isobutane	85	75	65	57
n-Butane	15	25	35	43
Pentanes				
Isopentane	95	85	78	71
n-Pentane	5	15	22	29
2,2-Dimethylpropane				
(neopentane)	0	0	0	0
Hexanes				
2,2-Dimethylbutane				
(neohexane)	57	38	28	21
2,3-Dimethylbutane	11	10	9	9
2-Methylpentane	20	28	34	36
3-Methylpentane	8	13	15	17
n-Hexane	4	11	14	17

CHAPTER 9

POLYMER CHEMISTRY

*P*olymers are large molecular compounds which are made up of repeating molecular units called monomers. They may be natural substances, such as proteins, starch and humic material, or synthetic substances such as nylon and plastics. The flexibility and toughness of polymers make them versatile for use in many different types of products. For instance, massive amounts of isoprene units of coiled chain will become rubber, an important polymer.

There are numerous new uses of polymers not only as flexible or high strength materials, but also electrical conductors or semiconductors, polymer based catalysts with transition metals at fixed sites, or polymers being antigens, whole cells or enzymes as therapeutic or slow drug releasers, even biopolymers used as potentially coding systems are in the developing stages.

In this chapter, we discuss molecular weights, polymerization process and control, solid-state properties such as crystallinity and grass transition. Lastly, we will address rheology namely, rubber elasticity, flow and dynamic mechanical testing.

9.1 MACROMOLECULAR CHEMISTRY

Large or giant molecules usually exhibit unique properties compared with ordinary molecules. Polymers are formed from a great number of monomers linking together. The "mers" is derived from Greek "meros" meaning part or repeating unit. For example, a linear polymer such as polyethylene (Fig. 9-1a) is made from ethylene (Fig. 9-1b), having a repeating unit as shown in Figure 9-1c, where n is called the **degree of polymerization**, DP.

$$CH_2-CH_2-CH_2-CH_2$$

(a)

$$H_2C = CH_2 \qquad\qquad -(CH_2\ CH_2)_n-$$

(b)   (c)

Figure 9-1. Linear polymer: (a) polyethylene, (b) ethylene, and (c) repeating unit of n.

If the repeating units are made of two different structures, usually the **co-polymer** is given. For three different units this is **tripolymer**, and so forth. For m different repeating units this is termed **multipolymer** (multipolymers will be discussed in Chapter 12 of this book). In order to differentiate two or more repeating units, polymers with only one repeating unit are often referred to as **homopolymers**. Depending on the arrangement of two repeating units, the copolymers can be classified as follows:

- **Alternating copolymer** – Fig. 9-2(a)

- **Random copolymer** – Fig. 9-2(b)

- **Block copolymer** – Fig. 9-2(c)

- **Graft copolymer** – Fig. 9-2(d)

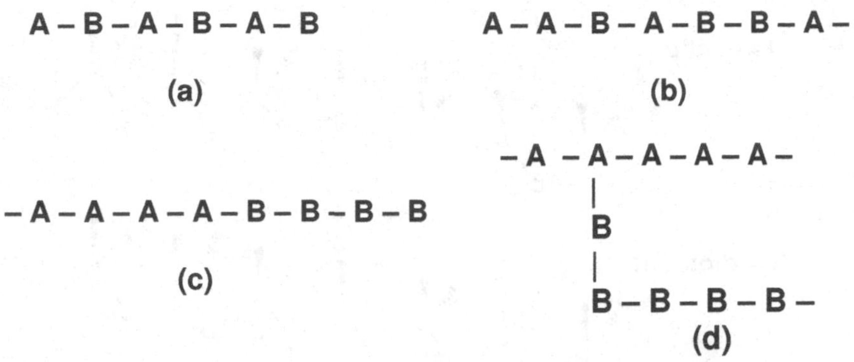

A – B – A – B – A – B   A – A – B – A – B – B – A –

(a)   (b)

                                         – A – A – A – A – A –
                                                         |
– A – A – A – A – B – B – B – B   B
                                                         |
(c)   B – B – B – B –
                                                (d)

Figure 9-2. Copolymer types. A and B represent the two types of repeating units. (a) Alternating copolymer (b) random copolymer (c) block copolymer (d) graft copolymer

If polyfunctional monomer instead of difunctional is used, a **network polymer** is formed. A network polymer is commonly referred to as a crosslinked polymer. Due to the diminished stability of the network, such crosslinked polymers cannot flow, melt, or be molded. Such polymers are said to be the **thermosetting** or **thermoset**. In the manufacturing process, the crosslinking must be disrupted to allow the polymer to flow. In contrast, the linear or branched polymer, which are not crosslinked, are said to be **thermoplastic**.

In addition to the sequential arrangement of monomers in the chain, the spatial arrangement of the substituents is also of importance. Besides conformation (geometrical arrangement) which has been discussed in Chapter 3, the configuration denoting the stereochemical arrangement of atoms cannot change without breaking the chemical bonds. The **isotactic** polymer has the substituent groups on the same side of the plane formed by the extended-chain backbone. If the substituent groups regularly alternate from one side to the other side of the plane, this is called a **syndiotactic** polymer (see Fig. 9-3 for iso- and syndiotactic polymers). In case the substituent groups have no preferred placement, then the polymer is termed **atatic** polymer.

Besides the network polymer as described with crosslinking (Fig. 9-4a), there are **star** polymers (Fig. 9-4b), **comb** polymers (Fig. 9-4c), **ladder** polymers (Fig. 9-4d), and **semiladder** or **stepladder** polymer (Fig. 9-4e).

Evidently, if one bond in the center of a linear polymer is cleaved, the molecular weight will decrease by one half, but not so in case of a network or ladder polymer. Finally, if the n value or DP is small, the low molecular weight polymer is termed **oligomer**. For most oligomers, n < 100.

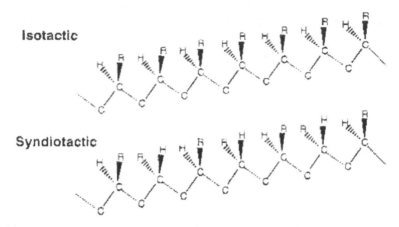

Figure 9-3. Two forms of stereochemical configuration of an extended-chain vinyl polymer having a substituent group R other than hydrogen.

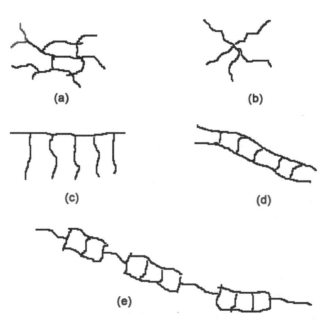

Figure 9-4.   Representation of (a) network, (b) star, (c) comb, (d) ladder, (e) semiladder or stepladder polymer.

9.1.1 Molecular Weight Distribution

Most synthetic polymers contain chains with a wide distribution of chain lengths. Only under certain laboratory conditions, monodispersed polymers can be prepared. For polymers with broad distribution, and average molecular weight $\overline{M}$ is used

$$\overline{M} = \frac{\sum_i N_i M_i^{\alpha}}{\sum_i N_i M_i^{\alpha-1}} \qquad [9\text{-}1]$$

Here, N_i indicates the number of moles of molecules with a molecular weight of M_i, and α is a weighting factor that defines a particular average of molecular weight distribution. The weight W_i of molecules with molecular weight N_i is then

$$W_i = N_i M_i \qquad [9\text{-}2]$$

When $i = n$, this is called **number average**

$$\overline{M}_n \quad (\alpha = 1) \qquad [9\text{-}3]$$

When $i = w$, this is called **weight average**

$$\overline{M}_w \quad (\alpha = 2) \qquad [9\text{-}4]$$

When $i = z$, this is called z-**average**

$$\overline{M}_z \quad (\alpha = 3) \qquad [9\text{-}5]$$

It follows that

$$\overline{M}_n = \frac{\sum_{i=1}^{N} N_i M_i}{\sum_{i=1}^{N} N_i} = \frac{\sum_{i=1}^{N} W_i}{\sum_{i=1}^{N} (W_i / M_i)} \qquad [9\text{-}6]$$

This expression can be put in an integration form

$$\overline{M}_n = \int_0^\infty NMdM \Big/ \int_0^\infty NdM \qquad\qquad [9\text{-}7]$$

Also, from Equation [9-1]

$$\overline{M}_w = \frac{\sum_{i=1}^N N_i M_i^2}{\sum_{i=1}^N N_i M_i} = \frac{\sum_{i=1}^N W_i M_i}{\sum_{i=1}^N W_i} \qquad\qquad [9\text{-}8]$$

It follows also

$$\overline{M}_w = \int_0^\infty NM^2 dM \Big/ \int_0^\infty NMdM \qquad\qquad [9\text{-}9]$$

Usually, a measure of breadth of molecular-weight distribution is given by **polydispersity index**, PDI

$$\text{PDI} = \overline{M}_w / \overline{M}_n \qquad\qquad [9\text{-}10]$$

For example, commercial polystyrene of $\overline{M}_n$ over 100,000 has a PDI between 2 – 5.

[Example 9-1] A polydisperse sample of polysytrene is prepared by mixing three *monodisperse* samples in the following proportions:

1 g   10,000 molecular weight

2 g   50,000 molecular weight

2 g   100,000 molecular weight

Using this information, calculate the number-average molecular weight, weight-average molecular weight, and PDI of the mixture.

From Eqs. [9-6] and [9-8]

$$\overline{M}_n = \frac{\sum\limits_{i=1}^{3} N_i M_i}{\sum\limits_{i=1}^{N} N_i} = \frac{\sum\limits_{i=1}^{3} W_i}{\sum\limits_{i=1}^{3} (W_i / M_i)} = \frac{1+2+2}{\dfrac{1}{10,000} + \dfrac{2}{50,000} + \dfrac{2}{100,000}} = 31,250$$

$$\overline{M}_w = \frac{\sum\limits_{i=1}^{3} N_i M_i^2}{\sum\limits_{i=1}^{N} N_i M_i} = \frac{\sum\limits_{i=1}^{3} W_i M_i}{\sum\limits_{i=1}^{3} W_i} = \frac{10,000 + 2(50,000) + 2(100,000)}{5} = 62,000$$

And from Eq. [9-10]

$$\text{PDI} = \frac{\overline{M}_w}{\overline{M}_n} = \frac{62,000}{31,250} = 1.98$$

There are a number of molecular weight distributions available. The single-parameter **Flory distribution** is given as

$$W(X) = X (\ln p)^2 p^x \qquad [9\text{-}11]$$

where X represents the degree of polymerization and p represents the fractional monomer conversion in a step-growth polymerization. The expressions for the number-average and weight-average degrees of polymerization are as follows:
First, the geometric series is given as

$$\sum_{k=1}^{\infty} ar^k = a + ar + ar^2 + ar^3 + \dots \frac{a}{1-r}, \text{ here } |r| < 1 \qquad [9\text{-}12]$$

By using Eqs. [9-6] and [9-8],

$$\overline{X}_n = \frac{\sum\limits_{x=1}^{\infty} W(X)}{\sum\limits_{x=1}^{\infty} W(X)/X} = \frac{(\ln p)^2 \sum\limits_{x=1}^{\infty} Xp^{x-1}}{(\ln p)^2 \sum\limits_{x=1}^{\infty} p^{x-1}} \qquad [9\text{-}13]$$

$$= \frac{\sum\limits_{x=1}^{\infty} Xp^{x-1}}{\sum\limits_{x=1}^{\infty} p^{x-1}} = \frac{1/(1-p)^2}{1/(1-p)} = \frac{1}{1-p}$$

$$\overline{X}_w = \frac{\sum_0^\infty XW(X)}{\sum_0^\infty W/X} = \frac{(\ln p)^2 \, p \sum_{x=1}^\infty X^2 p^{x-1}}{(\ln p)^2 \, p \sum_{x=1}^\infty X p^{x-1}} \qquad [9\text{-}14]$$

$$= \frac{\sum_{x=1}^\infty X^2 p^{x-1}}{\sum_{x=1}^\infty X p^{x-1}} = \frac{(1+p)/(1-p)^3}{1/(1-p)^2} = \frac{1+p}{1-p}$$

For step-growth polymers, Eqs. [9-13] and [9-14] are applicable. For the number-averagemolecular weight can be often determined using

$$\overline{M}_n = M_0/(1-p) \qquad [9\text{-}15]$$

where M_0 is the unit weight (molecular weight of the repeating unit). Another function, the **Schultz-Zimm molecular weight distribution**, is expressed as

$$W(X) = \frac{a^{(b+1)} X^b \exp(-aX)}{\Gamma(b+1)} \qquad [9\text{-}16]$$

where a and b are adjustable positive real numbers, the gamma function is

$$\Gamma(b+1) = \int_0^\infty t^n e^{-1} dt = b\Gamma(b) = b! \qquad [9\text{-}17]$$

The number average molecular weight $\overline{M}_n$ is the first moment of the molecular weight distribution, analogous to the center of gravity (the first moment of the mass distribution) in mechanics. The weight-average molecular weight $\overline{M}_w$, the second moment of the distribution corresponds to the radius of gyration in mechanics. Alternatively higher moments of distribution such as $\overline{M}_z$, the third moment, may be used.

9.2 POLYMERIZATION

There are two distinct methods of polymer synthesis. The **step-growth polymerization** is based on the random reactions between two molecules (monomers) which may have reactive groups on both ends of the molecule. On the other hand, the **chain-growth polymerization** will involve the attachment of a monomer to an "active" chain. This active end may be a free radical or an ionic site.

With some exceptions, most synthetic polymers can be derived from these two methods.

9.2.1 Stepgrowth Polymerization

A monomer molecule consisting of at least two functional groups can undergo step-growth polymerization either with another monomer molecule containing two different functional groups or with itself. For example, the condensation polymerization and non-condensation-polymerization as illustrated in Table 9-1 will illustrate this. Most commercial polymers such as nylon are produced in this manner. Also see Figures 9-5 and 9-6 for further illustration.

Table 9-1. Classification of Step-Growth Polymers

Classification	Monomer 1	Monomer 2
Condensation		
Polyamide	Dicarboxylic acid	Diamine
Polycarbonate	Bisphenol	Phosgene
Polyester	Dicarboxylic acid	Diol or polyol
Polyimide	Tetracarboxylic acid	Diamine
Polysiloxane	Dichlorosilane	Water
Polysulfone	Bisphenol	Dichlorophenylsulfone
Non-condensation		
Polyurethane	Diisocyanate	Diol or polyol
Poly(phenylene oxide)	2,6-Disubstituted phenol	Oxygen

9.2.2 Chaingrowth Polymerization

The bulk of the vinyl polymerization belongs to this category. Whether the chain is free radical or ionic, they all contain the **initiation, propagation**, and **termination** steps. Ionic polymerization usually involves monomers with the electron-donating groups belonging to cationic polymerization. In the same manner, monomers containing the electron-withdrawing groups belong to anionic polymerization.

For free radical polymerization and copolymerization, initiation consists of two steps. First, an initiator needs to dissociate to form free radical species, for example,

$$I_2 \xrightarrow{\;k_d\;} 2I \bullet \qquad\qquad [9\text{-}18]$$

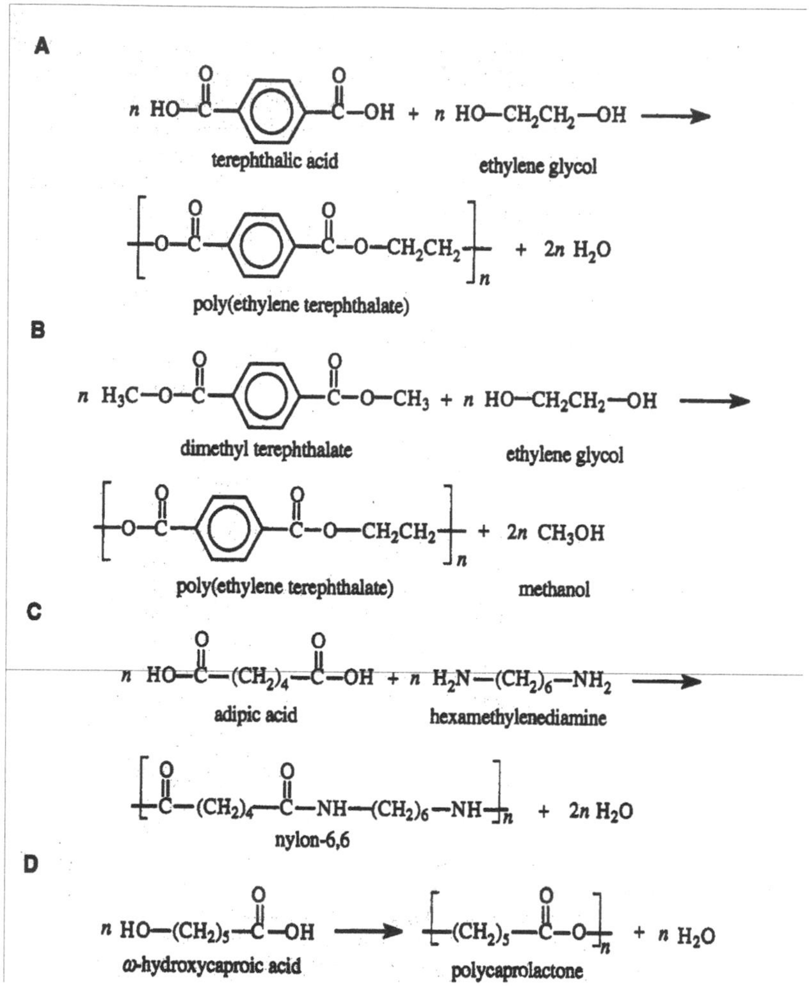

A

n HO—C(=O)—⟨benzene⟩—C(=O)—OH + n HO—CH$_2$CH$_2$—OH ⟶

terephthalic acid   ethylene glycol

$$\left[\!-O-\overset{O}{\overset{\|}{C}}-\text{⟨benzene⟩}-\overset{O}{\overset{\|}{C}}-O-CH_2CH_2\!-\right]_n \;+\; 2n\ H_2O$$

poly(ethylene terephthalate)

B

n H$_3$C—O—C(=O)—⟨benzene⟩—C(=O)—O—CH$_3$ + n HO—CH$_2$CH$_2$—OH ⟶

dimethyl terephthalate   ethylene glycol

$$\left[\!-O-\overset{O}{\overset{\|}{C}}-\text{⟨benzene⟩}-\overset{O}{\overset{\|}{C}}-O-CH_2CH_2\!-\right]_n \;+\; 2n\ CH_3OH$$

poly(ethylene terephthalate)   methanol

C

n HO—C(=O)—(CH$_2$)$_4$—C(=O)—OH + n H$_2$N—(CH$_2$)$_6$—NH$_2$ ⟶

adipic acid   hexamethylenediamine

$$\left[\!-\overset{O}{\overset{\|}{C}}-(CH_2)_4-\overset{O}{\overset{\|}{C}}-NH-(CH_2)_6-NH\!-\right]_n \;+\; 2n\ H_2O$$

nylon-6,6

D

n HO—(CH$_2$)$_5$—C(=O)—OH ⟶ $\left[\!-(CH_2)_5-\overset{O}{\overset{\|}{C}}-O\!-\right]_n$ + n H$_2$O

ω-hydroxycaproic acid   polycaprolactone

Figure 9-5. Examples of important polycondensations having a step-growth mechanism. (A) Polyesterification. (B) Ester-interchange polymerization. (C) Polyamidation. (D) Self-condensation of an A-B monomer. (After J. Fried, 2003)

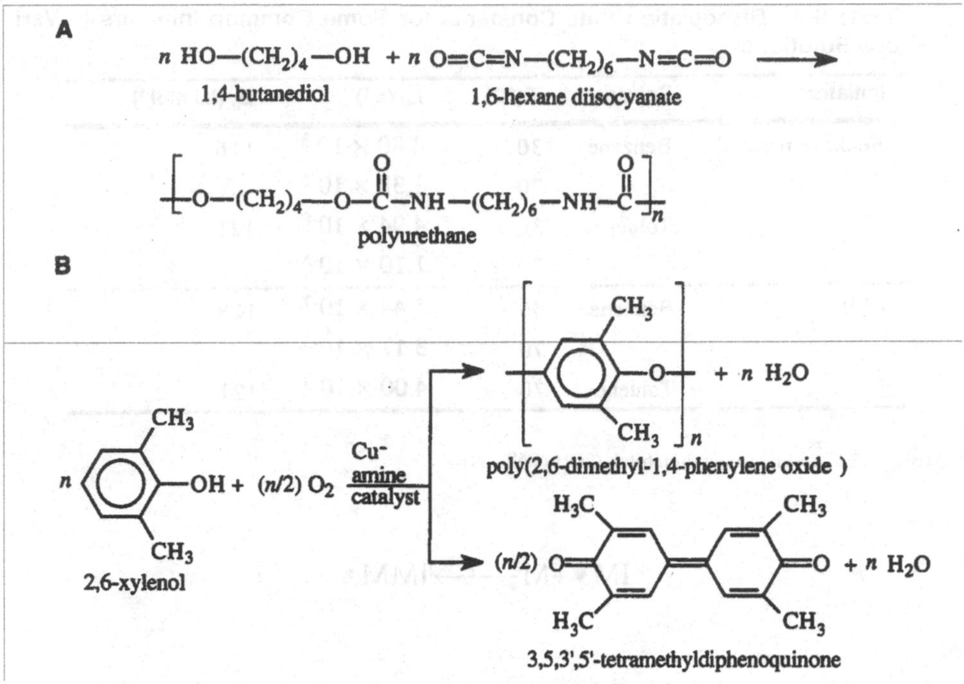

Figure 9-6. Two non-condensation step-growth polymerizations. (A) Addition polymerization of a polyurethane. (B) Oxidative-coupling polymerization of 2,6-xylenol to yield a high-molecular-weight polymer or a low-molecular-weight quinone as a by-product. (After J. Fried, 2003)

Here, k_d is the **dissociation rate constant**. It also follows the Arrhenius equation as discussed in Chapter 1.

$$k_d = A \exp(-E_a/RT) \qquad [9\text{-}19]$$

E_a is the activation energy for the dissociation of the initiatior. Commonly used initiators are listed in Table 9-2.

The second step of initiation is the attachment of the free radical species to the monomers and is known as monomer association.

$$I\bullet + M \xrightarrow{\quad k_a \quad} IM\bullet \qquad [9\text{-}20]$$

where k_a is the **rate constant of monomer association**.

Next, the propagation step involves the addition of a monomer to the initiated monomer species of Eq. 9-20.

Table 9-2. Dissociation Rate Constants for Some Common Initiators in Various Solutions

Initiator	Solvent	T (°C)	k_d (s^{-1})	E_a (kJ mol^{-1})
Benzoyl peroxide	Benzene	30	4.80×10^{-8}	116
		70	1.38×10^{-5}	
	Toluene	30	4.94×10^{-8}	121
		70	1.10×10^{-5}	
AIBN *	Benzene	40	5.44×10^{-7}	128
		70	3.17×10^{-5}	
	Toluene	70	4.00×10^{-5}	121

* AIBN is 2',2'-azobis (isobutylronitrite), $(2CH_3(CN)\!-\!N)\!\!=\!\!_n$

$$IM\bullet + M \xrightarrow{\ k_p\ } IMM\bullet$$

or

$$IM_x\bullet + M \xrightarrow{\ k_p\ } IM_{x+1}\bullet \qquad\qquad [9\text{-}21]$$

where k_p is the **propagation rate constant**. Finally, the termination step can be the combination of two propagating radical chains of degree of polymerization x and y,

$$IM_{x-1}M\bullet + \bullet MM_{y-1}I \xrightarrow{\ k_{tc}\ } IM_{x-1}M_2M_{y-1}I \qquad\qquad [9\text{-}22]$$

In this case, the terminal groups indicate that the sequence is head-to-head placement. It is also possible that disproportion may occur involving the transfer of a hydrogen atom to one propagatory chain speed such that

$$IM_{x-1}MH\bullet + \bullet HMM_{y-1}I \xrightarrow{\ k_{td}\ } IM_xH_2 + IM_y \qquad\qquad [9\text{-}23]$$

Chain transfer is another mode of termination, usually a hydrogen abstraction from other molecules such as an initiator, a monomer, an oligomer, or a solvent.

$$IM_{x-1}M\bullet + SH \xrightarrow{\ k_{tr}\ } IM_{x-1}MH + S\bullet \qquad\qquad [9\text{-}24]$$

Several rate constants are discussed here.

- k_{tc} **termination by combination**
- k_{td} **termination by disproportionation**
- k_{tr} **chain transfer**

Usually k_t signifies the total power of both k_{tc} and k_{td}, or

$$k_t = k_{tc} + k_{td} \qquad\qquad [9\text{-}25]$$

In kinetics, the overall rate of polymerization, R_o, of a free radical mechanism is simply the rate of chain propagation, R_p.

$$R_o \equiv R_p = k_p(\text{IM}_x\bullet)(\text{M}) \qquad\qquad [9\text{-}26]$$

Under steady conditions radicals formed in the initiation are equivalent to radicals consumed in the termination, or

$$R_i = R_t \qquad\qquad [9\text{-}27]$$

It follows that

$$\begin{aligned} R_i &= d(I\bullet)/dt \\ &= 2k_d(I) \end{aligned} \qquad\qquad [9\text{-}28]$$

For this equation, only a certain fraction, f, of the original initiator is effective in contributing to the chain propagation, so

$$R_i = 2fk_d(I) \qquad\qquad [9\text{-}29]$$

For the termination rate, if one only considers combination and disproportionation, k_t, then

$$R_t = -d(IM_x\bullet)/dt = 2k_t(IM_x\bullet)^2 \qquad\qquad [9\text{-}30]$$

Since

$$IM_x\bullet + IM_x\bullet \xrightarrow{\;k_t\;} P \qquad\qquad [9\text{-}31]$$

By steady state equating of Eqs. 9-29 and 9-30,

$$(IM_x\bullet) = (fk_d/k_t)^{1/2}I^{1/2} \qquad\qquad [9\text{-}32]$$

or

$$R_o = k_p(fk_d/k_t)^{1/2}(I)^{1/2}(M) \qquad\qquad [9\text{-}33]$$

Hence, in free radical polymerization, the rate is proportional to the monomer concentration and to the square root of initiator concentration.

Furthermore, the number average degree of polymerization can be estimated from the ratio of the rate of propagation to the rate of termination.

$$\overline{X_n} = R_p/R_t \qquad\qquad [9\text{-}34]$$

Using Eqs. 9-26, 9-30, and 9-32,

$$X_n = k_p(M)/2(k_t f\, k_d\, (I))^{1/2} \qquad\qquad [9\text{-}35]$$

The propagation step is important, since it also controls the kinetic chain length, υ, which is defined as the ratio of the rate of growth per effective radical or

$$\upsilon = R_p/R_i \qquad\qquad [9\text{-}36]$$

So far we only discussed the termination by combination and by disproportionation (the first two types). If the chain-transfer involves in the termination process, then Equation 9.27 should be:

$$\overline{X}_n = R_p/(R_{tc} + R_{td} + R_{tr}) \qquad\qquad [9\text{-}37]$$

Here R_{tr} is the rate of termination due to chain transfer, and it is expressed by Eq. 9-24 as

$$R_{tr} = k_{ts}\,(IM_x\bullet)(SH) \qquad\qquad [9\text{-}38]$$

Rearranging Eq. 9.37 and using Eq. 9.25,

$$\overline{X}_n^{-1} = \frac{(R_{tc} + R_{td})}{R_p} + \frac{k_{tr}(\text{Mx}\bullet)(\text{SH})}{k_p(\text{Mx}\bullet)(\text{M})}$$

$$= \overline{X}_{no}^{-1} + C(\text{SH})/(\text{M}) \qquad\qquad [9\text{-}39]$$

where $\overline{X}_{no}$ is the average degree in the absence of chain transfer, $\overline{X}_n$ is the total sum of chain transfer and other terminations, and C is the chain-transfer coefficient.

$$C = k_{tr}/k_p \qquad\qquad [9\text{-}40]$$

Some chain-transfer coefficient values are listed in Table 9-3. Evidently, the molecular weight will decrease with an increase in the concentration of chain-transfer agent in emulsion polymerization.

Table 9-3. Representative Values of Chain-Transfer Constants

Monomer	Chain-Transfer Agent	T(°C)	$C \times 10^4$
Styrene	Styrene	25	0.279
		50	0.35-0.78
	Polystyrene	50	1.9-16.6
	Benzoyl peroxide	50	0.13
	Toluene	60	0.125
Methyl metacrylate	Methyl metacrylate	30	0.117
		70	0.2
	Poly (methyl metacrylate)	50	0.22-1000
	Benzoyl peroxide	50	0.01
	Toluene	40	0.170

Some of the chain-modifiers can be prepared synthetically, such as substituted naphthyl disulfate, sometimes called polymerization modifiers.

9.3 SOLID STATE PROPERTIES

Some polymers such as polyolefins are highly crystalline materials having crystalline morphology consisting of chain folded lamella joined as spheralites. Others are crystalline or amorphous. For amorphous solid state, thermal energy can keep to long – range segments of each polymer chain to move in random micro-Brownian motions. As the polymer is cooled, a temperature is reached at which the long-large segment's characteristic motions cease. This temperature is called the **glass transition temperature** or T_g and varies quite widely with polymer structure. Below T_g, the only molecular motions that can occur are short range motion of several contiguous chain segments. Even before T_g, the chains assume their undisturbed dimension as they do in a solution under **θ-condition**. In the absence of excluded-volume effect, the characteristic ratio is its minimum size.

$$C_\infty = \langle r^2 \rangle_0 \Big/ nl^2 \qquad\qquad [9\text{-}41]$$

The ratio of unperturbed mean square end-to-end distance to the mean square of the end to end distance of freely joined model. The glass transition temperature is thought to be the "frozen out" of conditions 1 and 2. The molecular energy is only allowed for 4 and 5.

1.   Translational motion of the entire molecule which permit flow.
2.   Co-operative wiggling and jumping of the segments of molecules approximately 40-50 carbon in length.
3.   Motion of 5 to 6 carbon atoms along the main chain – **Schatzki model**
4.   Vibration of atoms about an equilibrium position in the crystal lattice.

9.3.1 Chain Entanglement and Repetition

Polymer chains that are sufficiently long can form stable, flow-restricting **entanglement** just as illustrated by Fig. 9-7. Entanglements usually effect viscoelastic properties as well as mechanical properties such as stress relaxation, creep, and craze. A minimum polymer chain length can be called the critical molecular weight, M_c. This quantity is for the formation of stable entanglement may be directly related to the characteristic ratio.

A related parameter M_e, which is the **molecular weight between entanglements**, represents values of both the M_c and M_e for a number of polymers are listed in Table 9-4. A rough relationship between M_c and M_e is

$$M_c \approx 2M_e \qquad\qquad [9\text{-}42]$$

Figure 9-7. Representation of a polymer chain intertwined in a network of entangled chains. (Adopted from Pierre-Gilles de Gennes: Scaling Concepts in Polymer Physics, 1979, Cornell University Press.)

Table 9-4. Entanglement Molecular Weights for Linear Polymers

Polymer	Mc	Me
Polycarbonate	4,800	2,490
Cis-Polyisoprene	10,000	5,800
Polyisobutylene	15,200	8,900
Polydimethylsiloxane	24,400	8,100
Poly (vinyl acetate)	24,500	12,000
Poly (methyl metacrylate)	27,500	5,900
Poly (2-methylstyrene)	28,000	13,500
Polystyrene	31,200	18,100

If individual chains entangled in the solid state, it is possible that by repetition, movement is possible so that heating can proceed through T_g and pass to the next stage (Fig 9-8).

Relaxation from the main chain has been proposed as the **Schatzki crankshaft rotation** model shown in Fig. 9-9. This limited non-cooperative motions of the contiguous bonds may occur at lower temperatures, for example at T_g or below.

Next, if two polymers are mixed together, an **interpenetrating polymer network** (IPN) is formed (Fig. 9-10). In this case, the entangling and twisting of the two polymers results in interaction between non-consecutive parts of each polymer (including the case that one of the polymers is cross-linked). Thus stability resembles that of a neutralized poly salt formation due to the amalgamation of a polycation and a polyanion forming a coacerate. Often the IPN can lead to the formation of a stable geopolymers such as bitumen or kerogen.

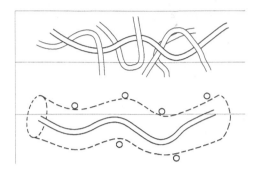

Figure 9-8. Reptation model of a polymer chain constrained in entangled network. A particular chain can be viewed as constrained to move within a virtual tube defined by neighboring entanglement sites. Circles pictured in the lower view represent cross sections of chains constituting the tube constraints. (Adapted from J.Klein, *Nature*, 271, 143(1978) with minor modifications)

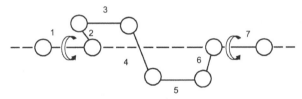

Figure 9-9. Schatzki model of crankshaft motion of a carbon-carbon backbone. The dashed line represents the virtual axis around which bonds 2-6 rotate.

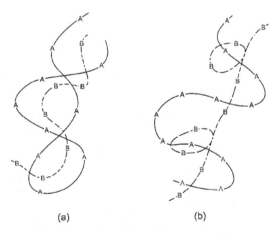

(a)   (b)

Figure 9-10. Interpenetrating polymer network (IPN): In (a), A is linked by a solid line, B linked by a dashed line. A,B are polymers. In (b), the polymer also has some degree of cross link.

[Example 9-2]   In Taiwan, people like to consume fish in their diet. However, fishes, in the marine environment, can adsorb mercury from water or aquatic organisms that they feed on. Chitosan is a type of positively charged biopolymer with functional group that can chelate mercury and remove it from human body. Biopolymers are also known to form interpenetrating polymer networks for stabilization. Negatively charged sodium alginate should be taken to neutralize chitosan at 1:1 mole ratio. Assume each person in Taiwan consumes two fishes per week with a mercury content of 3.5 ppmw. Assume that: 1) the average weight of one fish is 2 kg; 2) the maximum capacity of chitosan for mercury is 100 mg/g; 3) the net weight of the pill is 50 mg/tablet for both chitosan pill and sodium alginate pill and the purity are 20% and 50% respectively. Furthermore,

sodium alginate (unit weight = 268)

chitosan (uw = 217)

Determine how many tablets of chitosan and sodium alginate should be taken in order to remove all the mercury intaken by each person per week.

Mercury content in one fish = 3.5 ppmw × 2 kg/fish

$$= \frac{3.5}{10^6} \times 2 \times 10^6 \, \text{mg/fish} = 7 \, \text{mg/fish}$$

Mercury intake per person per week
$$= 7 \, \text{mg/fish} \times 2 \, \text{fishes/week}$$
$$= 14 \, \text{mg/week}$$

Chitosan pills needed to be taken per person

$$= \frac{14\text{mg/week}}{100\text{mg/g} \times 20\% \times 50\text{mg/tablet}}$$

$$= 14 \text{ tablets / week}$$

Sodium alginate pills needed to be taken per person

$$= \frac{\frac{14\text{mg/week}}{100\text{mg/g}} \times \frac{268 \text{ g/mol}}{217 \text{ g/mol}}}{50\% \times 50 \text{ mg/tablet}}$$

$$\approx 7 \text{ tablets/week}$$

Therefore, it is recommended to take 2 tablets of chitosan and 1 tablet of sodium alginate daily after meal.

9.3.2 The Glass Transition

Conventionally speaking, the temperature marking the transition from amorphous to the melt state is called glass-transition temperature T_g. Several phenomological models have been provided for some understanding:

- The **isoviscous state** – When the polymer is cooled from its melt state. Viscosity measured rapidly to a common maximum value. (ca·10^{12} Pa-s)
- The **isofree volume state** – Free volume V_f is defined as the difference between the specific (or actual) volume at a given temperature and its equilibrium volume at absolute zero, V_0,

$$V_f = V - V_0 \qquad [9\text{-}43]$$

- The **isoentropic state** – Gibbs and Dimarzio have suggested that there is a temperature T_2 at which the conformational entropy, S_c goes to zero.

Usually the equilibrium temperature lies approximately 52°C below T_g. T_g for several important polymers are given in Table 9-5. The values can very widely with the polymer chain for amorphous polymers.

Table 9-5. Glass-Transition Temperatures of Some Amorphous Polymers

Polymer	T_g (°C)
Polydimethylsiloxane	-123
Poly (vinyl acetate)	28
Polystyrene	100
Poly (methyl metacrylate)	105
Polycarbonate	150
Polysulfone	190
Poly (1,6-dimethyl-1,4-phenylene oxide)	220

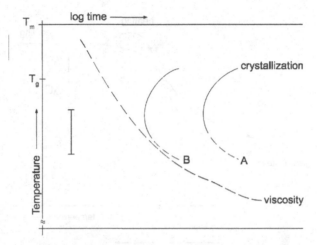

Figure 9-11. Donth's ttt diagram (time, temperature transition), which compares times from viscosity with times from a suitably defined crystallization rate. The bar indicates the forming interval for viscosity. The solid lines in A and B are the thermodynamic parts, and the dashed lines in A and B are the transport parts. (R. Khachatoorian et al., *Petroleum Science and Technology*, 41, 243,2003)

A changeover in behavior from that of a typical disordered solid to one more akin to a liquid occurs as the temperature is raised in a large number of amorphous materials.

The aim is usually to avoid crystallization in order to get a clear state glass. The situation can be usually expressed in **Donth's TTT** (time, temperature, transition) **diagram**, where a quantity such as viscosity can be compared with a crystallization rate after a suitable reduction of the two quantities to comparable time scales (Figure 9-11). For case A shown in the figure, the crystallization rate in the forming interval is slow so that it is easy to get a clear glass by sufficiently quick

action. For case B, the reciprocal crystallization rate at the 'node' is comparable
with the viscosity time scale in the forming interval so that a reasonable interplay
can be organized. Consider future physiochemical or biological aspects of glass
transition, e.g. formation of a glassy matrix of freezing of biological materials at
low temperatures. People involved in such activities used to find the glass transition by dynamic calorimetry with a given cooling or heating rate, e.g. by **DSC
(dynamic scanning calorimetry)**, as indicated by Figure 9-12. The glass transition is indicated by a continuous step in heat capacity. This calorimetric glass
transition is characterized by a typical temperature T_g, by a steep height ΔC_p and
surprisingly small transformation interval ΔT of order of 10K.

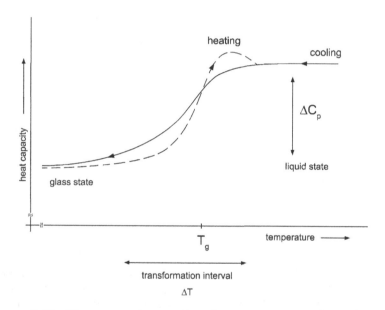

**Figure 9-12. Thermograph near the glass temperature T_g as determined by
DSC (After Khachatoorian et al., 2003, loc.cit.)**

Most polymers are semicryatalline and they may exhibit both a T_g corresponding to long-range segmental motions in amorphous regions and a crystal-
line-melting temperature T_m at which crystallite are destroyed. For many polymers, T_g is approximately one-half to two-thirds of T_m

$$\frac{T_g}{T_m} \approx 0.66 \text{ (Expressed in K)} \qquad [9\text{-}44]$$

The thermal transitions of some crystalline polymers are located in Table 9-6.

Table 9-6. Thermal Transitions of Some Semi-crystalline Polymers

Polymer	T_g (°C)	T_m (°C)
Poly caprolactone	-60	61
Pllyethylene (high-density)	-120	135
Poly (vinylidene fluoride)	-45	172
Polyoxymethylene	-80	198
Poly (vinyl alcohol)	85	258
Poly (hexamethylene adipamide) (nylon-6,6)	49	265
Poly (ethylene terephthalate)	69	265

Transition temperatures are extremely "structure-sensitive", partly due to steric effects and intra-inter-molecular interactions. Several researchers have proposed correlations between the chemical structure and the glass transition temperature of polymers. Their methods are usually based on the assumption that the structural groups in the repeating units provided weighted additive contributions to the T_g. In the case of ideal additivity, the combination of a given group is independent of adjacent groups. Although this ideal case is seldom encountered, additivity can be approximated by proper choice of structural groups.

An empirical approach developed by van Krevelen is that Tg.M (or Y_g) behaves in general as an additive function which is termed as molar glass-transition function.

$$Y_g = \sum Y_{gi} = T_g.M \qquad \text{[9-45]}$$

so that

$$T_g = Y_g/M = \sum Y_{gi}/M \qquad \text{[9-46]}$$

Equation 9-45 has been applied to available literature data on T_g's of polymers, in all nearly 600; from this study the correlation rules for Y_g have been derived. It appeared that the Y_{gi}-values of the relevant groups are not independent of some other groups present in the structural unit.

The group contributions and structural corrections of van Kervelen (For example. Table 9-7 lists the additive functions of Y_g) are used to calculate the values of Y_{gi} and M_i can be read and recorded for each functional group and by using Equation 9-46, the T_g values of the polymers are obtained.

Table 9-7(a). Partial List of Group Contributions (Increments) to Yg[K kg / mol] (Non-conjugating groups) based on van Krevelen D.W., 1990, *Properties of polymers*, 3rd ed.

Group		Y_{gi}	M_i	Group		Y_{gi}	M_i
–CH₂–	In main chains (gen.)	2.7		alicyclic ⟨H⟩	cis	(19)	82.1
	In hydrogen bonded ch.	4.3	14.0		trans	(27)	82.1
	In side chains	*		–CHF–		12.4	32.0
				C–	–CHCl–	19.4	48.5
–CHX–	–CH(CH₃)–	8.0	28.0	halide	–CFCl–	22.8	66.5
	–CH(I-propyl)–	19.9	56.1		–CF(CF₃)–	24	82.9
	–CH(ter-butyl)–	25.6	70.1		–CF₂–	10.5	50.0
	–CH(cyclopentyl)–	30.7	82.1		–CCl₂–	22.0	82.9
	–CH(cyclohexyl)–	41.3	96.2				
	–CH(C₆H₅)–	36.1	90.1		–O–	4	16.0
	–CH(p-C₆H₄CH₃)–	41.2	104.1		–NH–	(7)	15.0
					–S–	8	32.1
	–CH(OH)–	13	30.0		–SS–	16	64.2
	–CH(OCH₃)–	11.9	44.1	–Si(CH₃)₂–	free	7	58.2
	–CH(OCOCH₃)–	23.3	72.1		st. hind	16	58.2
	–CH(COOCH₃) –	21.3	72.1	–O–C(=O)–O–		20	60.0
	–CH(CN) –	17.3	39.0				
–CX₂–							
–CXY–							
	–C(CH₃)₂ – free	8.5	42.1				
	–C(CH₃)₃– st. hindered	15, ss26	42.1	C– hetero	–O–C(=O)–NH–	20	59.0
	–C(CH₃)(COOCH₃)–	35.1	86.1				
	–C(CH₃)(C₆H₅)–	51	104.1		–NH–C(=O)–NH–	20	58.0
	–C(C₆H₅)₂–	65	164.2				
	–C(CN)₂–	22	64.0				

* N = 9 $Y_g \sim Y_{g9}$
N < 9 $Y_g \sim Y_{g0} + [N(Y_{g9}-Y_{g0})/9]$
N > 9 $Y_g \sim Y_{g9}+7.5(N-9)$

9.3 Solid State Properties

359

Table 9.7(b). Partial list of group contributions to Yg[K. kg / mol] (Groups with potential mutual conjugation)

Double-bonded systems				Aromatic ring systems			
Group		Y_{gi}	M_i	Group		Y_{gi}	M_i
–CH=CH–	cis	3.8	26.0	(1,4-phenylene)	n	29.5	
	tr	7.4	26.0		c	35	76.1
–CH=CH(CH₃)–	cis	8.1	40.1		cc	41	
	tr	9.1					
				(1,2-phenylene)	n	25	
–C(CH₃)=CH(CH₃)–		16.1	54.1		c	29	76.1
–CH=CF–		9.9	44.0		cc	34	
–CH=CCl–		15.2	60.5				
–CH=CF–		20.3	62.0	(1,4-phenylene, x subst.)	x=CH₃	35	90.1
–C≡C–		11	24.0		x=C₆H₅		152.2
					x=Cl	51	110.5
–C(=O)–	n	9					
	c	14	28	(1,2-disubst. x)	x=CH₃	54	104.1
	cc	19			x=C₆H₅	118	228.3
–C(=O)–O–C(=O)–	n	22	72.0		x=Cl		145.1
					x=CH₃	30	90.1
–C(=O)–O–	n	12.5	44.1		x=C₆H₅		152.2
	c	13.5	114.0		x=Cl	(45)	110.5
	cc/1)	15					
–C(=O)–NH–	n	12.5		(naphthalene)		50	126.2
	c	13.5	114.0				
	cc/1)	15					
–S(=O)₂–	n	32.5		(naphthalene)		68	126.2
	c	36	64.1				
	cc	41					

n = non-conjugated (isolated in aliphatic chain).

c = one-sided conjugation with aromatic ring.

cc = two-sided conjugation/1) between two aromatic rings (rigid).

[**Example 9-3**]   Doi synthesized a series of five co-polymer polyesters, namely the polyhydroxybutyric acid P(3HB-Co-4HB) where the mole percentage of 4HB varies from 0 , 10, 20, 30 and 82. He also experimentally measured the T_g of this series and the values were 277, 270, 267, 263 and 247 in K. Use van Krevelen's method to support his T_g measurements.

From Table 9-7, computation for PHB are listed in Table 9-8 which is the case of P(3HB) only. The calculation as the increment of 4HB are:

Mole conc. of 4HB (%)	Calculated T_g (K)	Experimental T_g (K)
0	276	277
10	271	270
20	269	264
30	266	263
82	250	247

Table 9-8. Thermal Transitions of Some Semi-crystalline Polymers (for Example 9-3 only)

Functional Group	Y_{gi}	Y_{mi}
$CH(CH_3)$	8.0	13.0
(CH_2)	2.7	1.0
O	4.0	13.5
$-\overset{O}{\overset{\|}{C}}-$	9.0	12.0
M=86	Y_g=23.7	Y_m=39.5

$$T_g = 23.7\times10^3/86 = 276K$$
$$T_m = 39.5\times10^3/86 = 459K$$
$$T_g/T_m = 0.6$$

A plot is made to correlate the calculated T_g of Example 9-3 and Doi's experimental T_g. As seen in the figure, after the experimental T_g is plotted vs. the calculated T_g, the correlation is R^2=0.9772. The results also agree to Eq. 9-44 which states that the ratio of glass transition to melting is 0.6. A figure has also been constructed for the predicted T_{gs}.

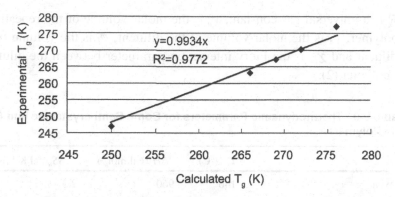

Figure for Example 9-3. Plot of experimental value of T_g versus calculated value of T_g by the additive group method. (y = 0.9934x, R^2=0.9772). All of the biopolymers used here are PHB type (3HB-Co-4HB). The mole percentages of 4HB of the sample from left to right are 0, 10, 20, 30, and 82, respectively. The experimental values of T_g for the five samples are respectively 277, 270, 267, 263, and 247 (Doi, 1990).

9.3.3 Crystallinity

From thermodynamics, the free energy per repeating unit of the polymer is

$$\Delta G_u = \Delta H_u - T\Delta S_u \qquad [9\text{-}47]$$

where ΔH_u and ΔS_u are the enthalpy and entropy per repeated unit. At the equilibrium melting temperature $(T=T_m{}^0)$

$$T_m^0 = \Delta H_u / \Delta S_u \qquad [9\text{-}48]$$

In general, the obsessed crystalline melting temperature T_m is always lower than the equilibrium value of $T_m{}^0$ (See Table 9-9). This can be due to crystalline size melting point depression, etc. Using the Flory-Hugguin theory, the melting point depression by a dilutent is

$$T_m^{-1} - T_m^0 = \left(\frac{R}{\Delta H_u}\right)\left(\frac{V_u}{V_1}\right)\left(\phi_1 - \chi_{12}\phi_1^2\right) \qquad [9\text{-}49]$$

where R is a universal gas constant, V_u is the molar volume of the repeating unit in the polymer, V_1 is the molar volume of the dilutent, ϕ_1 is the volume function of the diluent and χ_{12} is the Flory interaction parameter between the dilutent (1) and the polymer (2).

Table 9-9. Thermodynamic Parameters for Some Semi-crystalline and Crystalline Polymers

Polymer	T_m^0 (°C)	ΔH_u(cal mol^{-1})	ΔS_u(cal K^{-1} mol^{-1})
Polyethyene	146	960	2.3
Pllyoxymethylene	180	590	3.5
Polypropylene	200	1386	2.9
Poly (ethylene terephthalate)	280	6431	11.6
Polycarbonate*	335	6346	10.4

*Solvent-induced crystallization

Between T_g and T_m, there is a temperature at which the growth of crystals is maximum, T_e.

For example, (Polyethylene terephthalate) is such a case. (Fig. 9-13) An equation called **Avrami equation** is used to account for the functional crystalinity ϕ.

$$\phi = 1 - \exp(-kt^n) \qquad [9\text{-}50]$$

Where k is the growth rate parameter, n is the temperature-independent nucleation index, varying between 1 and 4. Crystalinity can be determined by a variety of methods. If the amorphous density ρ_a and crystalline density ρ_c (See Table 9-10) are known, then the functional crystalinity can be evaluated

$$\phi = \frac{(\rho - \rho_a)}{(\rho_c - \rho_a)} \qquad [9\text{-}50a]$$

9.4 RUBBER ELASTICITY

The unique and useful mechanical properties of natural and synthetic rubber are that they are capable of reverse extension of 600-700%. Due to vulcanization or crosslink, the bulk slippage of the molecule across one another is prevented.

Consider an element with dimension a×b×c as shown in Figure 9-14. If the first law of thermodynamics is applied,

$$dU = dq - dW \qquad\qquad [9\text{-}51]$$

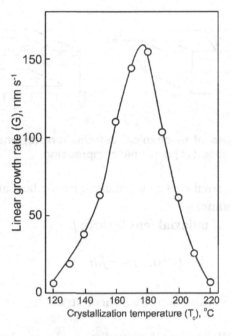

Figure 9-13. Plot of linear growth rate of spherulites in poly (ethylene terephthalate) (PET) as a function of temperature and at a pressure of 1 bar. The maximum growth rate is observed near 178°C. Values of T_g and T_m for PET are approximately 69° and 265°C, respectively. (Philips and Tsang, 1989)

Table 9-10. Amorphous and Crystalline Densities

Polymer	$\rho_a(\text{g cm}^{-3})$	$\rho_c(\text{g cm}^{-3})$
Nylon-6	1.09	1.12-1.14
Nylon-6,6	1.09	1.13-1.145
Poly (aryl ether ether ketone) (PEEK)	1.263	1.400
Poly (butylene terephthalate)	1.280	1.396
Poly (ethylene terephthalate)	1.335	1.515
Poly (vinyl chloride)	1.385	1.44-1.53
Polycarbonate	1.196	1.316
Poly (p-phenylene sulfide)	1.32	1.43
Poly (2,6-dimethyl-1,4-phenylene oxide)	1.06	1.31

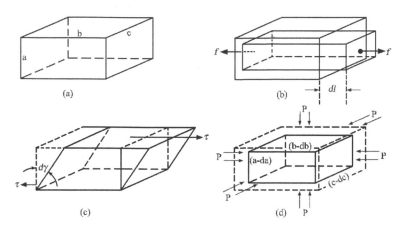

Figure 9-14. Types of mechanical deformation: (a) unstressed; (b) uniaxial tension; (c) pure shear; (d) isotropic compression.

The change of internal energy is balanced by the heat and work between the system and its surroundings.

For the first type of **uniaxial tensile force** f,

$$dW(tensile) = -fdl \qquad [9\text{-}52]$$

The second type is work done by a **shear stress** τ:

$$
\begin{aligned}
dW(\text{shear}) &= (\text{force})(\text{distance}) \\
&= -(\tau bc)(a\,d\gamma) \qquad [9\text{-}53]\\
&= -\tau V d\gamma
\end{aligned}
$$

where γ is the **shear strain**.

The third type is done by isotropic pressure **(isotropic compression)** in change the volume element.

$$
\begin{aligned}
dW(pressure) &= P(cb)da + P(ac)db + Pab(dc) \\
&= PdV
\end{aligned}
\qquad [9\text{-}54]
$$

If the process is reversible, then

$$dq = TdS \qquad [9\text{-}55]$$

where S is the system's entropy.

Combining the above equations, one can write **deformation** as

$$dU = TdS - PdV + fdl + V\tau d\gamma \qquad\qquad [9\text{-}56]$$

Now for the first type, uniaxial tension at constant volume and temperature

$$dV = \tau = 0 \qquad\qquad [9\text{-}57]$$

Dividing Eq. 9-56 by dl and solving for f,

$$f = \left(\frac{\partial U}{\partial l}\right)_{T,V} - T\left(\frac{\partial S}{\partial l}\right)_{T,V} \qquad\qquad [9\text{-}58]$$

For the second type, pure shear at constant volume and temperature, one obtains

$$dV = f = 0$$

$$\tau = \frac{1}{V}\left(\frac{\partial U}{\partial \gamma}\right)_{T,V} - \frac{T}{V}\left(\frac{\partial S}{\partial \gamma}\right)_{T,V} \qquad\qquad [9\text{-}59]$$

For the third type, the isotropic compression at constant temperature

$$P = \left(\frac{\partial U}{\partial V}\right)_{T} - T\left(\frac{\partial S}{\partial V}\right)_{T} \qquad\qquad [9\text{-}60]$$

It is difficult to carry out tensile experiments at constant volume. Most experiments are done at constant pressure (atmospheric). Fortunately Poisson's ratio for rubber is ca. 0.5; the change in volume is small.

The relationship of both deformation and temperature for rubber can best be illustrated by Fig. 9-15:

Based on thermodynamics,

$$-\left(\frac{\partial S}{\partial l}\right)_{T,V} = \left(\frac{\partial f}{\partial T}\right)_{l,V} \qquad\qquad [9\text{-}61]$$

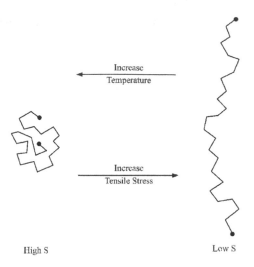

Increase
Temperature

Increase
Tensile Stress

High S   Low S

Figure 9-15. The effect of temperature on chain configurations.

Since experiments are conducted at constant pressure,

$$-\left(\frac{\partial S}{\partial l}\right)_{T,V} \approx \left(\frac{\partial f}{\partial T}\right)_{P,\alpha}$$
                                                              [9-62]

Here $\alpha = l \, / \, l_o$, the extension ratio or the ratio of stretched to un-stretched length at a particular temperature. Combining Eq. 9-61 and 9-58,

$$\left(\frac{\partial f}{\partial T}\right)_{l,V} = \frac{f}{T} - \frac{1}{T}\left(\frac{\partial U}{\partial l}\right)_{T,V}$$
                                                              [9-63]

Here both f and T are positive. For the first term on the right, the force increases with temperature. The second term products a relation of the tensile force with increasing temperature, since there is a negative sign in front of it. For rubbers at large values of f, the first entropy term becomes dominant. For ideal rubber,

$$\left(\frac{\partial U}{\partial l}\right)_{T,V} = 0$$
                                                              [9-64]

Hence, the interpretation of Eq. 9-63 becomes

$$f = (\text{constant})T \quad (\text{Ideal Rubber})$$
                                                              [9-65]

or
$$T\left(\frac{\partial f}{dT}\right)_{l,v} = f \qquad \text{for ideal rubbe} \qquad \text{[9-65a]}$$

or
$$T\left(\frac{\partial f}{dl}\right)_{T,v} = f \qquad \text{for ideal rubber} \qquad \text{[9-65b]}$$

This resembles the ideal gas law, P and T at constant V. Data in Figure 9-16 support this fact.

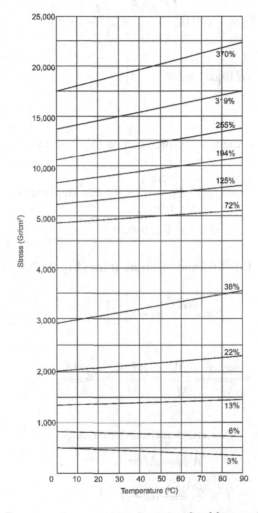

Figure 9-16. Force vs. temperature in natural rubber maintained at constant extension.

Furthermore, a standard treatment of the entropy is

$$\Delta S = S - S_0 = NR\ln\left(\frac{\Omega}{\Omega_0}\right) \qquad [9\text{-}66]$$

where Ω is the number of configurations available in N moles of network chains and R is the gas constant.

For constant volume stretching, it is possible to evaluate the Ωs as

$$S - S_0 = \frac{1}{2}NR\ln\left[\left(\frac{l}{l_0}\right)^2 + 2\left(\frac{l_0}{l}\right) - 3\right]$$

$$= \frac{1}{2}NR\ln\left[\alpha^2 + 2\alpha^{-1} - 3\right] \qquad [9\text{-}67]$$

For ideal rubber, the tensile force can be expressed as

$$f = -T\left(\frac{\partial S}{\partial l}\right)_{T,V} \qquad \text{for ideal rubber} \qquad [9\text{-}68]$$

$$\text{same as Eq. 9-65b}$$

Differentiating Eq. 9-67 and inserting it into Eq. 9-68,

$$f = \frac{NRT}{l_0}\left[\alpha + \alpha^{-2}\right] \qquad [9\text{-}69]$$

Also,

$$N = \frac{mass}{M_c} = \frac{\rho V}{M_c} = \frac{\rho l_0 A_0}{M_c} = \frac{\rho lA}{M_c} \qquad [9\text{-}70]$$

Here, ρ is density, A is cross-sectional area, and

$$A_0 l_0 \approx Al \text{ or}$$

$$f = \frac{\rho A_0 kT}{M_c}\left(\alpha - \alpha^{-2}\right) \qquad [9\text{-}71]$$

The **engineering tensile stress**, σ_e is defined as the tensile force over the initial cross section area, A_o and

$$\sigma_e = \frac{f}{A_0} = \frac{\rho RT}{M_c}\left(\alpha - \alpha^{-2}\right) \qquad [9\text{-}72]$$

The **true tensile stress** σ_t, the tensile force over the actual area A at length l is

$$\sigma_t = \frac{f}{A} = \frac{\rho RT}{M_c}\left(\alpha^2 - \alpha^{-1}\right) \qquad [9\text{-}73]$$

Since for the above,

$$\frac{A_0}{A} = \frac{l_0}{l} = \alpha^{-1} \qquad [9\text{-}74]$$

The **tensile strain** ε is defined as

$$\varepsilon = \frac{l - l_0}{l_0} = \alpha - 1 \qquad [9\text{-}75]$$

The slope of the stress-strain curve (**tangent Young's modulus**) is

$$E = \left(\frac{\partial \sigma_t}{\partial \varepsilon}\right)_T = \left(\frac{\partial \sigma}{\partial l}\right)_T \left(\frac{\partial l}{\partial \varepsilon}\right)_T$$

$$= \frac{\rho RT}{M_c}\left(2\alpha + \alpha^{-2}\right) \qquad [9\text{-}76]$$

And the initial modulus ($\alpha \to 1$) for an ideal rubber becomes

$$E(\text{initial}) = \frac{3\rho RT}{M_c} \quad \text{for ideal rubber} \qquad [9\text{-}76a]$$

This equation indicates that the force (or its modulus) in an ideal rubber held at a particular strain, is proportional to absolute temperature, but is inversely proportional to the molecular weight of the chain segments between the crosslinks. Therefore the increase in cross links is anticipated in stiff rubber.

For ideal rubber, equations 9-65 and 9-68 are useful. Also the exponent results of Fig 9-16 support this behavior.

9.4.1 Viscous Flow

Deformation and flow of polymeric systems include both melt and solution. Newton's law for flow and Hooke's law for elasticity are not sufficient. For basic definition, the viscosity of a material expenses to flow in terms of two parameters, the **shear stress**, τ and the **shear rate**, $\dot{\gamma}$. A simple viscometric flow can be described in Figure 9-17. A fluid layer at y moves with a velocity of $u = dx/dt$ in the x-direction while the layer at $y + dy$ has a velocity of $u + du$. The displacement gradient dx/dy is known as **shear strain** and expressed as

$$\gamma = \frac{dx}{dy} = \text{shear strain (dimensionless)} \qquad [9\text{-}77]$$

The **rate of shear strain** is

$$\dot{\gamma} = \frac{d\gamma}{dt} = \frac{d}{dt}\left(\frac{dx}{dy}\right) = \frac{d}{dy}\left(\frac{dx}{dt}\right) = \frac{du}{dy} \quad (time)^{-1} \qquad [9\text{-}78]$$

which is the velocity gradient.

also,
$$u = \frac{dr}{dt}$$

Next to shear stress is the force in the direction of flow per unit area normal to the Y-axis.

$$\tau_{yx} = \frac{F(\text{in x}-\text{direction})}{A(\text{in direction normal to y})} \qquad \left(\frac{dyne}{L^2}\right) \qquad [9\text{-}79]$$

Hence, the **viscosity**, η, is defined as the ratio of the shear stress to the shear rate.

$$\eta = \frac{\tau}{\dot{\gamma}} \qquad \left(dyne \cdot s \cdot cm^{-2}\right) \qquad [9\text{-}80]$$

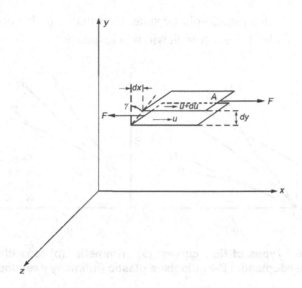

Figure 9-17. Definition of shear stress and shear rate.

Actually, viscosity can have the dimension as

$$dyne \cdot s \cdot cm^{-2} = P\,(\text{Poise})$$

$$= (1/10)\,Pa \cdot s \qquad\qquad [9\text{-}81]$$

$$= (1/10)\,N \cdot s \cdot m^{-2}$$

When Fig. 9-16 is used as a flow curve, the linear relation states Newton's flow. In reality,

$$\tau = \eta\dot{\gamma} \qquad\qquad [9\text{-}82]$$

For an arithmetic flow curve, where τ is plotted against $\dot{\gamma}$, a Newtonian fluid is characterized by a straight line through the origin with a slope of η (See Figure 9-18). However many liquids do not obey Newton's hypothesis. Both **dilatant** (shear-thickening) and **pseudoplastic** (shear-thinning) are observed with their slopes greater than or smaller than one, especially when the logarithmic plot is shown.

Molecular structure is found to be closely related to the range of shearing. Under low shearing field, the polymer is the most random and lightly entangled state as shown in Figure 9-19. As the shear is increased, entanglement and alignment can reach in a pseudo-plastic state. Eventually as shearing is increased, chain uncoiling will lead to a lower Newtonian response.

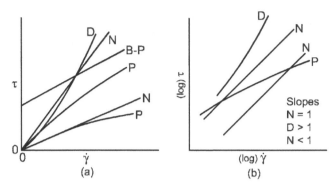

Figure 9-18. Types of flow curves: (a) arithmetic; (b) logarithmic. N. Newtonian; P, pseudoplastic; B-P, Bingham plastic (infinitely pseudoplastic); D, dilatant.

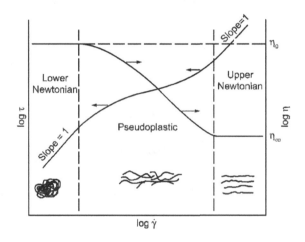

Figure 9-19. Generalized flow properties for polymer melts and solutions.

9.4.2 Linear Viscoelasticity

Single Spring

For viscoelastic response, it is better to use mechanical models for illustration. Figure 9-20(a) is used for linear elastic or **Hookean** solid which is illustrated by a spring.

$$\tau = G\gamma \qquad\qquad [9\text{-}83]$$

G here is defined as the constant **shear modulus** of stress, τ and strain, γ.

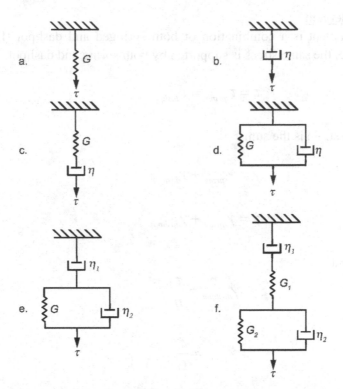

Figure 9-20. Linear viscoelastic models (spring and dashpot): (a) linear elastic; (b) linear viscous; (c) Maxwell element; (d) Voigt-Kelvin element; (e) three-parameter; (f) four-parameter.

Single Dashpot

Similarly, a linear viscous or Newtonian fluid is illustrated by a dashpot (Fig. 9-20b).

$$\tau = \eta \dot{\gamma} \qquad\qquad [9\text{-}84]$$

Note: This is for **shear deformation**. Notation is equally useful to **determine tensile deformation** as follows:

$$\text{tensile stress} = \sigma, \text{ shear stress} = \tau$$
$$\text{tensile strain} = \varepsilon, \text{ shear strain} = \gamma \qquad\qquad [9\text{-}85]$$
$$\text{elongational viscosity} = \eta_e, \text{ shear viscosity} = \eta$$

Maxwell Element

Maxwell element is a combination of both springed and dashpot (Fig. 9-20c). For example, the same stress is supported by both spring and dashpot

$$\tau = \tau_{spring} = \tau_{dashpot} \qquad [9\text{-}86]$$

but the total strain is the sum

$$\gamma = \gamma_{spring} + \gamma_{dashpot} \qquad [9\text{-}87]$$

Similarly,

$$\dot{\gamma} = \dot{\gamma}_{spring} + \dot{\gamma}_{dashpot} \qquad [9\text{-}88]$$

Realizing that

$$\dot{\gamma}_{dashpot} = \frac{\tau}{\eta} \qquad [9\text{-}89]$$

$$\dot{\gamma}_{spring} = \frac{\dot{\tau}}{G} \qquad [9\text{-}90]$$

Substituting back to Eq. 9-88 and after rearranging,

$$\tau = \eta\dot{\gamma} - \frac{\eta}{G}\dot{\tau} \qquad [9\text{-}91]$$

$$= \eta\dot{\gamma} - \lambda\dot{\tau}$$

Thus,

$$\lambda = \frac{\eta}{G} \qquad [9\text{-}92]$$

This quantity has the dimension of time and is known as the **relaxation time**.

A good example of Maxwell element is **creep testing** (Fig. 9-21).

$$\gamma(t) = \frac{\tau_0}{G} + \frac{\tau_0}{\eta t} \qquad [9\text{-}93]$$

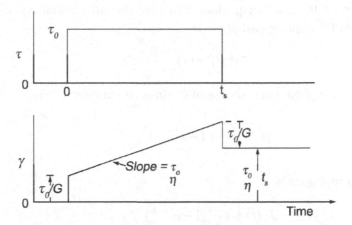

Figure 9-21. Creep response of Maxwell element.

Furthermore, a **creep compliance** $J_c(t)$ is defined as

$$J_c(t) = \frac{\gamma(t)}{J_0} = G^{-1} + \eta^{-1} \qquad [9\text{-}94]$$

For Maxwell element, the stress undergoes a first-order exponential decay.

$$\tau(t) = G\gamma_0 e^{-t/\lambda} \qquad [9\text{-}95]$$

In terms of **relaxation modulus**

$$G_r(t) = \frac{\tau(t)}{\gamma_0} = Ge^{-t/\lambda}. \qquad [9\text{-}96]$$

λ here is the relaxation time, which is a constant of the exponential decay.

Voigt-Kelvin Element

It follows that for Voigt-Kelvin model spring and dashpot are parallel (Fig. 9-20d)

$$\gamma = \gamma_{spring} = \gamma_{dashpot} \qquad [9\text{-}97]$$

$$\tau = \tau_{spring} + \tau_{dashpot} \qquad [9\text{-}98]$$

Combination of the above equations will give the differential equation for the deformation of the spring dashpot

$$\tau = \mu\dot{\gamma} + G\gamma \qquad\qquad [9\text{-}99]$$

When stress is applied, the response of strain is an exponential rise.

$$\gamma(t) = \frac{\tau_0}{G}(1 - e^{-t/\lambda}) \qquad\qquad [9\text{-}100]$$

The creep compliance is

$$J_c(t) = G^{-1}(1 - e^{-t/\lambda}) \qquad\qquad [9\text{-}101]$$

If the stress is removed, then the strain decays exponentially as

$$\gamma(t) = \frac{J}{G}e^{-t/\lambda} \qquad\qquad [9\text{-}102]$$

Three-Parameter model

The three parameter model (Fig. 9-20e) is a dashpot in series with a Voigt-Kelvin model. The differential equation can be expressed as

$$(1 + \lambda_1\frac{d}{dt})\tau = \eta_1(1 + \lambda_1\frac{d}{dt})\dot{\gamma} \qquad\qquad [9\text{-}103]$$

Where

$$\lambda_1 = \frac{(\eta_1 + \eta_2)}{G}, \ \lambda_2 = \frac{\eta_2}{G}$$

Four-Parameter model

Accordingly the four parameter model (Fig. 9-20f) is a series combination of a Maxwell element with a Voigt-Kelvin element. The differential equation can be

$$\ddot{\tau} + \left(\frac{G_1}{\eta_1} + \frac{G_1}{\eta_2} + \frac{G_2}{\eta_2}\right)\dot{\tau} + \left(\frac{G_1 G_2}{\eta_1 \eta_2}\right)\tau = G_1\ddot{\gamma} + \frac{G_1 G_2}{\eta_2}\dot{\gamma} \qquad\qquad [9\text{-}104]$$

The creep response and the resulting compliance is the sum of the creep response of the Maxwell and Vogit-Kelvin elements.

$$J_c(t) = G_1^{-1} + t\eta_1^{-1} + G_2^{-1}\left(1 - e^{-G_2t/\eta_2}\right)$$ [9-105]

In terms of molecular structures, the real situations can be drawn
- Dashpot 1 represents molecular slip and transnational motion.
- Spring 1 represents the elastic straining of bond angles and lengths Value of G_1, characterize the resistance from equilibrium value.
- Dashpot 2 represents the resistance of polymer chains to the uncoiling and coiling caused by chain entanglements (V-K)
- Spring 2 represents the restoring force by thermal agitation (Increasing in crosslinking)

Generalized Maxwell model
For a generalized Maxwell model G_1, G_2, ... G_n, and η_1, η_2, ...η_n, (Fig. 9-22a) the modulus of relaxation is

$$G_r(t) = \int_0^\infty G(\lambda)e^{-t/\lambda}d\lambda$$ [9-106]

where

$$\frac{\eta_i}{G_i} = \lambda_i$$

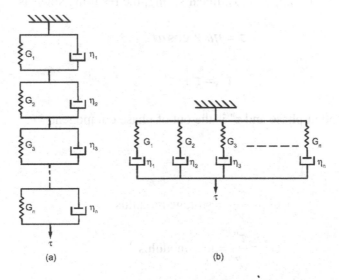

Figure 9-22. Generalized models. (a) Voigt-Kelvin; (b) Maxwell.

Generalized Voigt-Kelvin model

And for the generalized Voigt-Kelvin (Fig. 9-22b), G_1, G_2, ... G_n and η_1, η_2 ... η_n, the creep compliance is

$$J_c(t) = \int_0^\infty J(\lambda)\left(1 - e^{-t/\lambda}\right) d\lambda \qquad [9\text{-}107]$$

Where

$$\lambda_i = \frac{\eta_i}{G_i}$$

9.4.3 Dynamic Mechanical Testing

When viscoelastic material is tested in a sinusoidal varying stress and strain, it best interpreted by the complex notation of two vectors, τ^* and γ^* separated by a phase angle δ. A sinusoidal strain is shown in Fig. 9-23.

$$\gamma = \gamma' \sin \omega t \qquad [9\text{-}108]$$

Where ω is the angular frequency in radians/sec.

When this strain is applied to a linear spring, the resulting stress is

$$\tau = \eta \omega \gamma' \cos \omega t \qquad [9\text{-}109]$$

$$\tau^* = \tau' + i\tau'' \qquad [9\text{-}110]$$

where τ' is the in-phase and τ'' is the out of phase component.

Thus,

$$G' = \frac{\tau'}{\gamma'} = \text{storage modulus} \qquad [9\text{-}111]$$

$$G'' = \frac{\tau''}{\gamma} = \text{loss modulus} \qquad [9\text{-}112]$$

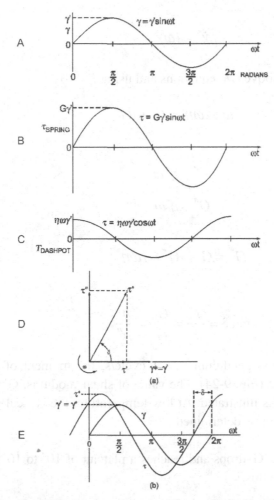

Figure 9-23. (A–C) Stress in a linear spring and dashpot in response to a sinusoidal applied strain, Quantities in dynamic testing: (D) rotating vector diagram; (E) stress and strain.

and G^* is the complex modulus

$$G^* = G' + iG'' = \frac{\tau' + i\tau''}{\gamma'} = \frac{\tau^*}{\gamma^*} \qquad [9\text{-}113]$$

and so is

$$\eta^* = \eta' + i\eta'' = \frac{\tau^*}{\gamma^*} \qquad [9\text{-}114]$$

and

$$\dot{\gamma}^* = i\omega\gamma^*$$ [9-115]

By combining the above two equations and using $i^2 = -1$,

$$\eta''\omega + i\omega\eta' = G' + iG''$$ [9-116]

Thus,

$$G' = \eta''\omega$$

$$G'' = \eta'\omega$$

Furthermore,

$$G^* = G' + iG'' = i\omega\eta^*$$ [9-117]

The loss tangent

$$\tan\delta = \frac{J''}{J'} = \frac{G''}{G'} = \frac{\eta'}{\eta''}$$ [9-118]

When using a torsion pendulum as a 1 cycle/s, the moment of polymethlyme-thacrylate was made (Fig. 9-24). The value of shear modulus, G' and $\approx$ damping $\pi\tan\delta$ vs. T ($^\circ$C) was illustrated. At low temperature, the typical glassy modulus of 10^{10} to 10^{11} dyn/cm^2 is determined.

Close to 110-130°C, G$^{'}$ drops and reaches a plateau of 10^6 to 10^7 dynes/cm^2 of a typical rubbery modulus.

9.4.4 The William-Landel-Ferry (WLF) Equation and Time-Temperature Superposition

There is a dimensionless **Deborah number**, which indicates the ratio of the characteristic material time to the time of deformation.

$$D_e = \frac{\lambda_c}{t_s}$$ [9-119]

If De $\gg$ 1, material will appear elastic. If $D_e \rightarrow 0$ or t_s is large, then material is viscous.

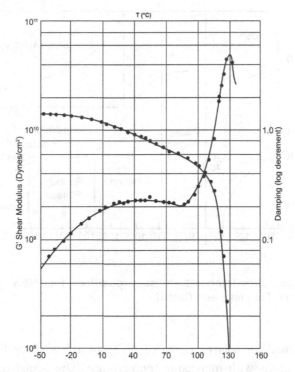

Figure 9-24. Dynamic mechanical properties of polymethyl methacrylate. The data were obtained with a torsion pendulum at about 1 cycle/sec.

In another way,

$$D_e = \lambda_c \omega \qquad [9\text{-}120]$$

Here polymer's characteristic time is a function of time. D_e number can be doubled by halving t_s or doubling ω in a dynamic test or by lowering the temperature enough to double the λ_c.

This **time-temperature superposition** is best illustrated by a tensile stress relaxation at various temperatures. For example, in Fig. 9-24 data for polyisobutylene are plotted in the form of a time dependent Young's modules tensile, $Er(t)$ vs. time on a log-log scale.

$$Er(t) = \frac{\sigma(t)}{\varepsilon_0} = \frac{f(t)/A}{\Delta l/l} \qquad [9\text{-}121]$$

When f(t) is the measured tensile force in the sample held at constant strain $\varepsilon_0 = \Delta l/l$ and A is the cross-sectional area.

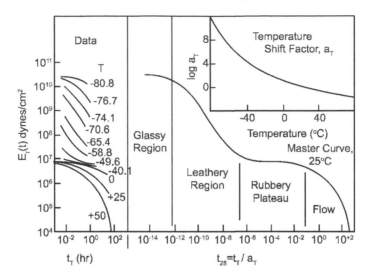

Figure 9-25. Time-temperature superposition for NBS polyisobutylene. Adapted from Tobolsky and Catsiff.

Data were obtained for a large range of tie from a few seconds to a couple of days. Modulus drops with increasing temperature. The actual data can be arbitrarily shifted to form a continuous curve. Using 25°C, a continuous curve is constructed with a shift factor, a_T along the X-axis as shown in Fig. 9-25.

$$a_T = \frac{t_T}{t_{T_0}} \quad \text{(for same response)} \qquad [9\text{-}122]$$

The master curve now represents stress relaxation at 25°C over 17 decades of time and also can be applicable to the prediction in creep or dynamic testing. If the polymer's glass – transition temperature is chosen as the reference temperature, the shift factors are given by the **William-Landel-Ferry (WLF) equation** in the range of

$$T_g < T < T_g + 100^o C$$

$$log\, a_T = \frac{-C_1(T-T^*)}{C_2+(T-T^*)} \quad \text{for } T_0 = T^* = T_g \qquad [9\text{-}123]$$

The "universal" constants C_1=17.44 and C_2=51.6 (with temperatures in K) give a rough fit for a wide variety of polymers.

[**Example 9-4**] The damping peak for polymethyleneacrylate is located at 130°C. Data were obtained at a frequency of 1 cycle/second. At what temperature would the peak be located if measurements were made at 1000 cycle/s. T_g for this polymer is 105°C.

$$a_T = \frac{\omega_{T_0}}{\omega_T}, \text{ since frequency is a reciprocal of rate. (See Fig. 9-24)}$$

using universal constants,

$$\log a_T = \log \frac{\omega_{T_g}}{\omega_T} = \frac{-17.44(T - T_g)}{51.6 + (T - T_g)}$$

$$= -5.69$$

$$\frac{\omega_{105°C}}{\omega_{130°C}} = 2.03 \times 10^{-6}$$

$$\omega_{105°C} = (1cps)(2.03 \times 10^{-6}) = 2.03 \times 10^{-6}$$

Now shifting from T_g to T,

$$\log \frac{2.03 \times 10^{-6}}{1000} = -8.69 = \frac{-17.44(T - 105)}{51.6 + (T - 105)}$$

Solving,

$$T = 156° C$$

REFERENCES

9-1.   Flory, Paul J., *Principles of Polymer Chemistry,* Cornell University Press, Ithaca, New York, 1953.

9-2.   Fried, Joel R., *Polymer Science and Technology*, 2nd ed., Prentice Hall, Upper Saddle River, NJ, 2003.

9-3.   R. Khachatoorian, I. G. Petrisor, and T. F. Yen, *Prediction of Plugging Effect of Biopolymers Using Their Glass Transition Temperatures, Petroleum Science and Engineering* **41** (4), 243-251 (2004).

9-4.   Kumar, Anil and Rakesh K. Gupta, *Fundamentals of Polymer Engineering,* Second Ed., Marcel Dekker, Inc., New York, 2003.

9-5.   Ravve, A., *Principles of Polymer Chemistry,* Second Ed., Kluwer Academic / Plenum Publishers, New York, 2000.

9-6.   Rosen, S. L. , *Fundamental Principals of Polymeric Materials*, 2nd ed., John Wiley and Sons, New York, 1993.

9-7.   Stevens, Malcolm P., *Polymer Chemistry: An Introduction*, Second Ed., Oxford University Press, New York 1990.

9-8.   Wei, W., C. A. Bennett, et al., "Detailed Kinetic Models for Catalytic Reforming," *Prepr. ACS Div Petr. Chem* 2005 **50** (4) 386-387.

9-9.   Van Krevelen, D.W., *Properties of Polymers:Their Correlation with Chemical Structure; Their Numerical Estimation and Prediction from Additive Group Contributions,* 3rd Ed., Elsevier, Amsterdam, 1990.

9-10. Doi, Y. *Microbial Polyesters,* VCH Publishers, New York, 1990.

PROBLEM SET

1. Determine the storage moduli at 200 °C and 1 rad/sec of a polystyrene high MW fraction, using the time-temperature superposition ($\sim 2.5 \times 10^4$ dyne/cm^2) as below:

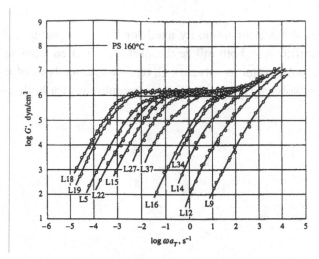

Master curve of G' for a different MW fraction of polystyrene at reference temperature of 160 °C low MW L<12, intermediate MW 14<L<22, high MW L>27. Data from S.Onogi, Macromolecules 3, 109 (1970)

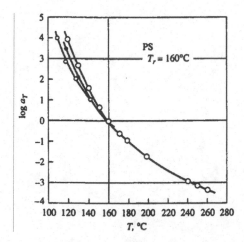

The logarithm of the shift fact a$_T$ plotted against the temperature for polystyrene. Open circles for high MW, closed circles for intermediate MW and small open circles for low MW. Data from S. Onogi, Macromolecules 3, 109 (1970)

2. Biopolymers such as Xanthan, Guar and Xanthan-Guar mixture can crosslink with borax to form a gel. If the network modules after gel formation are determined as 1.784, 0.3982 and 7.759 Pa, can you predict which one of those are densely crosslinked? Also rank them. (After Hsuan Jing Lai, 2005).

    Hint: $G_N = C \, (N_A/M_C) \, k_B \, T$

3. Polymer membranes are commonly used for barrier separation. Reverse osmosis and ultrafiltration both utilize the pressure gradient causing separation of solution (usually water as solvent). Give a sample analysis of transport salt (species 1) and solvent (species 2) through membranes for both cases.

CHAPTER 10

CEMENT CHEMISTRY

Chemistry of constructive materials remains one of the most important aspects in civil engineering. Portland cement, the admixtures for special use of cement, and the resulting concrete should be examined and studied via scientific methods. The mechanisms of various stages of intermediates discovered from studying raw chemicals are worthy of exploration from a chemistry viewpoint. The basic instrument for studying the structure of concrete has been explored. Chemicals for admixture for concrete improvement have been analyzed. Also, different concrete problems and issues have been evaluated based on chemistry.

10.1 PORTLAND CEMENT AND ITS CONSTITUENT PHASES

The ordinary **Portland cement** is prepared in a two-step process. The first step is the high-temperature mixing and processing of limestone, sand, and clay starting materials to produce a cement powder. Heating of the starting materials results in **clinkers** (or cement precursors) after cooling and release of CO_2 and H_2O. The second step involves the **hydration**, mixing, and setting of the powder into a final cement product (**concrete**).

Generally, when a mixture of limestone, sand, and clay is heated to 1450°C, partial fusion occurs and nodules of clinker are produced. The clinker is mixed with small amounts of gypsum and is finely ground to produce the so-called cement. The composition is usually 67% CaO, 22% SiO_2, 5% Al_2O_3, 3% Fe_2O_3, and is balanced by other compounds.

A special nomenclature is used in this field of cement chemistry. The appropriate abbreviations are as follows:

$$C = CaO$$
$$S = SiO_2$$
$$A = Al_2O_3$$
$$F = Fe_2O_3$$
$$M = MgO$$
$$K = K_2O$$
$$N = Na_2O$$
$$T = TiO_2$$
$$P = P_2O_5$$
$$H = H_2O$$

There are two overhead bars to indicate gaseous species:

$$\overline{C} = CO_2$$
$$\overline{S} = SO_2$$

For example,

$$C_3S = 3\,CaO \cdot SiO_2 = Ca_3SiO_3$$

Commonly used abbreviations are listed in Table 10-1.

Table 10-1. Common Cement Component Names, Compositions, Formulae, and Abbreviations (Maclaren and White, 2003)

Component Name	Composition	Empirical Formula	Abbreviation
Calcium oxide (lime)	CaO	CaO	'C'
Silicon dioxide (silica)	SiO_2	SiO_2	'S'
Aluminum oxide	Al_2O_3	$AlsO_3$	'A'
Iron (III) oxide	Fe_2O_3	Fe_2O_3	'F'
Dicalcium silicate	$2CaO.SiO_2$	$CaSiO_4$	'C2S'
Tricalcium silicate	$3CaO.SiO_2$	Ca_3SiO_5	'C3S'
Tricalcium aluminate	$3CaO.Al_2O_3$	$Ca_3Al_2O_6$	'C3A'
Tetracalcium aluminoferrate (Brown millerite)	$4CaO.Al_2O_3.Fe_2O_3$	$Ca_4Al_2Fe_2O_{10}$	'C4AF'
Caclium silicate hydrate gel	$(CaO)_x.SiO_2.yH_2O$ with x < 1.5 in solid solution with $Ca(OH_2)$	(variable)	'CSH'
Calcium silicate (Wollastonite)	$CaO.SiO_2$	$CaSiO_3$	'CS'
Calcium silicate (Rankinite)	$3CaO.2SiO_2$	Ca_3SiO_7	'C3S2'
Calcium aluminium silicate (Gehlenite)	$2CaO.Al_2O_3.SiO_2$	$Ca_2Al_2SiO_7$	'C2AS'
Aluminium silicate (Mullite)	$3Al_2O_3.2SiO_2$	$Al_6Si_2O_{13}$	'A3S2'
Calcium aluminium silicate (Anorthite)	$CaO.Al_2O_3.2SiO_2$	$CaAl_2Si_2O_8$	'CAS2'
Aluminium silicate	$2Al_2O_3.2SiO_2$	$Al_4Si_2O_{10}$	'A2S2'
Calcium aluminate	$CaO.Al_2O_3$	$CaAl_2O_4$	'CA'
Calcium dialuminate	$CaO.2Al_2O_3$	$CaAl_4O_7$	'CA2'
Dodecacalcium septaluminate	$12CaO.7Al_2O_3$	$Ca_{12}Al_{14}O_{33}$	'C12A7'
Calcium hexaliminate	$CaO.6Al_2O_3$	$CaAl_{12}O_{19}$	'CA6'

Normally, Portland cement clinker contains four major phases as listed below as well as some other minor phases:

- **Alite**

About 50-70% of Portland cement clinker is C_3S. It also undergoes a series of reversible transitions, which are detected by differential thermal analysis (DTA) and X-ray diffraction (XRD).

$$T1 \underset{}{\overset{620}{\rightleftharpoons}} T2 \underset{}{\overset{929}{\rightleftharpoons}} T3 \underset{}{\overset{980}{\rightleftharpoons}} M1 \underset{}{\overset{990}{\rightleftharpoons}} M2 \underset{}{\overset{1000}{\rightleftharpoons}} M3 \underset{}{\overset{1070}{\rightleftharpoons}} R$$

(all numbers are °C, T=triclinic, M=monoclinic, R=rhombohedral)

It is possible to evaluate the different crystalline transitions by XRD. A good example is illustrated in Figure 10-1. For investigation, both high temperature XRD and high temperature light microscopy are indispensable.

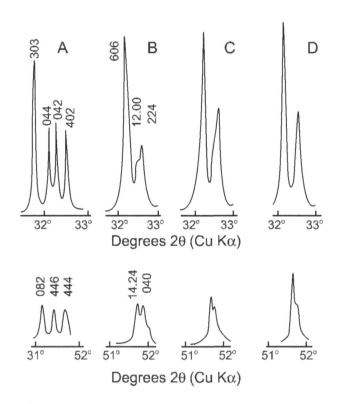

Figure 10-1. Portions of XRD powder patterns of (A) the T_1 modification of C_3S at 605° C; (B), (C) and (D) clinkers containing, respectively, M_3, ($M_3 + M_1$), and M_1 alites. Indexing of T_1 and M_3 patterns is based on the axes and calculated intensities. (After H. F. W. Taylor). The lower half is the continuation of the upper portion of the graph for (A) through (D).

- **Belite**

    About 15-30% of Portland cement clinker is C_2S. It has five polymorphs at ordinary pressures:

$$\alpha \underset{\longleftarrow}{\overset{1425}{\longrightarrow}} \alpha'_H \underset{\longleftarrow}{\overset{1160}{\longrightarrow}} \alpha'_C \underset{\longleftarrow}{\overset{630\text{-}680}{\longrightarrow}} \beta \underset{\longleftarrow}{\overset{<500}{\longrightarrow}} \gamma$$

$$780\text{-}860$$

(numbers indicate °C, H=high, L=low)

- **Aluminate phase**

    The aluminate phase constitutes 5-10% of most Portland cement clinkers. This phase can be represented by C_3A and does not exhibit polymorphism. However, C_3A can incorporate Na^+ by gradual substitution of Ca^{2+} into other structures, as shown in Table 10-2.

Table 10-2. Modifications of the C_3A Structure of General Formula $Na_{2x}Ca_{3-x}Al_2O_6$

Approximate Na_2O (%)	Compositional range, x	Designation	Crystal system	Space group
0-1.0	0-0.04	C_I	Cubic	Pa3
1.0-2.4	0.04-0.10	C_{II}	Cubic	$P2_13$
2.4-3.7	0.10-0.16	$C_{II} + O$	-	-
3.7-4.6	0.16-0.20	O	Orthorhombic	Pbca
4.6-5.7	0.20-0.25	M	Monoclinic	$P2_1/a$

- **Ferrite phase**

    Usually the ferrite phase makes up 5-15% of the Portland cement clinkers. It is essentially C_4AF and can be modified by Al/Fe ratio and in corporation of foreign ions such as Mn^{3+}, Mg^{2+}, Ti^{4+}, and so forth.

These four phases are the major constituent of the Portland cement clinker. Occasionally due are minor phases such as alkali sulfates and calcium oxide.

10.1.1 High Temperature Phase Diagram for the Clinkers

The composition of two or three component systems where reaction takes place at high temperature is best to study under the change of polymorphisms as a function of temperature. The following phase diagram illustrates this principle. For the CaO-SiO_2 system and for the CaO-Al_2O_3 system, the following two-phase diagrams in Figures 10-2 and 10-3 demonstrate good behavior. Notice the formation of various compounds in different compositions at different temperature ranges.

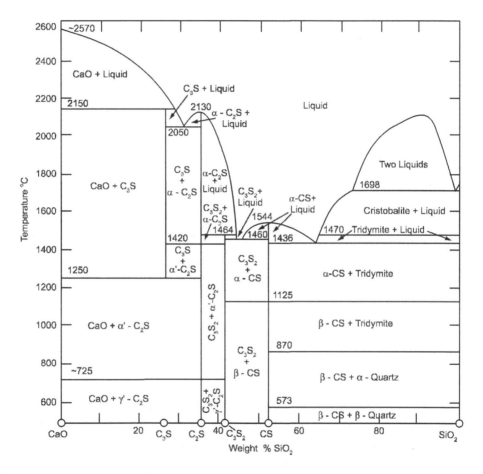

Figure 10-2. The system CaO-SiO_2. (After Day et al. with later modifications)

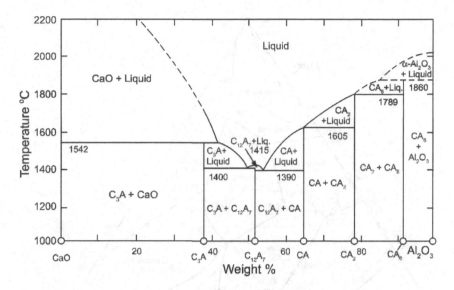

Figure 10-3. The system CaO-Al$_2$O$_3$ modified by the presence of small amounts of H$_2$O and O$_2$, and thus representing the behavior in air of ordinary humidity. (After Rankin and Wright with later modifications).

The phase diagram of CaO-Al$_2$O$_3$-SiO$_2$ system is very important and can best be illustrated with a tertiary diagram as in Figure 10-4. The phase field of C$_3$S, C$_2$S can be visualized. A portion of the tertiary diagram is important to depict the formation of cement clinker. For example, parts X and P are boundaries for the 1500°C isotherm (Fig. 10-5).

Minor phases such as the quaternary phase CaO-C$_2$S-C$_{12}$A$_7$-C$_4$AF can also be evaluated under a diagram such as that shown in Fig. 10-6.

The raw materials for manufacturing Portland cement clinker are generally utilized by blending a calcareous material, typically limestone, with a small amount of an argillaceous one, typically clay or shale. These are intimately mixed and heated to a temperature of 1450°C. Reaction below about 1300°C is the decomposition of calcite, also called calcing, and the decomposition of clay materials:

$$CaCO_3 (\text{calcite}) \rightarrow CaO + CO_2(g), \quad \Delta H = \frac{1782 kJ}{kg} \qquad [10\text{-}1]$$

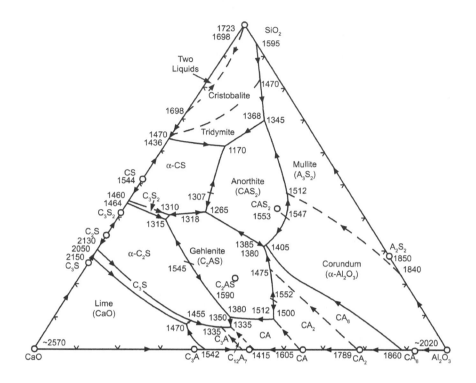

Figure 10-4. The system CaO-Al₂O₃-SiO₂. For the detailed CaO corner, see Fig. 10-5. (After Muan and Osborn, with later modifications).

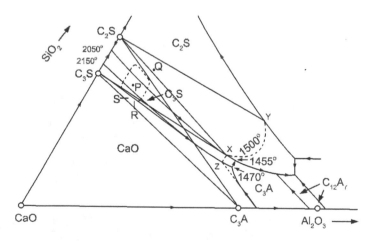

Figure 10-5. Part of the system CaO-Al₂O₃-SiO₂ illustrating the formation of Portland cement clinker. (This portion is from Fig. 10-4).

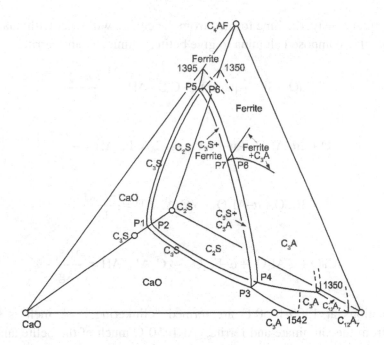

Figure 10-6. The pseudosystem CaO-C₂S-C₁₂A₇-C₄AF, showing the primary phase volume of C₃S. (After Lea and Parker, with later modifications).

$$AS_4 (\text{pyrophyllite}) \rightarrow \alpha-A\ell_2O_3 + 4SiO_2(\text{quartz}) + H_2O(g), \quad \Delta H = \frac{224\text{kJ}}{\text{kg}} \quad [10\text{-}2]$$

$$AS_2H_2 (\text{kaolinite}) \rightarrow \alpha-Al_2O_3 + 2SiO_2(\text{quartz}) + 2H_2O(g),$$

$$\Delta H = \frac{538\text{kJ}}{\text{kg}} \quad [10\text{-}3]$$

$2FeO \cdot OH$ (goethite)

$$\text{or} \quad FH \rightarrow \alpha-Fe_2O_3 + H_2O(g), \quad \Delta H = \frac{254\text{kJ}}{\text{kg}} \quad [10\text{-}4]$$

$$2CaO + SiO_2(\text{quartz}) \rightarrow \beta-C_2S, \quad \Delta H = \frac{-734\text{kJ}}{\text{kg}} \quad [10\text{-}5]$$

Simultaneously, the lime formed from the calcite will react with quartz and clay mineral decomposed alumina to give belite, aluminate, and ferrite.

$$3CaO + SiO_2(quartz) \rightarrow C_3S, \ \Delta H = \frac{-495kJ}{kg}$$
                                                                                        [10-6]

$$6CaO + 2\alpha\text{--}Al_2O_3 + \alpha\text{--}Fe_2O_3 \rightarrow C_6A_2F, \ \Delta H = \frac{-157kJ}{kg}$$
                                                                                        [10-7]

$$3CaO + \alpha\text{--}Al_2O_3 \rightarrow C_3A, \ \Delta H = \frac{-27kJ}{kg}$$
                                                                                        [10-8]

$$4CaO + \alpha \ Al_2O_3 + \alpha\text{--}Fe_2O_3 \rightarrow C_4AF, \ \Delta H = \frac{-105kJ}{kg}$$
                                                                                        [10-9]

Reactions at 1300-1450°C are termed "clinkering."   A melt is formed, usually from the aluminate and ferrite. At 1450°C much of the belite and nearly all of the lime react to form alite. The resulting material modulizes after cooling the remaining liquid aluminate and ferrite.   The overall enthalpy changes in forming the clinker and is dominated by the strong endothermic decomposition of calcite.   The theoretical heat required to produce 1 kg of clinker is about 1750kJ. Additional heat is required for the entire process.

The total process of cement-clinker formation is summarized in Figure 10-7, which shows the main component as a function of temperature. Limestone, quartz, clay raw materials, and water are combined and heated.   As the temperature rises, the gas components are lost to form a calcium aluminate phase (erringite). One of the reasons to add minor components is that they help to flux the system to a lower temperature.   Typical contents of phase in a Portland cement clinker are shown in Figure 10-8. Table 10-3 illustrates that small amounts of outer minerals can lower the temperature of the liquid phase.

In cement chemistry, there are a few parameters that can characterize the clinker formation.

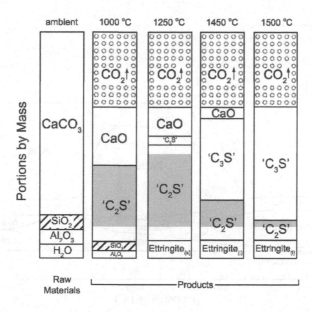

Figure 10-7. A schematic view of the components of cement-clinker formation, their reactions, and the products formed as the temperature of the mixture is raised. Calcium carbonate decomposes to form calcium oxide and carbon dioxide. Calcium oxide reacts with silica to form dicalcium silicate at temperatures below 1250°C, which converts to tricalcium silicate at temperatures above 1250°C. Formation of a liquid aluminate, Ettringite, phased at about 1450°C facilitates the conversion of dicalcium silicate to tricalcium silicate. (Jackson P.J., Chemistry of Cement and Concrete, 1998)

Table 10-3. Clinker Components in the 'C$_3$S' Cement Phase Field and Their temperatures of Liquid Formation (After Bogue, 1947)

Components	Temperature of Liquid Formation (°C)
CaO-SiO$_2$	2065
CaO-SiO$_2$-Al$_2$O$_3$	1455
CaO-SiO$_2$-Al$_2$O$_3$-Na$_2$O	1430
CaO-SiO$_2$-Al$_2$O$_3$-MgO	1375
CaO-SiO$_2$-Al$_2$O$_3$-Fe$_2$O$_3$	1450
CaO-SiO$_2$-Al$_2$O$_3$-Na$_2$O-MgO	1365
CaO-SiO$_2$-Al$_2$O$_3$-Na$_2$O-Fe$_2$O$_3$	1315
CaO-SiO$_2$-Al$_2$O$_3$-MgO-Fe$_2$O$_3$	1300
CaO-SiO$_2$-Al$_2$O$_3$-Na$_2$O-MgO-Fe$_2$O$_3$	1280

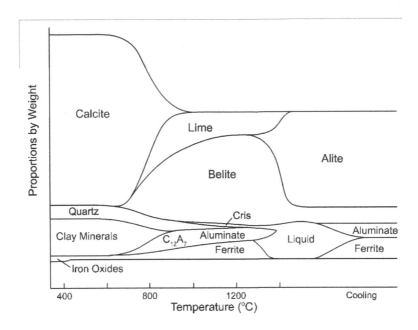

Figure 10-8. Schematic diagram showing the variations in typical contents of phases during the formation of Portland cement clinker. (Loosely based on a figure by Wolter).

$$\textbf{Lime Saturation Factor (LSF)} = \frac{CaO}{\left(2.8SiO_2 + 1.2Al_2O_3 + 0.65Fe_2O_3\right)} \qquad [10\text{-}10]$$

$$\textbf{Silica Ratio (SR)} = \frac{SiO_2}{\left(Al_2O_3 + Fe_2O_3\right)} \qquad [10\text{-}11]$$

$$\textbf{Alumina Ratio (AR)} = \frac{Al_2O_3}{Fe_2O_3} \qquad [10\text{-}12]$$

For all of the above equations, the chemical formulae denote the weight percent in the cement. For LSF, typical values are approximately 0.92 – 0.98. The SR and HR are referred to as silica modules and alumina modules, respectively. Usually $SR \cong 2.0 - 3.0$, and $AR \cong 1.0 - 4.0$. An approximate method for potential phase composition is termed **Bogue** calculation. The calculation is as follows:

- Assume that the compositions of the four major phases are C_3S, C_2S, C_3A, and C_4AF.
- Assume that the Fe_2O_3 occurs as C_4AF.
- Assume that the remaining Al_2O_3 occurs as C_3A.
- Deduct from the CaO content the amounts attributed to C_4AF, C_3A, and free lime, and solve two simultaneous equations to obtain C_3S and C_2S.

Actually, the free lime can be determined by chemical titration; therefore,

$$C_4AF = 3.0432\,Fe_2O_3 \qquad [10\text{-}13]$$

$$C_3A = 2.6504\,Al_2O_3 - 1.6920\,Fe_2O_3 \qquad [10\text{-}14]$$

$$C_2S = -3.0710\,CaO + 8.6024\,SiO_2 + 5.0683\,Al_2O_3 + 1.0785\,Fe_2O_3$$
$$= 2.8675\,SiO_2 - 0.7544\,C_3S \qquad [10\text{-}15]$$

$$C_3S = -4.0710\,CaO + 7.6024\,SiO_2 - 6.7187\,Al_2O_3 - 1.4297\,Fe_2O_3 \qquad [10\text{-}16]$$

[Example 10-1] Free lime in a cement clinker is determined at 50%. Assuming the LSF is 0.90 and SR and AR are respectively 5.0 and 2.0, evaluate the composition of the phases for the clinker.

From Eqs. [10-10] to [10-12]:

C = 50
A / F = 2
A = 2F and S(A + F) = 5
S = 7.5A
LSF = 0.90 = 50 / (2.8S + 1.2A + 0.65F)

Solving A = 2.466
$$F = 1.233$$
$$S = 18.49$$

Using Bogue calculation by Eqs. [10-13] to [10-16]

$C_4AF = 4.360\%$ = ferrite phase

$C_3A = 4.111\%$ = aluminate phase

$C_2S = 19.56\%$ = belite

$C_3S = 44.36\%$ = alite

10.1.2 Hydration of Cement

When anhydrous cement or its constituent phases are mixed with water, chemical reactions take place into the corresponding hydrates. A mixture of water and cement in such proportions that setting and hardening occurs is termed a paste. For a paste, the water/cement (w/c) or water/solid weight ratio typically is 0.3-0.6. **Setting** refers to stiffening without development of compressive strength; **hardening**, on the other hand, is significant development of compressive strength. **Curing** usually means storage under conditions such that **hydration** occurs. The hydration of C_3S or C_2S often results in CH or CSH. CH is calcium hydroxide. CSH is $CaO{\bullet}SiO_2{\bullet}H_2O$. There is the expression C-S-H, which refers to the amorphous or poorly crystalline calcium silicate hydrate. The dash here indicates that no particular composition is implied.

The rate of hydration of C_3S in a Portland cement is shown in Fig. 10-9. Upon contact with water, C_3S undergoes an intense, short-lived reaction. In the preinduction period, the rate is as high as 5 day^{-1}. The dissolution of C_3S proceeds as

$$O^{2-}(\text{lattice}) + H^+ (aq) \rightarrow OH^- (aq) \qquad [10-17]$$

$$2OH^- (aq) + Ca^{2+} (aq) \leftrightarrow Ca(OH)_2 (aq) \qquad [10-18]$$

Also, silicate from C_3S lattice enters:

$$SiO_4^{4-} (\text{lattice}) + n\,H^+ (aq) \rightarrow H_nSiO_4^{(4-n)-} (aq) \qquad [10-19]$$

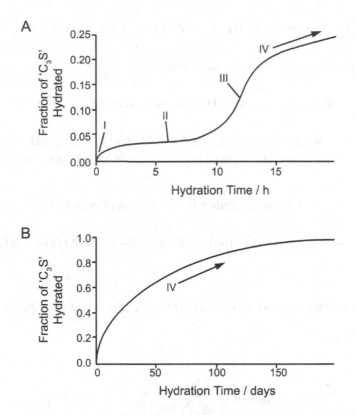

Figure 10-9. A graphical representation of the rate of consumption of tricalcium silicate ('C3S') as a function of hydration time: (A) changes in hydration rates in the first few hours as a result of (I) pre-induction, (II) induction, (III) acceleration, and (IV) develeration processes. (B) An expanded view showing the length of time required for complete cement hydration.

The dissolved compounds combine to form the calcium silicate hydrate CSH gel, and amorphous, two-compound solid solution.

$$2(3CaO \cdot SiO_2)_{(l)} + 6H_2O_{(l)} \rightarrow 3CaO \cdot 2SiO_2 \cdot 2SiO_{2(s)} + 3H_2O_{(s)} + 3Ca(OH)_{2(aq)} \quad [10\text{-}20]$$

Most cement powders have gypsum added. Gypsum acts to slow down the pre-induction period to avoid rapid setting of cement. It reacts with C_3A to form various phases which are collectively referred to as ettingite phases.

$$3CaO \cdot Al_2O_{3(s)} + 3CaSO_{4(s)} + 32H_2O_{(l)} \rightarrow 3CaO \cdot Al_2O_3 + 3CaSO_4 \cdot 32H_2O_{(s)} \quad [10\text{-}21]$$

$$3CaO \cdot Al_2O_{3(s)} + 3CaSO_{4(s)} + 12H_2O_{(l)} \rightarrow 3CaO \cdot Al_2O_3 + 3CaSO_4 \cdot 12H_2O_{(s)} \qquad [10\text{-}22]$$

C_3A and C_4A_I can also hydrate independently.

$$3CaO \cdot Al_2O_3\,(s) + 6H_2O\,(l) \rightarrow 3CaO \cdot Al_2O_3 \cdot 6H_2O\,(s) \qquad [10\text{-}23]$$

$$
\begin{aligned}
4CaO \cdot Al_2O_3 \cdot Fe_3O_3\,(s) \quad &\rightarrow \quad 3CaO \cdot Al_2O_3 \cdot 6H_2O\,(s) \\
+\, 2Ca(OH)_2 + 10H_2O\,(l) \quad & \qquad + 3CaO \cdot Fe_2O_3 \cdot 6H_2O\,(s)
\end{aligned}
$$
$$[10\text{-}24]$$

After acceleration stages, C_2S is also hydrated:

$$2(2CaO \cdot SiO_2)\,(s) + 4H_2O\,(aq) \rightarrow 3CaO \cdot 2SiO_2 \cdot 3H_2O\,(s) + Ca(OH)_2\,(aq)$$
$$[10\text{-}25]$$

A tentative phase diagram of C_1S and H can be illustrated as in Figure 10-10.

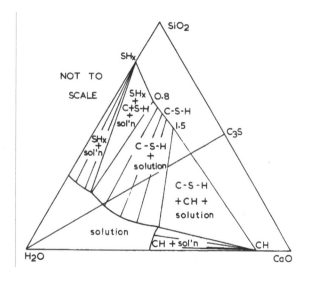

Figure 10-10. The system $CaO\text{-}SiO_2\text{-}H_2O$ at ordinary temperatures (schematic).

Actually, the C-S-H system is very complex. Table 10-4 will illustrate this point.

Table 10-4. Water/Calcium Ratios and Observed Densities (kg m⁻³) ᵃ

Material	H_2O/Ca	Density
C-S-H gel (saturated)	2.3	1850-1900
Plombierite (0.8 CaO•SiO₂•xH₂O; saturated)	2.5	2000-2200
C-S-H gel (11% RH)	1.2	2180
	?	2430-2450
Plombierite (11% RH)	1.64	?
Jennite	1.22	2320
1.4-nm tobermorite	1.80	2200
Tacharanite (Ca₁₂Al₂Si₁₈O₆₉H₃₆)	1.50	2360
C-S-H gel (110 C)	0.85	2600-2700
Plombierite (110 C)	1.07	?
Metajennite	0.78	2620
1.1-nm tobermorite	1.00	2440
Oyclite (Ca₂₀Si₁₆B₄O₈₃H₅₀)	1.25	2620

ᵃValues are in some cases calculated from data given. (After H.F.W. Taylor).
The hydrated aluminate, ferrite, and sulfate phases are more complex. The term AFm is not only $Al_2O_3-Fe_2O_3$-mono but denotes $C_3(A,F) \cdot CaX_2 \cdot yH_2O$ or $C_4(A,F)X_2 \cdot yH_2O$ where X represents a singly charged anion or half unit of a doubly charged anion. A suggested ground form of C-A-S-H system is sketched in Fig. 10-11.

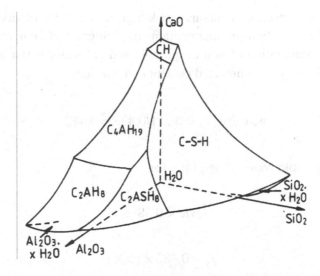

Figure 10-11. Suggested general form of the metastable equilibria in the CaO-Al₂O₃-SiO₂-H₂O system at ordinary temperature, showing solubility surfaces. Modified from Dron.

Reaction of C_3A with water in the presence of calcium sulfate will evolve intermediates before AF in phase. For example,

$$C_3A + 3C\overline{S}H_2 + 26H_2O \rightarrow C_6A\overline{S}_3H_{32}$$

$$2C_3A + C_6A\overline{S}_3H_{32} + 4H_2O \rightarrow 3C_4ASH_{12}$$

$$C_3A + CH + 12H_2O \rightarrow C_4AH_{13}$$

During hydration of cement, the development of the microstructure of C-S-H can be changed as a 'silicate garden' according to time (see Fig. 10-12). For complete hydration, a minimum **water to cement weight ratio** is required, usually expressed as w/c and w/c > 0.38. Here, w indicates the initial state. If w_e is the evaporable water and w_t is the total water added in cement,

$$w_e/c = w_t/c - 0.277$$

since the hydration product contains per kilogram 0.227 kg of non-evaporable water. Also, the hydration product contains per kilogram of cement 0.211 kg of gel water. The summation of non-evaporable and gel water is 0.438 kg. Pastes with w/c < 0.38 consist of unreacted cement and will have

$$(w_e/c)/(w_t/c) = 0.211/0.438 = 0.482$$

fraction of evaporable water. Therefore, for

$$w/c < 0.38$$

$$w_e/c = 0.482 \times (w_t/c)$$

Thus, it is concluded that most pastes are made above 0.38, the transition point.

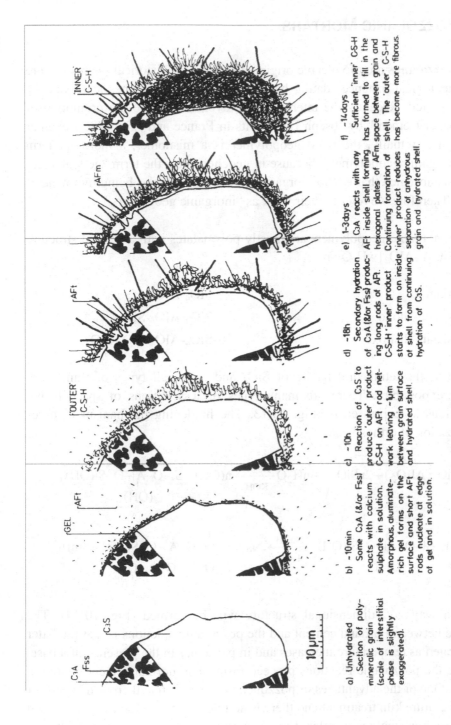

Figure 10-12. Development of microstructure during the hydration of Portland cement from 10 min. to 14 days. (After Scrivener, Ph.D. Dissertation, University of London, 1984).

a) Unhydrated
Section of poly-mineralic grain (scale of interstitial phase is slightly exaggerated).

b) ~10min
Some C₃A (&/or Fss) reacts with calcium sulphate in solution. Amorphous, aluminate-rich gel forms on the surface and short AFt rods nucleate at edge of gel and in solution.

c) ~10h
Reaction of C₃S to produce 'outer' product C-S-H on AFt rod network leaving ~1μm between grain surface and hydrated shell.

d) ~18h
Secondary hydration of C₃A (&/or Fss) producing long rods of AFt. C-S-H 'inner' product starts to form on inside of shell from continuing hydration of C₃S.

e) 1-3days
C₃A reacts with any AFt inside shell forming hexagonal plates of AFm. Continuing formation of 'inner' product reduces separation of anhydrous grain and hydrated shell.

f) ~14days
Sufficient 'inner' C-S-H has formed to fill in the space between grain and shell. The 'outer' C-S-H has become more fibrous.

10.2 POZZOLANIC MORTARS

Natural **pozzolanas** are of volcanic origin and remain of historical interest. There are archaeological instances dating back to ancient Roman times and even the Egyptian period. Recently the interest has rejuvenated due to the promotion of "Geopolymer Cement" by Joseph Davidovits in France and J.S.J. Van Deventer of Australia. Actually, the word geopolymer is a misnomer, a more apt term being "Inorganic geopolymer"; because in geochemistry, the term "geopolymer" refers to a substance that is 100% organic in nature, examples being bitumene or kerogen. Therefore we refer to their work as "inorganic geopolyers".

The inorganic geopolymers are mostly polysialates with a general structure of $Mn\{(SiO_2)_2\text{-}AlO_2\}wH_2O$ or

Polysialte	$-SiO_4\text{-}AlO_4-$
Polysiliate-siloxo	$-SiO_4\text{-}AlO_4\text{-}SiO_4-$
Polysialate-disiloxo	$-SiO_4\text{-}AlO_4\text{-}SiO_4\text{-}SiO_4-$

Due to the tetrahedral nature of SiO_4 and AlO_4, all types of framework models are possible. There are many types of tessellations of space filling combinations as expressed by Fig. 10-13. The hardening mechanism involves polymerization of orthosilicate.

$$n(SiO_5, Al_2O_2) + 2nSiO_2 + 4nH_2O \xrightarrow[\text{KOH}]{\text{NaOH}} n(OH)_3\text{-}Si\text{-}O\text{-}\overset{\ominus}{Al}\text{-}O\text{-}Si(OH)_3 \atop (OH)_2$$

$$n(OH)_3\text{-}Si\text{-}O\text{-}\overset{\ominus}{Al}\text{-}O\text{-}Si(OH)_3 \atop (OH)_2 \xrightarrow[\text{KOH}]{\text{NaOH}} Na,K^{(+)}\text{-}\underset{O}{Si}\text{-}O\text{-}\overset{\ominus}{\underset{O}{Al}}\text{-}O\text{-}\underset{O}{Si}\text{-}O + 4nHCl$$

In such a way, a 3-dimensional structure will be formed (Fig. 10-14). The difference between Portland cement and the pozzolanic reactions is that the latter is accelerated as temperature increases and in particular in the presence of a base. Therefore the pozzolanic reactions proceed faster than normal cement hydration reactions. One of the advantages of pozzolanic mortars is that there is no need for high temperature kiln treatment and there is no release of greenhouse gas. Natural minerals such as zeolites or related are formed in natural pozzolonas, including analcime, leucite, chabazite, phillipsite and dinopltilolite. A number of natural pillolanas and their compositions are listed in Table 10-5.

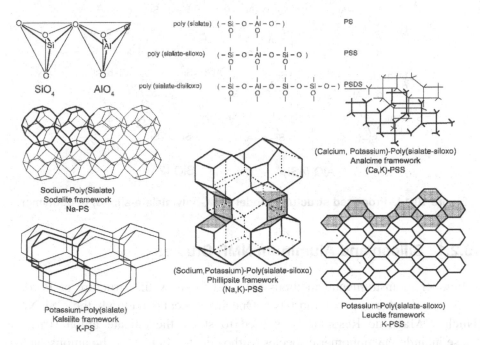

Figure 10-13. Computer molecular graphics of polymeric M_n-(-Si-O-Al-O)$_n$, poly(sialate), and Mn-(-Si-O-Al-O-Si-O)$_n$ poly(sialate-siloxo), and related frameworks.

Table 10-5. Chemical Compositions of Some Natural Pozzolanas

	Na_2O	MgO	Al_2O_3	SiO_2	K_2O	SO_3	CaO	Fe_2O_3	Loss	Total
Sacrofano (Italy)	-	-	3.05	89.22	-	2.28	0.77	4.67	-	99.99
Bacoli (Italy)	3.08	1.2	18.20	53.08	0.65	7.61	9.05	4.29	3.05	100.24
Segni (Italy)	0.85	4.42	19.59	45.47	6.35	0.16	9.27	9.91	4.03	100.05
Santorin Earth	3.8	2.0	13.0	63.8	2.5	-	4.0	5.7	4.8	99.6
Renish Trass	1.48	1.20	18.29	52.12	-	5.06	4.94	5.81	11.10	100.00
Rhyolite Pumice	4.97	1.23	15.89	65.74	1.92	-	3.35	2.54	3.43	99.07

Some heat treated pozzolanas, including clays such as surkhi (India), qaize (France), moler (Denmark) are often used. The most cited example is **pulverized fuel ash (pfa)** including fly ash activated by alkali.

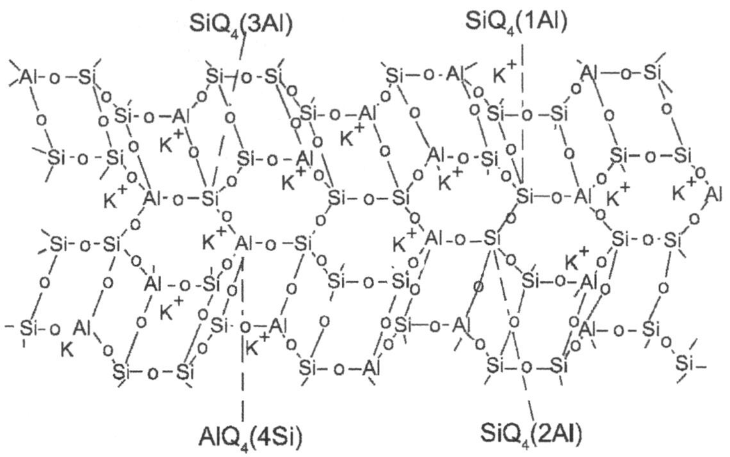

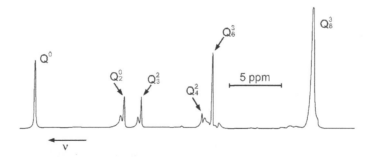

Figure 10-14. Proposed structural model for K-poly(sialate-siloxo) geopolymer.

10.2.1   Silicon and Aluminum NMR Studies

A number of instrumental methods are used to study the structure of Portland cement and other type of mortars. One instrumental method, the ^{29}**Si NMR** (Nuclear Magnetic Resonance) is used to study the silicate anion structure. These include the monomeric species (orthosilicate, SiO_4^{4-}), cyclic anions such as $Si_3O_9^{6-}$ and $Si_6O_{18}^{12-}$ and infinite chain anion such as $(SiO_3^{2-})_n$. Many of the species can be directly verified by the ^{29}Si NMR spectra. To illustrate this point a ^{29}Si NMR spectra is shown in Figure 10-15.

Figure 10-15. ^{29}Si NMR at 79.5 MHz and 5°C of alkaline aqueous tetramethyl-ammonium silicate solution with atomic ration Na:Si = 1.0 and concentration ~2M in Si. Sample was enriched in ^{29}Si to 95.3%. Peak assignments are indicated. Peak due to cubic octamer is not taken to its full height. Measurements show this peak is ca. 12 times as intense as that of monomer. (After Harris et al., 1982)

Both **magic angle spinning (MAS)** ^{27}Al NMR and ^{29}Si NMR are of interest.

The spectrum taken in an alkaline aqueous solution of tetraethylammonium silicate at 5^{0}C with Na/Si = 1.0 and Si is 2M. The shift of ppm with the monomer and other oligomers are summarized in Table 10-6.

Table 10-6. Chemical Shift of Monomeric Orthosilicate and Related Oligomers Species

	^{29}Si/ppm
Monomer, Q^0	0.00
Dimer, Q_2^1	-8.62
Cyclic trimer, Q_3^2	-10.19
Cyclic tetramer, Q_4^2	-16.10
Prismatic hexamer, Q_6^3	-17.21
Cubic octamer, Q_8^3	-27.55

The structure of the silicate anions can be best designated as Q_j^i where j is the number of Si atoms and i is the number of attached silahexane bridges, as shown in Figure 10-16.

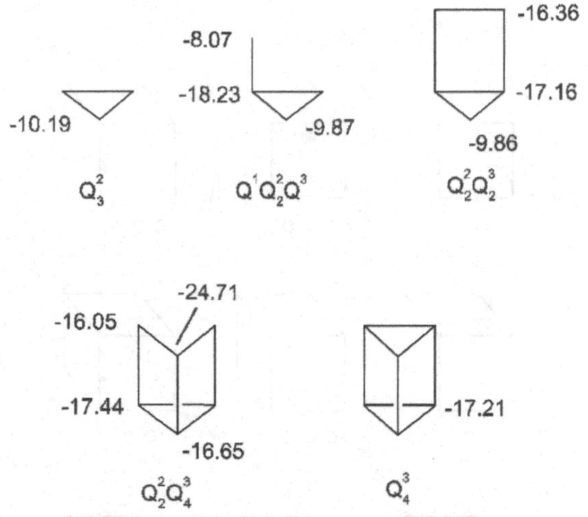

Figure 10-16. Identified series of species containing three-membered rings in solution. Vertices in diagrams indicate positions of silicon atoms, which are bridged by oxygens. Additional oxygens (or OH groups) are present as required for valency fulfillment. Chemical shifts are given in ppm with respect to signal due to monomer resonance on high-frequency-positive convention, as obtained for solution of alkaline aqueous potassium silicate with atomic ratio K:Si = 1.0 and concentration 0.65 M in Si.

The transition of cyclic trimer to a prismatic hexamer can be expressed as

$$Q_3^{\ 2} \rightarrow Q^1 Q_2^{\ 2} Q^3 \rightarrow Q_3^{\ 2} Q_2^{\ 3} \rightarrow Q_2^{\ 2} Q_4^{\ 3} \rightarrow Q_6^{\ 3}$$

Similarly the transformation of a cyclic tetramer to a cubic octomer can take place:

$$Q_4^{\ 2} \rightarrow Q^1 Q_3^{\ 2} Q^3 \rightarrow Q_2^{\ 1} Q_2^{\ 2} Q_2^{\ 3} \rightarrow Q_2^{\ 3} Q_4^{\ 2} \rightarrow Q_3^{\ 3} Q_3^{\ 2} Q^1 \rightarrow Q_4^{\ 3} Q_3^{\ 2}$$
$$\rightarrow Q_5^{\ 3} Q_2^{\ 2} Q^1 \rightarrow Q_6^{\ 3} Q_2^{\ 2} \rightarrow Q_8^{\ 3}$$

This is illustrated in Fig. 10-17. Also additional species can also be present such as in Fig. 10-18.

As a rule, the highly symmetric species are more likely to be present than the asymmetric species.

For individual compounds, the ^{29}Si NMR shift from TMS can be tabulated for Si(4Al), Si(3Al), Si(2Al), Si(1Al) and Si(0Al) in Table 10-7.

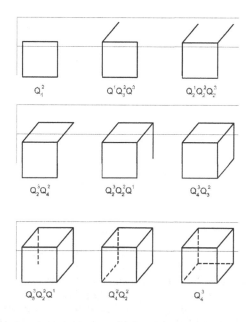

Figure 10-17. Series of square to cube intermediates. (T.F. Yen, 1994)

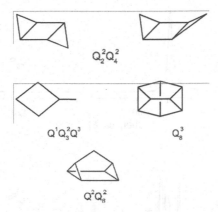

$Q_2^2Q_4^2$

$Q^1Q_3^2Q^3$ Q_8^3

$Q^2Q_8^2$

Figure 10-18. Additional species that may be present in solution from a spectrum. Evidence for existence of these species is not definitive.

Table 10-7. ^{29}Si MAS-NMR spectra of zeolitic species (exerpt from Klinowski J, *Prog in NMR Spectroscopy*, 16, 237 (1984)

		^{29}Si chemic shifts (ppm from TMS)					
Zeolite	Formula	Si/Al	Si(4Al)	Si(3Al)	Si(2Al)	Si(1Al)	Si(0Al)
Zeolite A	$NaAlSiO_4.H_2O$	1.0	-88.9	-	-	-	-
Analcime	$NaAlSi_2O_6.H_2O$	2.0	-	-92.0	-96.3	-101.3	-108.0
Leucite	$KAlSi_2O_6$	2.0	-81.0	-85.2	-91.6	-97.4	-101.0
Sodalite	$NaAlSiO_4.H_2O$	1.0	-84.8	-	-	-	-

For polysilicate, if the silicon tetrahedral are joined individually through oxygen bridges, for $Q^0 = -70$ppm (monomer), for linear dimmer, Q^1 -80ppm, for linear trimer, $Q^2 = -88$ppm. An iso (or tertiary) structure with four Si atoms, $Q^3 = -98$ppm. A quaternary structure with five Si atoms, $Q^4 = 110$ ppm. In the above discussion, the superscript numeral attached to Q referred to the number of $(SiO)^{4-}$ units. In the CSH formation during the hydration process, Q^0, Q^1 and Q^2 have been observed (see Figure 10-19 a, b and c). The formation of the CSH gel has been often inferred to as the formation of the tobermorite and jennite minerals with layers of Ca^{+2} or $Ca(OH)_2$ (Fig. 10-20).

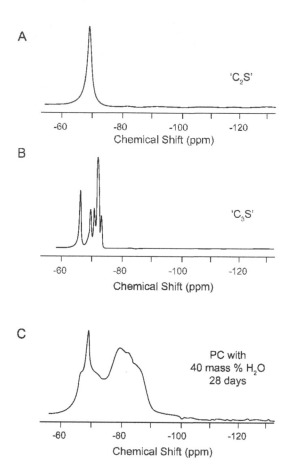

Figure 10-19. ^{29}Si NMR examination of cement hydration: (A) pure dicalcium silicate, (B) pure tricalcium silicate, (C) Portland cement sample hydrated at 40% by mass of water for 28 days. (A) and (B) show Q^0 29Si NMR signals (~ -70ppm). The addition of Q^1 and Q^2 Si signals (-80ppm to -90ppm is seen upon hydration. The slow hydration rate of sicalcium silicate is shown by a large peak of unreacted pure metal at about -70ppm). (Black, et al, *Journal of Material Chemistry*, vol 33, 859, 1998.)

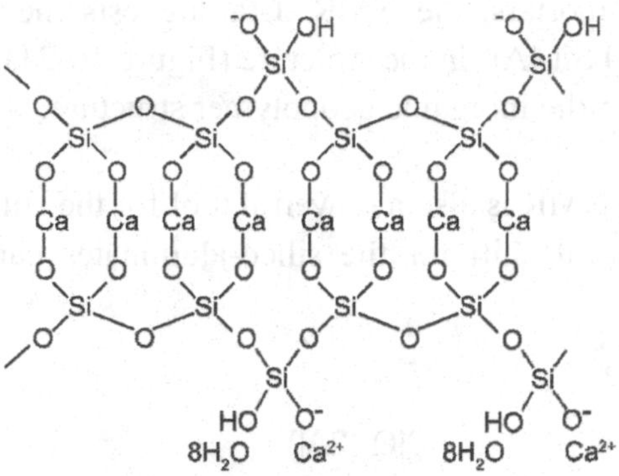

Tobermorite
$[Ca_4(SiO_3O_9H)_2]Ca_2.8H_2O$

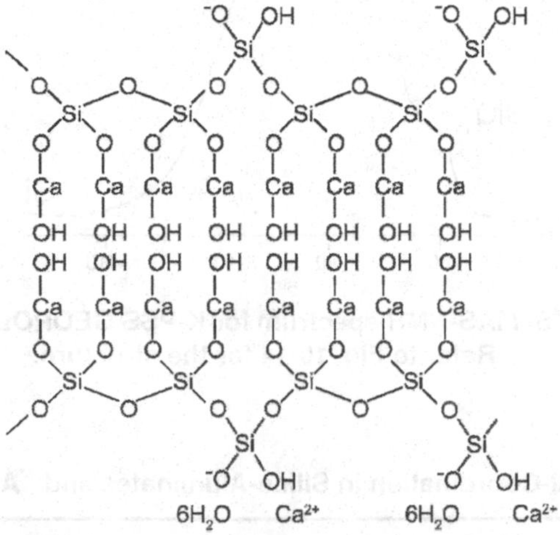

Jennite
$[Ca_8(SiO_3O_9H)_2(OH)_8]Ca_2.6H_2O$

Figure 10-20. Cement paste is believed to closely resemble the minerals tobermorite and jennite. These minerals are characterized by layers of polymerized silicon dioxide cross-linked with calcium oxide or calxium hydroxide.

For the pozzolanic mortars, the NMR data suggests the presence of Si(1Al), Si(2Al), Si (3Al) and Si(4Al) in the structure (Figure 10-21). The structure can be seen in Fig. 10-14 for the inorganic geopolymer structure.

Similarly, ^{27}Al NMR is also a powerful tool for the study of these materials. The observed chemical shift for the silico-aluminates can be found in Table 10-8.

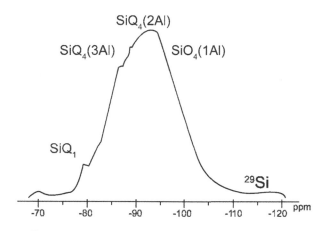

Figure 10-21. ^{29}Si MAS-NMR spectrum for K-PSS GEOPOLLYMITE® binder. Refer to Fig. 10-14 for the structure.

Table 10-8. Al-Coordination in Silico-Aluminates and [27]Al Chemical Shift

Name	Formula	Coordination	Chemical shift (ppm)
K-Feldspar	$KAlSi_3O_8$	4	54
Muscovite	$KAl_3Si_3O_{11}.H_2O$	6 , 4	-1 , 63
Biotite	$K(Mg,Fe)_3AlSi_3O_{11}.H_2O$	4	65
Calcium aluminate	$Ca_3Al_4O_7$	4	71
Sodium aluminate	$NaAlO_2$	4	76
Philipsite	$(K,Ca)AlSi_2O_6.H_2O$	4	55
K-Poly(sialate-siloxo)	$KAlSi_2O_6.H_2O$	4	55

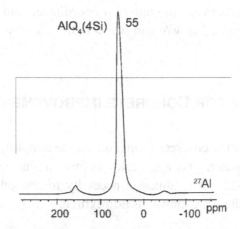

Figure 10-22. ^{27}Al MAS-NMR spectrum for K-PSS GEOPOLLYMITE® binder.

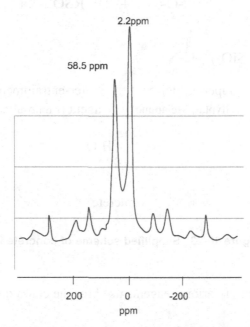

Figure 10-22(a). ^{27}Al MAS-NMR spectrum for a Fly ash synthesized inorganic geopolymer.

A MAS of the ^{27}Al-NMR or polysilicate-polysiloxo, the inorganic geopolymers given in Figure 10-22, the spectra indicate the chemical shift of 55ppm for $[Al(H2O)_6]^{3+}$ and the Al is tetrahedrally coordinated. For reference the ^{27}Al-NMR also studied for fly ash synthetic inorganic geopolymer (Fig. 10-22a). This sample indicated that there is a mixture of 4-coordinated and 6-coordinated Al is required, usually expressed as w/c and w/c > 0.38. Here, w indicates the initial state.

10.3 ADMIXTURE FOR CONCRETE IMPROVEMENT

The entire process for the concrete formation can be simplified as the addition of hydraulic cements, water, and aggregates before or during mixing as shown generally in Fig. 10-23. It is almost a necessity to add other materials to the essential ingredient; these are known as **admixture**.

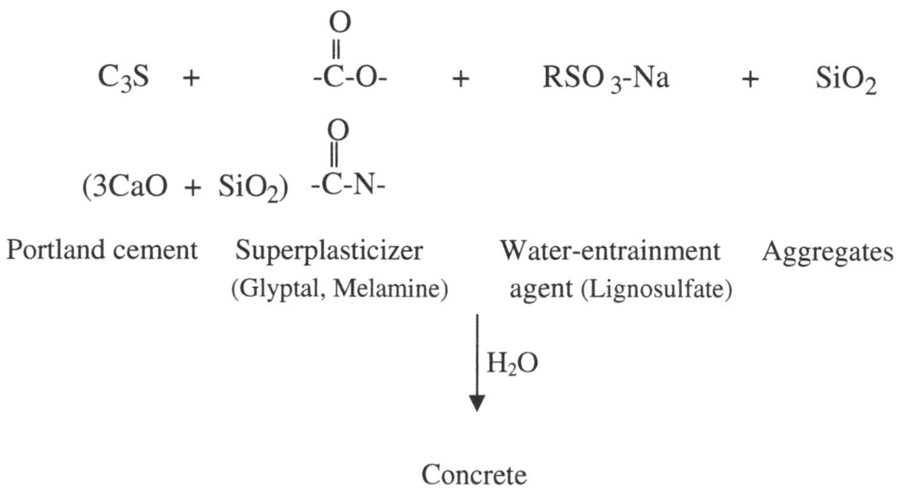

$$C_3S \quad + \quad \overset{\overset{O}{\|}}{-C-O-} \quad + \quad RSO_3\text{-Na} \quad + \quad SiO_2$$

$$(3CaO + SiO_2) \quad \overset{\overset{O}{\|}}{-C-N-}$$

Portland cement   Superplasticizer   Water-entrainment   Aggregates
                   (Glyptal, Melamine)   agent (Lignosulfate)

$$\downarrow H_2O$$

Concrete

Figure 10-23. Simplified scheme of concrete formation.

The following is a list of various reagents used for the common admixture.

*   **Retarders**

    Sugars which have the ability to stabilize CH are good inhibitors. Maltose, lactose, and cellulose are good inhibitors, while trehalose and α-methyl glucoside are not effective. These facts are supported by solubility findings. Evidently, the adsorption of sugars with functional groups of $-CO \cdot C(OH) =$ on clinker phase or hydration products is anticipated to slow down the hydration reaction. Studies with sucrose are illustrated in Fig. 10-24 to explain the retardation mechanism. The concentration of calcium, silicon, aluminum, iron, and hydroxyl in the solution phases of slurries of cement with w/c = 2 with or without the addition of sucrose were analyzed at ages up to 7 hours. In the absence of a retarder, the concentrations of Si, Al, and Fe are very low. In the presence of 50m mole L^{-1} of sucrose, the Ca and OH$^-$ were increased (Fig. 10-24a), but the Si, Al, and Fe were increased in the order of 1000 (Fig. 10-24b). This fact definitely points to the functional group of sucrose having been adsorbed by the hydration products, thereby preventing further settling.

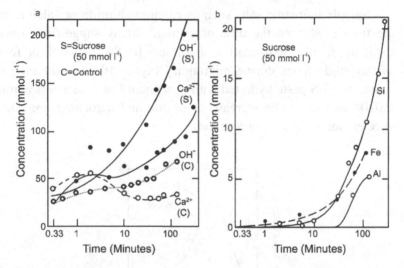

Figure 10-24. Concentrations in the aqueous phase of Portland cement hydrated at w/c = 2.0 in water or in sucrose solutions. (After Thomas and Birchall). For the control of Si, Fe, and Al, the concentrations are too small to be included in the scale.

In a typical case, addition of 0.1% of sucrose based on cement may increase the initial set from 4 h to 14 h, while a 0.25% may delay it to 6 days. Calcium or sodium salts of lignosulfonates may also be used as functional retarders. These are paper and pulp wastes that are three-dimensional random polymers with structural elements shown in Fig. 10-25.

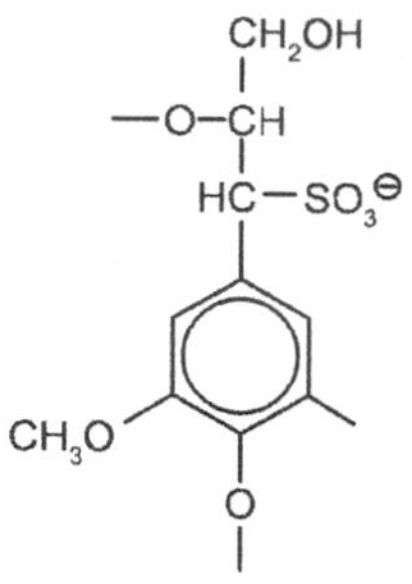

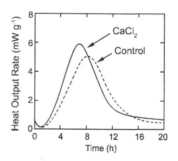

Figure 10-25. Typical element of structure from a lignosulfonate anion.

- **Accelerators**

    Sample inorganic salts such as calcium chloride or calcium formate are used to shorten the time of setting. Many suggestions have been made to provide the formation of surfaces for the nuclei of the formation of the solid phases during migration. Figure 10-26 is the case of heat output for C_3S paste hydrated with or without $CaCl_2$ in a concentration of 0.0204 mol L^{-1}. The increase of setting and hardening are due to the acceleration of hydration reaction.

Figure 10-26. Rates of heat output from C_3S pastes hydrated at 25°C and a w/c ratio 0.6 with and without $CaCl_2$ in a concentration of 0.0204 mol L^{-1}. (After Brown *et al.*)

    Other salts with both cations and anions also contribute to the accelerating effects. For example,

$$Ca^{2+} \rangle Sr^{2+} \rangle Ba^{2+} \rangle Li^+ \rangle k^+ \rangle Na^+ \cong Cs^+ \rangle Rb^+$$
$$Br^- \cong Cl^- \rangle SCN^- \rangle I^- \rangle NO_3^- \rangle ClO_4^-$$

The ranking series is broadly similar to the **Hofmeister series** in coagulation. (see T.F.Yen, *Chemical Processes for Environmental Engineers*, 2006, Imperial College Press)

- **Air-entraining agents**

    The ability of concrete to resist damage from freezing of pore solution is increased by introducing voids, which can later be expanded. Air is trapped during normal mixing, but only a limited amount because the voids are too large and unevenly spaced. The use of cationic surfactants will improve the workability. As discussed in Chapter 5, Surface Chemistry, there are many types of surfactants. In general, salts of fatty or alkyl aryl sulfonic acids of 0.05% (based on cement) can be used.

- **Grinding aids**

    In order to achieve a fineness of grinding or throughput for clinker grinding, amines or polyhydric alcohols are most effective. Generally the amount used is 0.01 – 0.1% of the weight of clinker. Again, they are surfactants.

- **Water-reducers**

    The purpose is to add an agent in such a way as to achieve workability at a lower w/c ratio. Usually addition of 0.2% of the weight of cement, salts, hydroxy carboxylic acids, hydrolyzed carbohydrates, or hydrolyzed proteins will allow w/c to be decreased by 5-15%.

- **Superplasticizers**

    If a water-reducer can allow the w/c to be decreased by 30%, then it can be categorized as a superplasticizer. Salts of sulfonated melamine formaldehydes (SMF) and salts of sulfonated naphthalene formaldehydes (SNF) are two important types used in industry. The sulfonated groups are used to tender the polymer water-soluble. Both polymers are inexpensive due to the fact that they are derived from industrial wastes. These condensation polymers are readily formed (Fig. 10-27A and B).

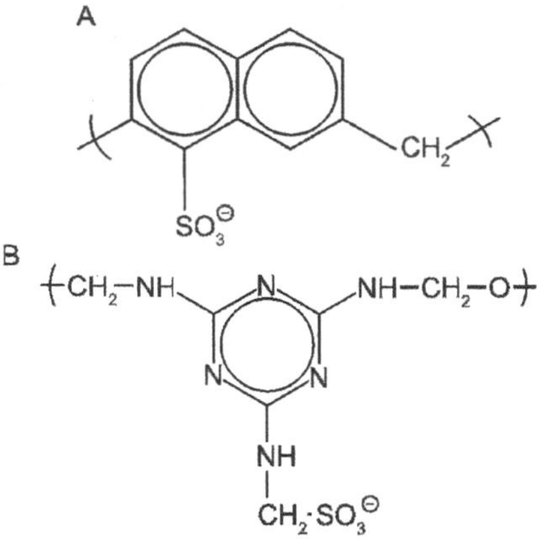

Figure 10-27. Repeating units of the structures of superplastizer anions: (A) naphthalene formaldehyde condensate; (B) melamine formaldehyde condensate. Occasionally structure A will contain CH_2O- as a repeating unit.

Subsidiary agents are often used in order to increase the interaction between aggregates in the system and superplasticizers.

* **Subsidiary agent**

For superplasticizers, we often adopt the Bakelite resin type of random polymerization. For example, phenol-fermaldehyde or arene-formaldehyde will give the resin, as indicated in Figure 10-28.

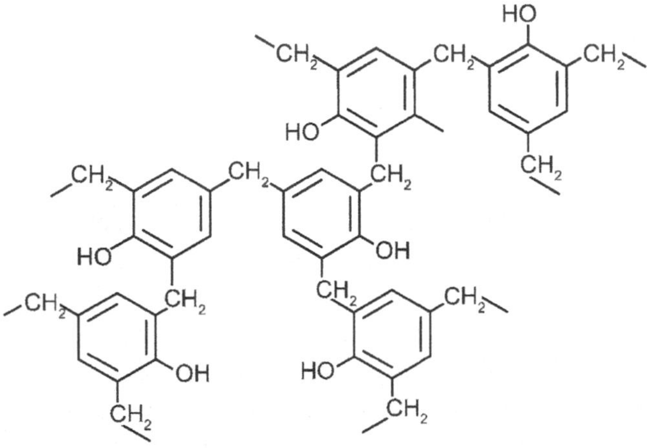

Figure 10-28. Phenol-formaldehyde resin.

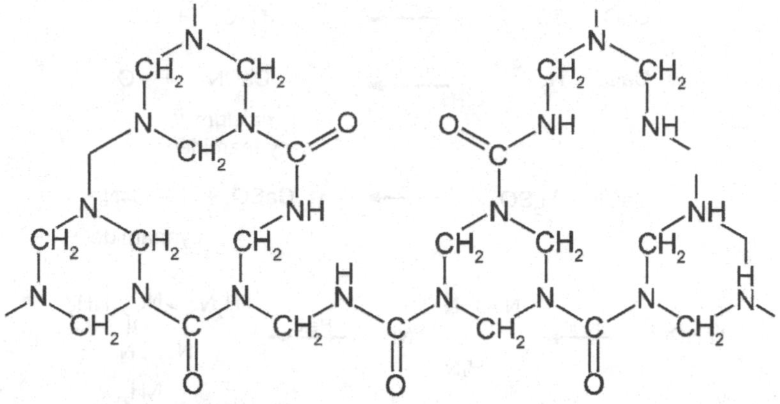

Figure 10-29. Urea-formaldehyde resin.

Another inexpensive polymer is the urea-formaldehyde resin, such as

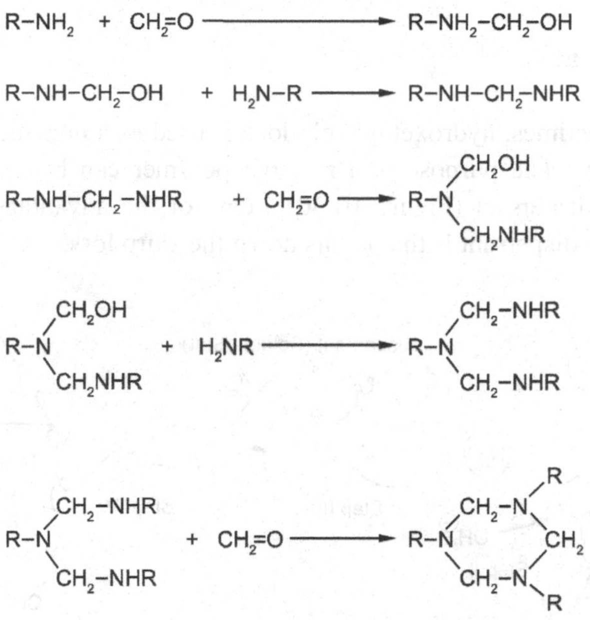

$$R-NH_2 \ + \ CH_2=O \longrightarrow R-NH_2-CH_2-OH$$

$$R-NH-CH_2-OH \ + \ H_2N-R \longrightarrow R-NH-CH_2-NHR$$

$$R-NH-CH_2-NHR \ + \ CH_2=O \longrightarrow R-N\begin{smallmatrix} CH_2OH \\ \\ CH_2NHR \end{smallmatrix}$$

$$R-N\begin{smallmatrix} CH_2OH \\ \\ CH_2NHR \end{smallmatrix} \ + \ H_2NR \longrightarrow R-N\begin{smallmatrix} CH_2-NHR \\ \\ CH_2-NHR \end{smallmatrix}$$

$$R-N\begin{smallmatrix} CH_2-NHR \\ \\ CH_2-NHR \end{smallmatrix} \ + \ CH_2=O \longrightarrow R-N\begin{smallmatrix} CH_2-N \\ \\ CH_2-N \end{smallmatrix}\begin{smallmatrix} R \\ \\ CH_2 \\ \\ R \end{smallmatrix}$$

The resultant polymer will be a three-dimensional polymer as shown in Fig. 10-29 above. Usually melamine is synthesized from carbon, calcium and nitrogen at very high temperatures as follows:

$$CaO + 3C \xrightarrow{2000°} CaC_2 + CO$$

$$CaC_2 + N_2 \xrightarrow{1000°} CaCN_2 + C$$

calcium
cyanamide

$$CaCN_2 + H_2SO_4 \longrightarrow CaSO_4 + H_2N-C\equiv N$$

cyanamide

$$H_2NC\equiv N \xrightarrow{Heat} \underset{\substack{H_2N}}{\overset{\substack{H \\ N\equiv C-N}}{\underset{C=NH}{}}} \xrightarrow{Heat}$$

dicyanamide

2, 4, 6-triamino-
1, 3, 5-triazine
(melamine)

The triazine type is useful as a subsidiary agent, since the formation is sensitive to heat.

- **Dispersant**

Sometimes, hydroxethyl cellulose is used as a non-segregation agent in water. The purpose is a reactive polymer can be converted into a soluble dispersant (Figure 10-30). One of the advantages of using an adequate dispersant is that it cuts down the slurp loss.

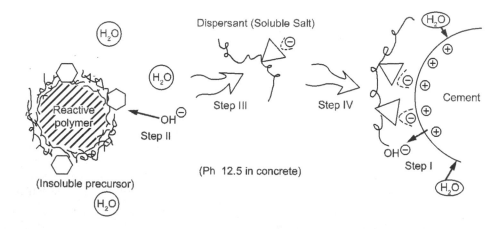

Figure 10-30. Dispersing mechanism of reactive polymers.

10.4 CONCRETE PROBLEMS AND RELATED ISSUES

10.4.1 Corrosions

Crystalline calcium hydroxide makes up about 10% of the volume of used common cement, and some **physical corrosion** results from the leaching of calcium hydroxide. A process called efforescence occurs when opaque white materials appear to ooze out of concrete walls. Often water containing dissolved $Ca(OH)_2$ has the concrete leached out and left a layer of calcium hydroxide to react with carbon dioxide and form calcium carbonate.

$$Ca(OH)_{2\,(s)} + CO_{2\,(g)} \rightarrow CaCO_{3\,(s)} + H_2O_{\,(l)}$$

Actually, CO_2 can diffuse into cement.

$$CO_{2\,(g)} + H_2O_{\,(l)} \rightarrow H_2CO_{3\,(aq)}$$
$$H_2CO_{3\,(aq)} + CaCO_{3\,(s)} \rightarrow Ca(HCO_3)_{2\,(aq)}$$
$$Ca(HCO_3)_{2\,(aq)} + Ca(OH)_{2\,(s)} \rightarrow 2CaCO_{3\,(s)} + 2H_2O$$

Another problem, especially in a marine environment, is the action of sulfates such as ammonium sulfate or magnesium on concrete. These salts often react with calcium hydroxide to form calcium sulfate.

$$(NH_4)_2SO_{4\,(aq)} + Ca(OH)_{2\,(g)} \rightarrow CaSO_{4(s)} + 2NH_{5\,(aq)} + H_2O_{\,(l)}$$
$$Mg(SO_4)_{\,(aq)} + Ca(OH)_{2\,(s)} \rightarrow CaSO_{4\,(s)} + Mg(OH)_{\,(aq)}$$

These reactions deplete cement paste of calcium hydroxide which are quite destructive due to large volume of sulfate salts formed. For example, the 74.2 mL/m of formed versus the original volumes removed, 33.2 mL/mole, is almost double; this leads to the stress of the acceleration of cracks formed.

10.4.2 Behavior of Water in Cement

According to Powers and Brownyard, water in saturated, mature cement can be capillary water, gel water, and also non-evaporable water. If one understands the movement and distribution of water in the pores of cement, one will know the corrosion of cement. In this aspect, **magnetic resonance imaging (MRI)** will be able to contribute valuable information. In cement gels, MRI has shown that water freezes in two steps. The first step occurs between 0 and -2° C, where free bulk water and water is the capillary preference. Owing to the volume expansion of water, the resulting internal pressure will force the migration of water from the gel pore to large pore regions, which will form ice. This results in a secondary freezing point at about -40 to -45°C as illustrated in Figure 10-31.

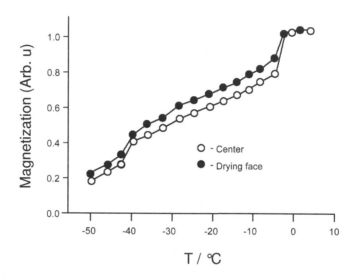

Figure 10-31. Magnetization signal for evaporable water in a sample of Portland cement mixed with 14-mm diameter graded quartz aggregate measured using APRITE as a function of temperature. Freezing of water is associated with a decrease in signal intensity. P. J. Pordo, et al., Cement and Concrete Research, 13,93, (1998).

10.4.3 Steel Reinforcement

Usually, reinforced concrete is used in lodge decks, sidewalks, and roads. A common occurrence is the attack of the chloride ion, which is transported by water to the rebar of the **steel reinforcement**, causing corrosion within. High pH is important for maintaining formation of a passive oxide layer on the surface of metal.

$$2Fe_{(s)} \rightarrow 2Fe^{2+}_{(aq)} + 4e^-$$
$$O_{2(g)} + 2H_2O_{(l)} + 4e^- \rightarrow 4OH^-_{(aq)}$$
$$2Fe^{2+}_{(aq)} + 4Cl^-_{(aq)} \rightarrow 2FeCl_{2(aq)}$$
$$2FeCl_{2(aq)} + 4OH^-_{(aq)} \rightarrow 2Fe(OH)_2 + 4Cl^-_{(aq)}$$
$$2\,Fe(OH)_2 + 1/2O_{2(g)} \rightarrow Fe_2O_3\,(s) + 2H_2O_{(l)}$$

$$2Fe_{(s)} + 3/2\,O_{2(g)} \rightarrow Fe_2O_{3(s)}$$

In this manner, peeling can create localized corrosion cells between the rebar in the existing chloride-concentrated concrete and also in the new chloride-free patch. An approach to treating chloride regress is when **electrochemical chloride extraction (ECE)** in chloride ions are effectively pulled from the concrete using the arrangement illustrated in Figure 10-32.

A DC circuit is set up using the steel rebar surface as the anode and an electrolyte gel packed on to create surface as the cathode. When a DC potential of 10 to 30 kV is applied, water is hydrolyzed at the anode, replenishing the hydroxide content of the systems. The negatively charged steel rebar repels chloride ions to the surface of the anode and into the electrolyte gel. After sealing with a water-proof coating, the reinforced anode is effectively protected.

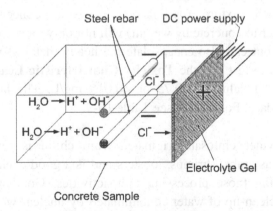

Figure 10-32. Schematic of electromechanical chloride extraction (ECE). A dc voltage of 10-30,000 V is applied between the steel rebar (anode) and an electrolyte gel (Cathode) on the surface of concrete. Hydrolysis of water takes place at the rebar, and chloride ions are repelled from the concrete into the electrolyte gel. Diothimore et al., *Civil Engineering*, 1999, Jan 26.

10.4.4 Solidification and Stabilization of Hazardous Waste

As the restrictions on landfilling become stronger and wastes are banned from land disposal, **solidification** and **stabilization** could potentially play an important role.   The cement-based solidification and stabilization approach always uses cement as an aborting agent.   Current uses are not only limited to Portland type, such as with the pozzolanic mortars, the inorganic geopolymers, the pulverized fused ashes (pfa), the fly ashes, wasted glasses, and even cathode ray tube (CRT) glasses.   One of the criteria for this type of **immobilization** or **encapsulation** is by the use of a leaching test.   Some of the leaching tests are only of marginal success.

Polymers have been known to be used as mortar and concrete modifiers for improvement of certain properties such as fracture toughness, impermeability, durability etc. In certain cases to make the concrete lighter in densities, the consequence is that the compression strength suffers greatly. In some other cases, the addition of adequate amount of water-soluble polymers as water-dispersible power, for example, hydroxyethyl cellulose (HEC) vinyl acetate-ethylene copolymer is helpful.

A successful example is the use of minute amounts (ppm level) of biopolymer with proper crosslinking. In the case of xanthan and guar gum (Fig. 10-33) or their mixture crosslinked by borax at a concentration of 18-100 mg/kg (biopolymer to concrete by weight) will not only increase the compressive strength by 30% but can also encapsulate the heavy metal. It was evaluated for lead by the leaching test that the **Toxicity Characteristic Leaching Procedure (TCLP)** indicates a value of <0.75 or <0.015 mg/L. The latter is below the groundwater standard. For details, see Table 10-9.

The biopolymer chitosan originated from chitin is found in shells of crustaceans such as crabs, shrimps, lobsters and is a good chelator for a number of metals including those processing radioactivities. Good results have been obtained for the clean-up of water contaminated by nuclear waste. Owing to the biodegradation of chitosan, the problem is still not solved.

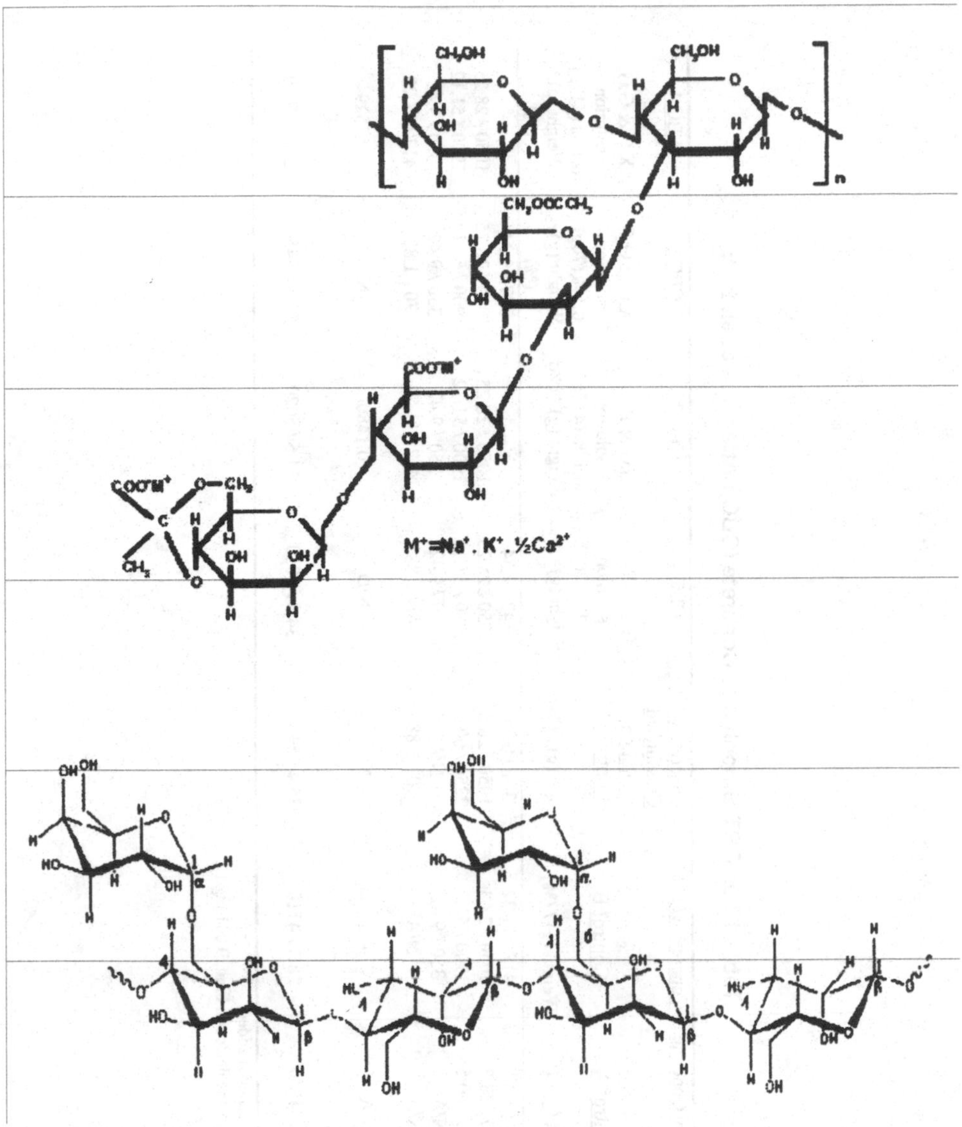

Figure 10-33. Structure of xanthan and guar gum.

Table 10-9. CRT-Biopolymer Concrete (CBC). After Kim et al, 2005

	Ordinary Concrete	Concrete w/ CRT	Concrete w/ Crosslinked solution	CBC 1	CBC 2	CBC 3	CBC 4
	Control A	Control B	Control C	G.G & Boric solution 0.1% (70g)	G.G & Boric solution 0.1% (417g)	X.G & G.G solution 0.1% (70g)	X.G & G.G solution 0.1% (417g)
	Weight (g) / (%)	Weight (g) / (%)	Weight (g) / (%)	Weight (g) / (%)	Weight (g) / (%)	Weight (g) / (%)	Weight (g) / (%)
Water	417 / 11.2	417 / 11.22	347 / 9.34	347 / 9.34	—[a]	347 / 9.34	—[a]
Cement	1050 / 28.3	1050 / 28.3	1050 / 28.3	1050 / 28.25	1050 / 28.25	1050 / 28.25	1050 / 28.25
Sand	2250 / 60.5	1900 / 51.1	2250 / 60.5	1900 / 51.12	1900 / 51.12	1900 / 51.12	1900 / 51.12
CRT glass	N/A	350 / 9.42	N/A	350 / 9.42	350 / 9.42	350 / 9.42	350 / 9.42
Crosslinked solution	N/A	N/A	70 / 1.88	70 / 1.88	417 / 11.22	70 / 1.88	417 / 11.22
TCLP (mg/L)	N/A	4.6	N/A	N/D	0.1480	N/D	0.2529
Compressive strength (MPa/psi)	28.5 / 4140	28.3 / 4110	35.4 / 5130	34.8 / 5050	34.6 / 5020	37.6 / 5450	21.2 / 3070

N/A: Not applicable or no material added.
N/D: Non-detectable (below the detection limit, 0.001 mg/L).
[a] No water added.

The concept of **geopolymers** will help. Geopolymers such as bitumen, asphaltene or kerogen have been formed in the lithosphere over a geological time scale. They can produce oil, gas and carbon, which as the precursors for petroleum. A modern analogy of geopolymers is the humus materials, which are resistant to biodegradation. A sample geopolymer formation can be visualized through the transformation scheme as shown in Fig. 10-34.

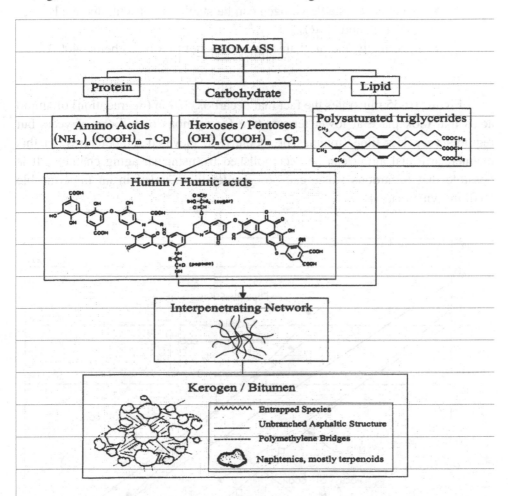

Figure 10-34. Formation of geopolymer (kerogen) from biomass. Notice that some biopolymers are the precursors of geopolymers (Kim et al, 2006).

The conversion time of geopolymers can be shortened by appropriate temperature treatment as a tradeoff can be made between time (geological age) and temperature. The advantages of geopolymers are as follows:

- Metals cannot be leached out in aqueous solution of various pH levels.
- Metals are coordinated through gaps or holes of the carbonized condensed aromatic rings or in association in the π-system.
- Radiation stable-asphaltene is a free radical acceptor and has extremely low G-yield.
- Chemically stable-kerogen can be stable in concentrated acids, e.g., HF, HCl and H_2SO_4.
- Thermally and mechanically stable- not pyrolyzed below 400°C.

Figure 10-35 illustrates the fact that decarboxylation (degradation) of amino acids that took place at lower temperature for millions or billions of years but requires very less time (minutes or seconds) at elevated temperatures. In this manner, artificial aging can be accomplished by using an aging chamber. It is possible that hazardous waste can be completely included in an impermeable medium (vitrification).

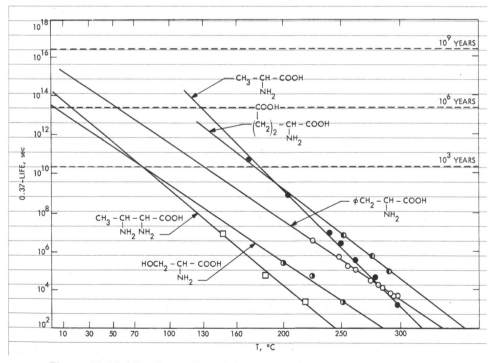

Figure 10-35. Kinetic studies of decarboxylation of some amino acids.

REFERENCES

10-1. *Chemistry of Cement: Proceedings of the Fourth International Symposium*, 1960, Vol. II, U.S. Department of Commerce, National Bureau of Standards, 1962.

10-2. International RILEM Symposium, *Admixtures for Concrete: Improvement of Properties*, Chapman and Hall, London, 1990.

10-3. Taylor, H. F. W., *Cement Chemistry*, Academic Press Inc., San Diego, 1990.

10-4. Vázquez, E., ed., *Admixtures for Concrete: Improvement of Properties*, Proceedings of the International RILEM Symposium, Chapman and Hall, New York, 1990.

10-5. T. F. Yen, R. D. Gilbert, and J. H. Fendler, eds., *Advances in the Applications of Membrane-Mimetic Chemistry*, Plenum Press, New York, 1994.

10-6. D.Kim, I.G.Petrisor and T.F.Yen, *Evaluation of Biopolymer-Modified Concrete Systems for Disposal of Cathode Ray Tube Glass*, J.Air Waste Management Assoc., **55**, 961-969, 2005.

10-7. J.Davidovitz, *From Ancient Concrete to Geopolymers*, Arts Meliers Mag., **180**, 8-16, 1993.

10-8. D.Kim, H.J.Lai, G.V.Chilingar and T.F.Yen, *Geopolymer Formation and its Unique Properties*, Environ. Geo, **51**, 103-111, 2006.

PROBLEM SET

1. A converted graph of w/c = 4 which contains C_3AH_6 and AH_3 was cured at 7 days at 70 °C. What is its compressive strength? What is the porosity value? How would these values change if w/c = 3?

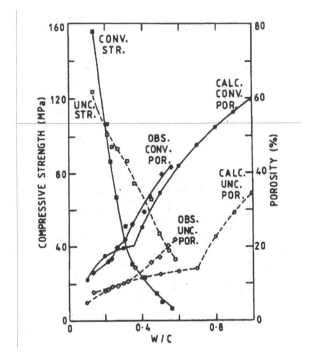

Compressive strengths and observed and calculated porosities for pastes of Ciment Fondu of varying w/c values. The unconverted paste, which contained CAH_{10}, was cured for 7 days at 10 °C. The converted paste, which contained C_3AH_6 and AH_3, was cured for 7 days at 70 °C. Data from C.M.George, *Structures and Performance of Cements* (ed. P. Barnes) P.415, Applied Science Publishers, London.

2. During clinker formation, the overall enthalpy change is dominated by the endothermic decomposition of calcite, $\Delta H = 2138$ kJ/kg. If a plant has a capacity of 20,000 ton/K what is the emission of greenhouse gas? What is the energy requirement?

3. According to many discussions, mortars from many concrete structures are still in slow transformation. It was verified that the plaster mortar of medium Egyptian pyramids (2600 BC) are composed of porous argillaceous matrix

with gypsum aggregates with the inclusion of halite. Also for the Unas pyramid (2250 BC), the mortar found is still in evolution with a slow anhydrite-gypsum conversion and also the exudation of halite. (see M.Regourd et al., *Cement Research*, **18**, 81-90, 1998). In your judgment, which pyramid is stable?

MATERIALS CHEMISTRY

Smart materials, or advanced engineering materials, cover a wide range of applications, such as aerospace, automobile, sporting goods, and electronic packaging materials; or, carbon electrodes, such as high strength, high modulus, low density, high temperature resistance. They often consist of a number of composites, including carbon fiber composites, with a polymer matrix, metal matrix, carbon matrix, ceramic matrix, hybrid matrix, or concrete matrix. Engineering often emphasizes on the macro scale of structure and ignores the micro scale or molecular orders of the structure. One should understand the structure is continuous, and without the microstructure, the macrostructure can never be completely understood and only partial or fragmental information can be achieved. Since most books on composites for engineers only speak about the mechanics for the macro approach, the following sections are devoted to the micro approach related to the molecular levels.

First, carbon and carbon bonded compounds such as graphite and diamond are discussed. Carbon fibers and their application as composites are illustrated. Highly interesting molecules such as fullerene and its homologs form the foundations of nanotechnology. Finally, ceramics, especially electronic ceramics, leading to useful materials such as piezoelectrics and electrooptics are explained.

11.1 GRAPHITE AND DIAMONDS

Carbon can undergo sp^3–hybridization to form a 6-membered nog form of benzene and numerous polynuclear aromatic hydrocarbons (PNA) as described in Chapter 3. Certainly the tessellation of forming the infinite net of hexagons of "chicken wire" or "bathroom tile" is of interest. Many regular or semi-regular **tessellations** can lead to very stable molecules. The two-dimensional space-filling regular tessellations and seven semi-regular tessellations are shown in Figure 11-1.

Depending on their resonance energy or the energy of certain geometric shapes of PNA, some are more stable than others. A simple example is that polyphenylenes are much more stable than polyacenes. Also the z-shape strap is of outstanding stability. Figure 11-2 illustrates the important classes of stable PNA. Some important stable PNA molecules are also shown in Fig. 11-3.

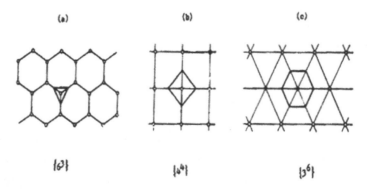

Figure 1-1(A). Regular tessellations. This graph shows all three types of regular tessellations; (a) fused hexagons {6³}, (b) fused squares {4⁴}, and (c) fused triangles {3⁶}.

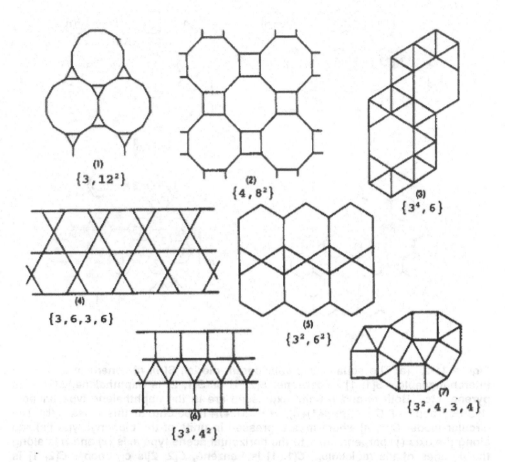

Figure 11-1(B). All seven semi-regular tessellations; (1) and (2) give a 3-connected planar network; (4) and (5) give a 4-connected planar network; and (3), (6), and (7) give a 5-connected planar network. The common ladder polymer belongs to the 3-connected planar type network. (T. F. Yen, 1972).

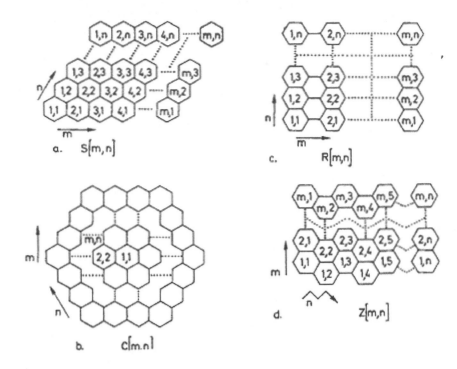

Figure 11-2. (a) The square or parallelogram model, S[m, n], where m and n are interchangeable. S[1, 1] is benzene, S[2, 1] or S[1, 2] is naphthalene, S[2, 2] is pyrene, etc. Both m and n units expressed are in the naphthalene type linkage. Hydrocarbons of $C = 2[m(n+1)+n]$, $H = 2(m+n+1)$ belong to this class. (b) The circular model C[m, n] where m is expressed in units of the biphenyl type linkage along the axis (y) perpendicular to the horizontal acene type axis (x) and n is along the 4-edges of the molecule. C[1, 1] is benzene, C[2, 2]is coronene, C[2, 1] is pyrene, C[1, 2] is naphthalene, C[2, 3] is ovalene, etc. Hydrocarbons of $C = 2m(m+2n)$ and $H = 2(2m+n)$ belong to this class. (c) The rectangular model R[m, n] originated by Coulson. Index m indicates units of the biphenyl type linkage along the x-axis and n, the naphthalene type linkage along the y-axis. R[1, 1] is benzene, R[2, 1] is naphthalene, R[1, 2] is biphenyl, R[2, 2] is perylene, etc. Hydrocarbons of $C = 2m(2n+1)$ and $H = 2(2m+n)$ belong to this model. (d) The skew strip model is Z[m, n]. Index m is units of naphthalene type linkage along the zig-zag path of the x-axis and n is that along the straight-line path of the y-axis. Z[1, 1] is benzene, Z[2,1] is naphthalene, Z[3, 1] is phenanthrene, Z[2, 2] is pyrene, Z[3,2] is 1,12-benzoperylene, etc. (After T.F. Yen, 1971).

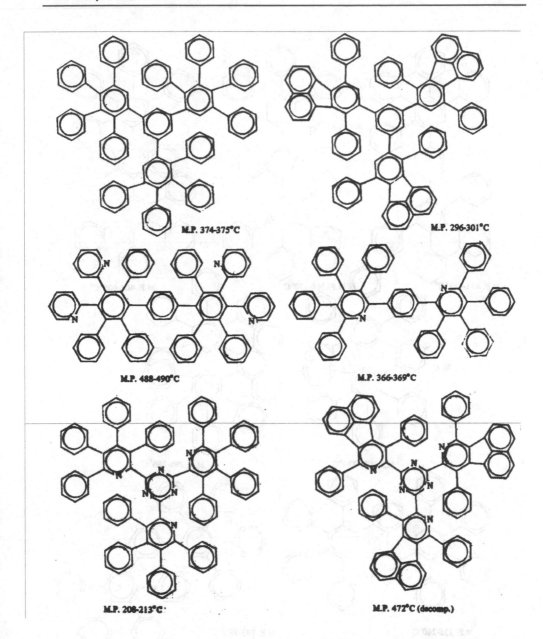

Figure 11-3. Melting point of a number of linked cluster-type polyphenylenes.

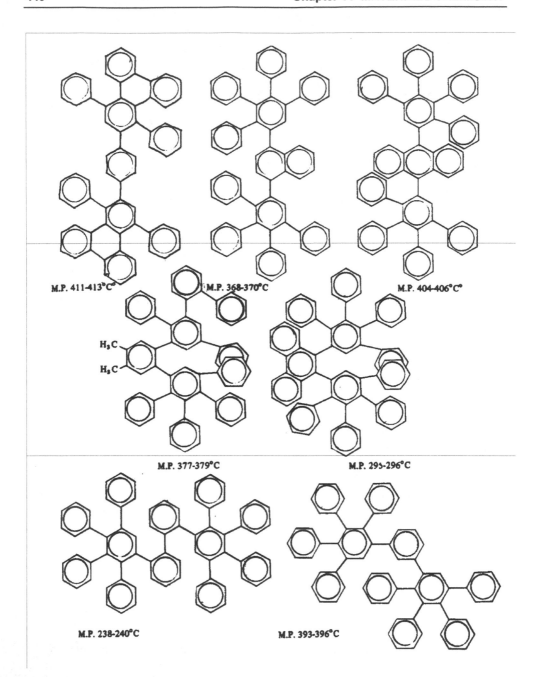

M.P. 411-413°C   M.P. 368-370°C   M.P. 404-406°C°

M.P. 377-379°C   M.P. 295-296°C

M.P. 238-240°C   M.P. 393-396°C

Figure 11-3. (Continued)

The unit cell of graphite consists of the axial direction L_a, layer diameter, and the L_c, the perpendicular stack height direction. The axial layer, which is the x, y –direction, can lude an infinite number of hexagons, either $R[m, n]$ or $C[m, n]$ model where $m \rightarrow \infty$. The graphite crystal is highly unisotropic, that is, there is only Vander Waals bonding along the L_c direction. However, the inplane covalent bonding, which provides the overlap of sp^2-hybridized orbitals, can give metallic bonding as shown in the crystallite in Figure 11-4.

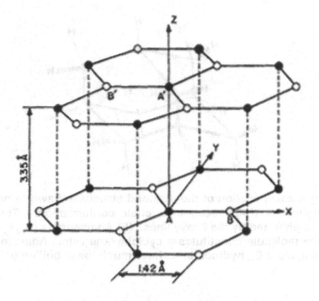

Figure 11-4. The crystal structure of graphite.

Another carbon **allotropic** form is diamond. Here, one carbon is bonded tetrahedrally with another carbon through sp^3-hybridization. One has to recall that diamond is produced under high pressure and high temperature conditions. This compact structure can produce the hardest material – diamond.

Diamond and graphite are allotropic forms of carbon. In diamond, the carbon atoms form a three-dimensional network with each carbon covalently bound to other three atoms that are located at the corners of a tetrahedron in sp3 hybridized bonds. The structure may be visualized as a three-dimensional array of fused cyclohexane rings, each ring in a chair configuration with bonds staggered (Figure 11-5a). The structure of admixture may be taken as a port repent unit for the geomixture arrangement of carbon atoms in diamond crystals.

a.

b.

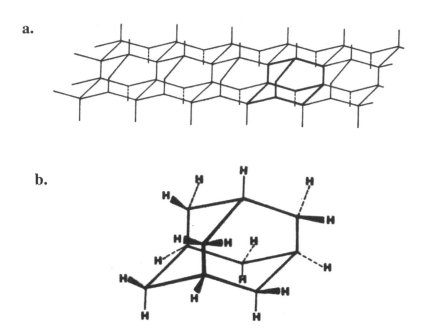

Figure 11-5. **(a) A small section of the diamond structure drawn to better illustrate the fused cyclohexane rings, each in a chair conformation. The adamantine relationship is emphasized by the heavy lines. (b) Adamantane ($C_{10}H_{16}$). Each of the four faces of the molecule constitutes a cyclohexane ring. Adamantane melts at 269°C; in contrast, most C_{10} hydrocarbons have much lower boiling points.**

The first number of diamond series can be viewed as cyclohexane (n = 1), adamantane (Figure 11-5b) (n = 2), or dimentane (n = 3). Further growth of trimentane (n = 4), etc., a polymentane is equivalent to diamond

$$C_6H_{12} \rightarrow C_{10}H_{16} \rightarrow C_{14}H_{20} \rightarrow C_{2(2n+1)}H_{4(n+2)}$$

Each chair conformation presents one lattice face. The polymentane will contain (1 + 3[n – 1]) faces or (3n – 2) faces as n increases.

Diamond undergoes a rapid transformation to graphite when heated above 1200°C at atmospheric pressure in an inert environment. One version of phase diagram of carbon is illustrated in Fig. 11-6.

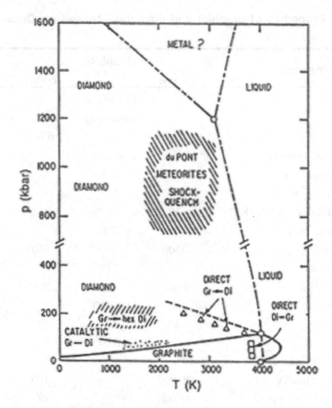

Figure 11-6. Phase diagram of carbon according to Bundy. Di and Gr are respectively diamond and graphite. DuPont, meteorites, and shock-quench may be related to carbynes. Liquid carbon has been studied at low pressure and light temperature.

Properties of graphite and diamond are listed in Table 11-1. Diamond and graphite (in place) exhibit the highest thermal conductivity and melting point. Diamond also has the highest density of any organic solid, which explains its extreme hardness and abrasion resistance.

Table 11-1. Properties of graphite and diamond. (Based on Dresselhaus, et al., 1995).

Property	Graphite a-axis	Graphite c-axis	Diamond
Lattice structure	Hexagonal		Cubic
Space group	$P6_3/mmc(D_{6h}^4)$		
Lattice constant[a] (Å)	2.462	6.708	3.567
Atomic density (C atoms/cm^3)	1.14×10^{23}		1.77×10^{23}
Specific gravity (g/cm^3)	2.26		3.515
Specific heat (cal/g · K)	0.17		0.12
Thermal conductivity[a] (W/cm · K)[b]	30	0.06	~25
Binding energy (eV/C atom)	7.4		7.2
Debye temperature (K)	2500	950	1860
Bulk modulus (GPa)	286		42.2
Elastic moduli (GPa)	1060[c]	36.5[c]	107.6[d]
Compressibility (cm^2/dyn)	2.98×10^{-12}		2.26×10^{-13}
Mohs hardness[e]	0.5	9	10
Band gap (eV)	-0.04[f]		5.47
Carrier density (10^{18}/cm^3 at 4 K)	5		0
Electron mobility[a] (cm^2/Vsec)	20,000	100	1800
Hole mobility[a] (cm^2/Vsec)	15,000	90	1500
Resistivity (Ωcm)	50×10^{-6}	1	$\sim10^{20}$
Dielectric constant[a] (low ω)	3.0	5.0	5.58
Breakdown field (V/cm)	0	0	10^7 (highest)
Magnetic susceptibility (10^{-6} cm^3/g)	-0.5	-21	--
Refractive index (visible)	--	--	2.4
Melting point (K)	4450		4500
Thermal expansion[a] (/K)	-1×10^{-6}	$+29 \times 10^{-6}$	$\sim1 \times 10^{-6}$
Velocity of sound (cm/sec)	$\sim2.63 \times 10^5$	$\sim1 \times 10^5$	$\sim1.96 \times 10^5$
Highest Raman mode (cm^{-1})	1582	--	1332

[a] Measurements at room temperature (300K).
[b] Highest reported thermal conductivity values are listed.
[c] In-plane elastic constant is C_{11} and c-axis value is C_{33}. Other elastic constants for graphite are $C_{12} = 180$, $C_{13} = 15$, $C_{44} = 4.5$ GPa.
[d] For diamond, there are three elastic constants, $C_{11} = 1040$, $C_{12} = 170$, $C_{44} = 550$ GPa.
[e] A scale based on values from 0 to 10, where 10 is the hardest material (diamond) and 1 is talc
[f] A negative band gap implies a band overlap, i.e., semimetallic behavior.

11.1.1 Carbon Fibers

The non-crystalline portion or the amorphous portion of carbon does not exist. In the chapter on asphalt density, the **mesomorphic carbon** is described in conjunction with asphaltene. The proportion of graphite in a carbon fiber can range from 0 to 100%. When the proportion is high, the fiber is referred to as graphitic or graphite-fiber. However, **graphite-fiber** is always poly-crystalline when compared to a **graphite whisker** since the whisker is of a single crystal nature. There are numerous ceramic fibers and whiskers, especially compared to some common polymer fibers (Table 11-2).

Table 11-2. Properties of various fibers and whiskers.

Material	Density[a] (g/cm³)	Tensile strength[a] (GPa)	Modulus of elasticity[a] (GPa)	Ductility (%)	Melting temp.[a] (°C)	Specific modulus[a] (10^6 m)	Specific strength[a] (10^4 m)
E-glass	2.55	3.4	72.4	4.7	<1 725	2.90	14
S-glass	2.50	4.5	86.9	5.2	<1 725	3.56	18
SiO_2	2.19	5.9	72.4	8.1	1 728	3.38	27.4
Al_2O_3	3.95	2.1	380	0.55	2 015	9.86	5.3
ZrO_2	4.84	2.1	340	0.62	2 677	7.26	4.3
Carbon (high-strength)	1.50	5.7	280	2.0	3 700	18.8	19
Carbon (high-modulus)	1.50	1.9	530	0.36	3 700	36.3	13
BN	1.90	1.4	90	1.6	2 730	4.78	7.4
Boron	2.36	3.4	380	0.89	2 030	16.4	12
B_4C	2.36	2.3	480	0.48	2 450	20.9	9.9
SiC	4.09	2.1	480	0.44	2 700	12.0	5.1
TiB_2	4.48	0.10	510	0.02	2 980	11.6	0.3
Be	1.83	1.28	300	0.4	1 277	19.7	7.1
W	19.4	4.0	410	0.98	3 410	2.2	2
Polyethylene	0.97	2.59	120	2.2	147	12.4	27.4
Kevlar	1.44	4.5	120	3.8	500	8.81	25.7
Al_2O_3 whiskers	3.96	21	430	4.9	1 982	11.0	53.3
BeO whiskers	2.85	13	340	3.8	2 550	12.3	47.0
B_4C whiskers	2.52	14	480	2.9	2 450	19.5	56.1
SiC whiskers	3.18	21	480	4.4	2 700	15.4	66.5
Si_3N_4 whiskers	3.18	14	380	3.7	—	12.1	44.5
Graphite whiskers	1.66	21	703	3.0	3 700	43	128
Cr whiskers	7.2	8.90	240	3.7	1 890	3.40	12

[a] based on D. R. Askeland, 1989

The processing of carbon fibers has to go through a mesophase. The mesophase molecules are the intermediate stages of the polyaromatic. The two major sources of the process are polyacylonitrate (PAN) or pitch. Pitch is the given fraction isolated from petroleum. The sequence of reaction during the thermal-oxidative stabilization of PAN is outlined in Figure 11-7. Different modes of **cross-linking** during carbonization to yield PNA are also listed in Fig. 11-8(a), (b), and (c).

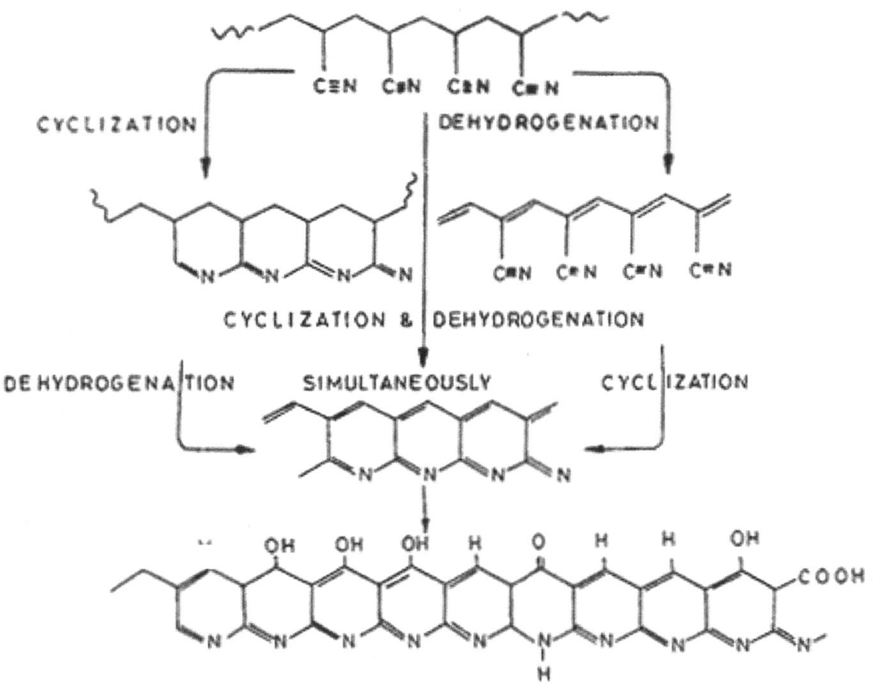

Figure 11-7. Sequence of reactions during thermooxidative stabilization of PAN. (From A. R. Gupta, et al., 1991).

a.

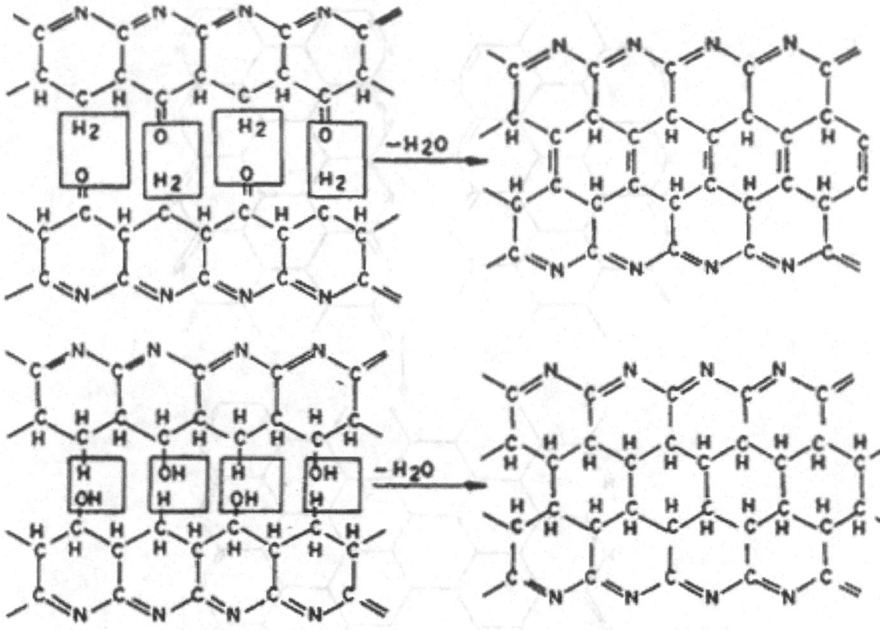

Figure 11-8(a). Intermolecular crosslinking of stabilized PAN fibers during carbonization through oxygen-containing groups. (From A.R. Cupta, et al., 1991).

b.

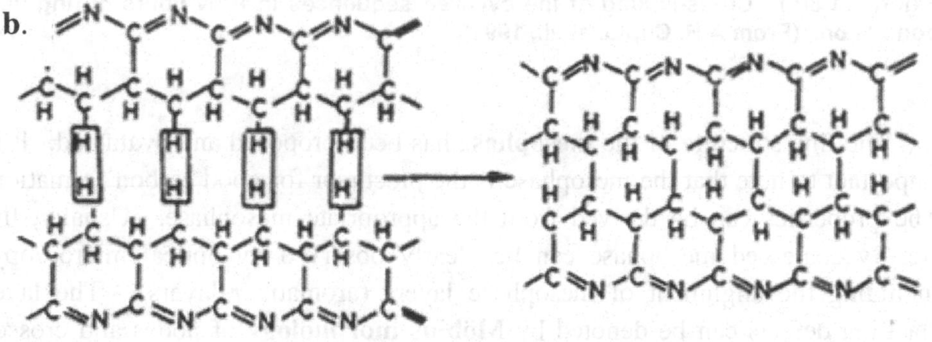

Figure 11-8(b). Intermolecular cross-linking of stabilized PAN fibers during carbonization through dehydrogenation. (From A.R. Cupta, et al., 1991).

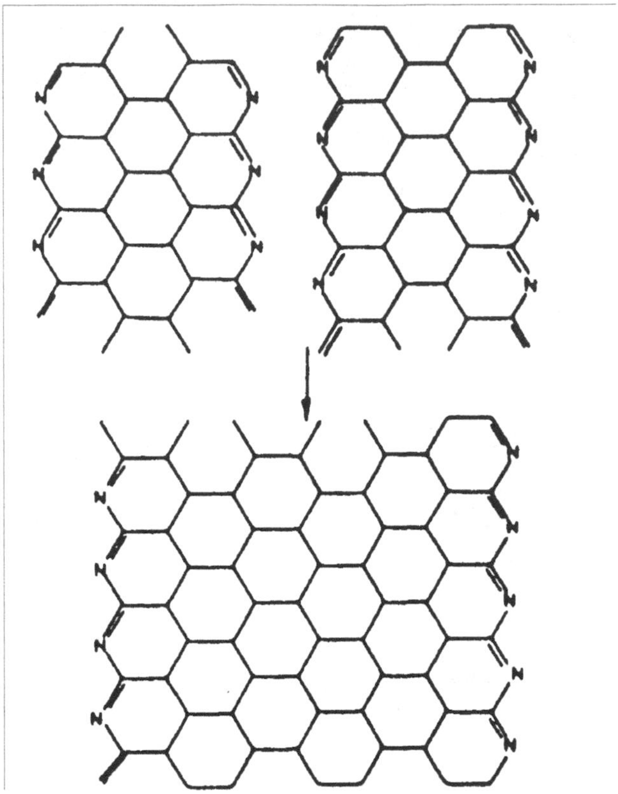

Figure 11-8(c). Cross-linking of the cyclized sequences in PAN fibers during car-bonization. (From A.R. Cupta, et al., 1991).

The fine structure of the **mesophase** has been proposed and evaluated. It is important to note that the mesophase is the precursor for good carbon formation. The properties can be derived from the appropriate mesophase. Usually, the freshly coalesced mesophase can be clearly observed by optical microscopy, including the alignment of mesophase layers (aromatic π-layers). The layer stacking defects can be denoted by **Möbius morphology** of nodes and crosses based on Honda's nomenclature:

u	co-rotating node
y	counter-rotating node
o	co-rotating cross
x	counter-rotating cross

Evidently, the polarized light response by light microscopy of the **lamelliform structure** of the four common types of the layer stacking defects (disclinations) is essential to characterizing the precursor material during carbonization. For a given microconstituent, the spacing of the extinction contour offers a measure of the **bend, twist, splay** and **disclinations** densities. An example is illustrated in Fig. 11-9(a-e).

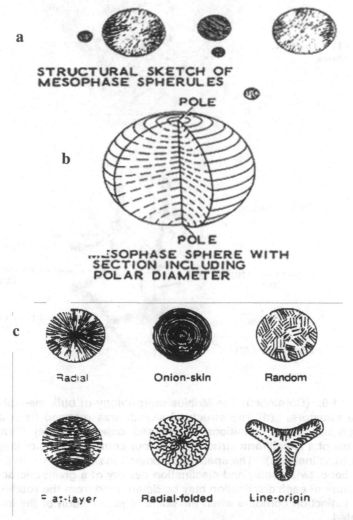

Figure 11-9. Brooks-Taylor spherules. (a) Alignment of mesophase layer; (b) spherules with sections; (c) cross-sectional microstructures of mesophase carbon fibers.

d

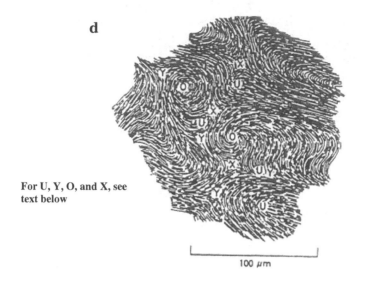

For U, Y, O, and X, see
text below

100 μm

e

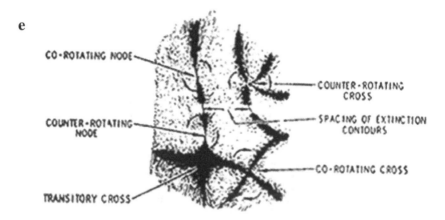

Figure 11-9. (Continued) The Mobius morphology of bulk mesophase with a coarse structure. (d) The structural sketch was mapped from a series of micrographs under conditions of crossed polarizers. (e) Polarized light response of a lamelliform structure with four common types of layer stacking defect (disclinations). The spacing of extinctions contours offers a measure of the bend, twist, splay and disclination density of a given microconstituent. The nature of each disclination may be determined from the rotation direction of the extinction contours when the plane of polarization of the incident light is rotated.

Other than pitch and PAN, petroleum resin, and asphaltene fractions, oil shale bitumen fractions and different coal liquids can serve as a precursor material for mesophase. For various source materials, the temperature of treatment is important, as shown in Fig. 11-10.

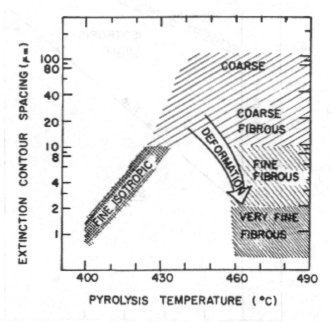

Figure 11-10. Pattern of formation of types of mesophase microconstituents with temperature of solidification (After White and Zimmer, 1976).

Both time and temperature are required for the large aromatic sheet of the mesophase. The temperature range for sphere nucleation is usually between 380-430°C. The large molecules must form and the viscosity of the formation must be low enough for molecular mobility and ordering of the mesophase. After the sphere is formed, increasing the heating time or temperature will cause more sphere to grow. Mesophase, which remains plastic above 430°C, is extensively deformed to give coarse mosaic or coarse deformed structures. When mesophase remains plastic to 460°C, the microstructure becomes finer due to deformation by bubble percolation. The fine fibrous structure is desirable because it behaves like needle coke morphology and tends to have good thermal and tensil properties. Heat treatment in 400-500°C is important because the precursor leaching leads to good morphology of carbon fiber material. The effect of aromaticity, oxygen content, and sulfur content has a bearing on mesophase microstructure (for example, see Figs. 11-11 and 11-12).

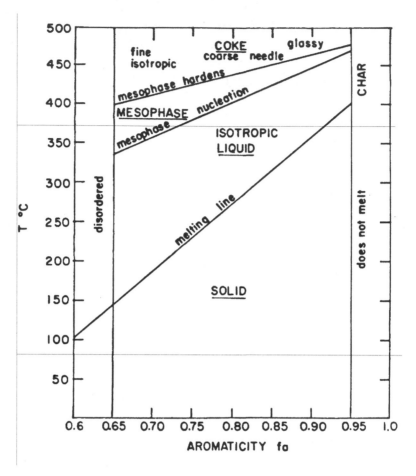

Figure 11-11. Effect of aromaticity on microstructure.

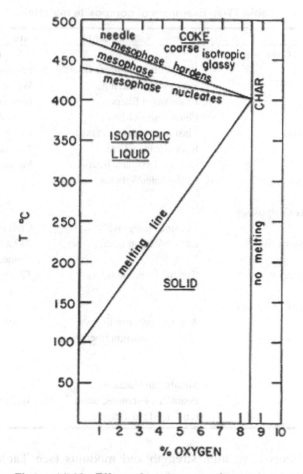

Figure 11-12. Effect of oxygen on microstructure.

11.1.2 Composites

The broad definition of **composites** implies that the material be made of two or more different parts. Often, the performance characteristics of the whole are better than that of the individual. Also, composites consist of one or more discontinuous phases distributed in one continuous phase. Examples of composite materials are illustrated in Table 11-3. The continuous phase is referred to as matrix, whereas the discontinuous phase is referred to as reinforcement. In the case of a composite in which the reinforcement is made of fibers, the orientation of the fibers determines the anisotrophy of the composite.

Table 11-3. Examples of composite materials.

	Constituents	Areas of Application
1. *Organic Matrix Composites*		
Paper, cardboard	Resin/fillers/cellulose fibers	Printing, packaging
Particle panels	Resin/wood shavings	Woodwork
Fiber panels	Resin/wood fibers	Building
Coated canvas	Pliant resins/cloth	Sports/building
Impervious materials	Elastomers/bitumen/textiles	Roofing, earthworks, etc.
Tires	Rubber/canvas/steel	Automobiles
Laminates	Resin/fillers/glass fibers	Multiple areas
Reinforced plastics	Resins/microspheres	
2. *Mineral Matrix Composites*		
Concrete	Cement/sand/gravel	Civil engineering
Carbon-carbon composites	Carbon/carbon fibers	Aviation, space, sports
		biomedicine, etc.
Ceramic composites	Ceramic/ceramic fibers	Thermomechanical items
3. *Metallic Matrix Composites*		
	Aluminum/boron fibers	Space
	Aluminum/carbon fibers	
4. *Sandwiches*		
Skins	Metals, laminates, etc.	
Cores	Foam, honeycombs, balsa	Multiple areas
	reinforced plastics, etc.	

Carbon fibers have high strength and modulus (see Table 11-4). These properties make carbon fibers useful as reinforcement for polymers, metals, carbons, and ceramics, even though the carbon fibers are brittle. The mechanisms of fiber-matrix bonding include chemical, Vander Waals, and mechanical interlocking. Polymers used often belong to either a thermoset or a thermoplast. The former class usually includes epoxy, phenolics, and furfuryl resin. The processing temperature is about 200°C. The thermalplasts, such as polyimide (PI), polyethersulfone (PES), polyetheretherketone (PEEK), polyetherimide (PEI) and polyphenyl sulfide (PPS), require processing temperatures ranging from 300-400°C. The most commonly used polymer is epoxy resin, which has the structure

Table 11-4. Tensile modulus, strength, and strain to failure of carbon fibers.

Manufacturer	Fiber	Modulus (Gpa)	Strength (Gpa)	Strain to failure (%)
PAN-based, high modulus (low strain to failure)				
Celanese	Celion GY-70	517	1.86	0.4
Hercules	HM-S Magnamite	345	2.21	0.6
Hysol Grafil	Grafil HM	370	2.75	0.7
Toray	M50	500	2.50	0.5
PAN-based, intermediate modulus (intermediate strain to failure)				
Celanese	Celion 1000	234	3.24	1.4
Hercules	IM-6	276	4.40	1.4
Hysol Grafil	Apollo IM 43-600	300	4.00	1.3
Toho Beslon	Sta-grade Besfight	240	3.73	1.6
Amoco	Thornel 300	230	3.10	1.3
PAN-base, high strain to failure				
Celanese	Celion ST	235	4.34	1.8
Hercules	AS-6	241	4.14	1.7
Hysol Grafil	Apollo HS 38-750	260	5.00	1.9
Toray	T 800	300	5.70	1.9
Mesophase pitch-based				
Amoco	Thronel P-25	140	1.40	1.0
	P-55	380	2.10	0.5
	P-75	500	2.00	0.4
	P-100	690	2.20	0.3
	P-120	820	2.20	0.2

Usually the curing of expoxy resin requires a crosslinking agent. The epoxy or hydroxyl groups are the reactive sites for cross-linking. An example is given in Fig. 11-13. The structure of common polymers serving as a matrix is given in Figure 11-14.

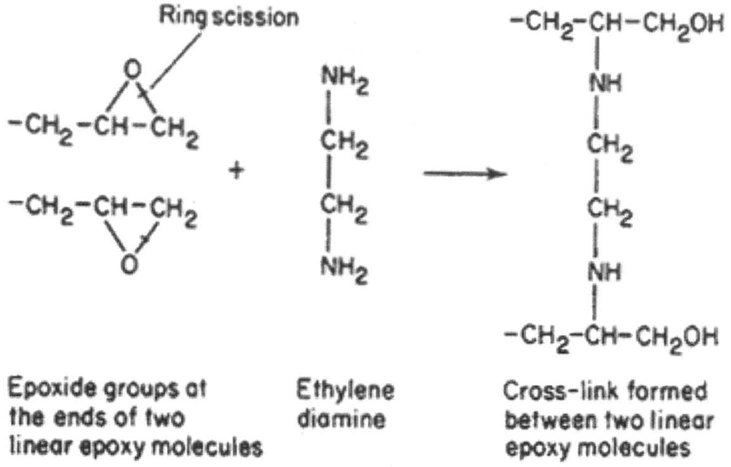

Figure 11-13. The crosslinking of expoxy resin.

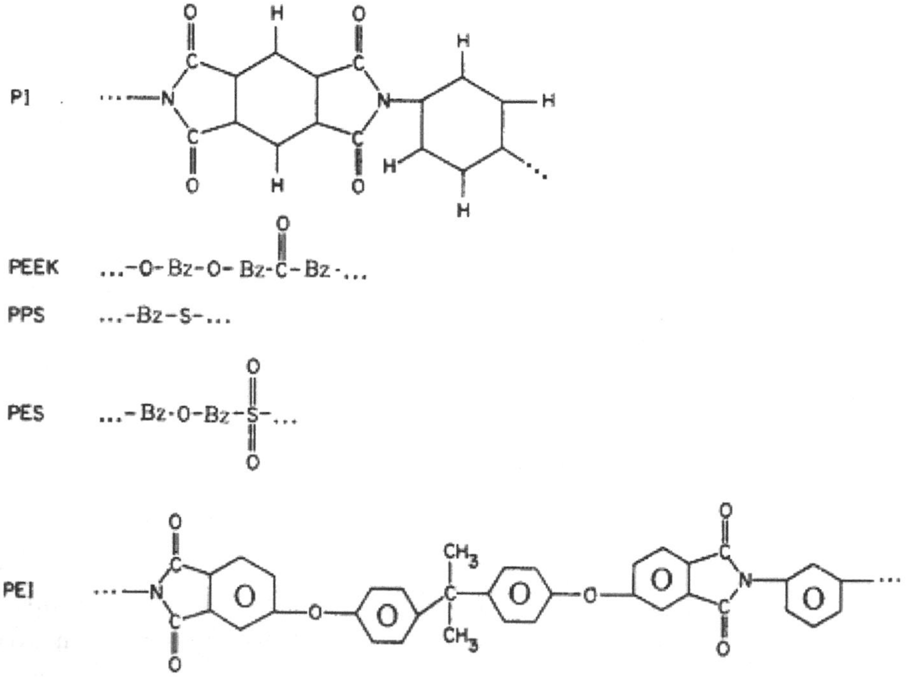

Figure 11-14. Typical thermoplasts used for matrices for carbon fibers.

Mechanical properties of common carbon fiber-polymer matrices are listed in Table 11-5. Fibers are usually made with diameters of a few microns and are gathered together with a bundle called a **strand**. A single fiber is termed a **monofilament**. Continuous or discontinuous strands, or **yarns**, are characterized by their linear density, which is the weight per unit length. The fineness of strands or yarns can be measured in terms of linear density or the **tex** number

$$1 \text{ tex} = 1 \text{ g} / \text{km} \qquad\qquad [11\text{-}1]$$
$$= 10^{-6} \text{kg} / \text{m}$$

For a two-dimensional or orthogonal woven fabric, there are two sets of interlaced strands or yarns. The lengthwise set is called the **worp** and the crosswise set is the **weft** or **till**.

Besides polymer as matrix, other matrices such as metal, carbon, and ceramic are known to form composites. If more than one type of filler or more than one type of matrix is used for the composite, they are called **hybrid composites**. For example, to fill the pores in carbon-carbon composite, a second type of matrix can be SiC, TiC, or other ceramic materials. This is often performed by **chemical vapor infiltration** (CVI) through codeposition at elevated temperatures.

Table 11-5. Properties of thermoplasts for carbon fiber polymer-matrix composites.

	PES	PEEK	PEI	PPS	PI
T_g (°C)	230[a]	170[a]	225[a]	86[a]	256[b]
Decomposition temperature (°C)	550[a]	590[a]	555[a]	527[a]	550[b]
Processing temperature (°C)	350[a]	280[a]	350[a]	316[a]	304[b]
Tensile strength (MPa)	84[c]	70[c]	105[c]	66[c]	138[b]
Modulus of elasticity (GPa)	2.4[c]	3.8[c]	3.0[c]	3.3[c]	3.4[b]
Ductility (% elongation)	30-80[c]	50-150[c]	50-65[c]	2[c]	5[b]
Izod impact (ft lb/in.)	1.6[c]	1.6[c]	1[c]	<0.5[c]	1.5[c]
Density (g/cm³)	1.37[c]	1.31[c]	1.27[c]	1.3[c]	1.37[b]

[a]From Whang and Lin, 1990
[b]From Sherman et al. 1988
[c]From Askeland, 1989

11.1.3   Reinforcement Mechanism

There are a number of fiber-reinforced composites. For example, matrix materials such as thermoplastic or thermosetting resins, reinforced with carbon, glass, organic polymers and other fibers.

When there is little or no chemical interaction between matrix and reinforcement material the relative density, ρ_c, of such composite materials can be calculated by the 'rule of mixtures'. Then:

$$\rho_c = \rho_r V_r + \rho_m (1 - V_r)$$ [11-2]

Where V_r is the volume of the reinforcement fraction and ρ_r and ρ_m the relative densities of reinforcement and matrix respectively. The following is an example for such a composite.

[**Example 11-1**] In a fiber-reinforced composite the continuously aligned cylindrical fibers are packed as closely as possible in a single direction. Given that the relative densities of fiber material and matrix material are 2.41 and 1.28 respectively, calculate the maximum relative density of the resultant composite assuming pure ingredients. (After Higgins)

The closest packing of cylindrical fibers is achieved in a hexagonal pattern as indicated in the following cross sectional diagram:

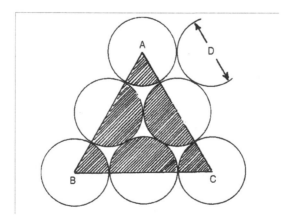

Figure 11-15. Elements in close hexagonal packing.

Consider the surface area of the cross-section within the triangular element, ABC.

The length of the side of triangle ABC is:

$$D + 2\left(\frac{D}{2}\right) = 2D$$

Therefore, the area is:

$$\frac{2D \times D\sqrt{3}}{2} = D^2\sqrt{3}$$

The number of circles within the triangle is:

$$\left(3 \times 3 \ \frac{1}{2}\right) + \left(3 \times 3 \ \frac{1}{6}\right) = 2 \quad \text{radius D/2}$$

Packing density is given by:

$$\frac{\text{Area of 2 circles}}{\text{Area of } \triangle \text{ABC}} = \frac{2\pi D^2 / 4}{D^2\sqrt{3}}$$

$$= \frac{\pi}{2\sqrt{3}}$$

$$= .907$$

Let ρ_c be the relative density of the composite, then using

$$\rho_c = \rho_r V_r + \rho_m(1 - V_r):$$

$$\rho_c = (2.41 \times 0.907) + 1.28(1 - 0.907)$$

$$= 2.19 + 0.119 = 2.31$$

The relative density of the composite is 2.31.

Usually, the reinforced fibers contribute the individual mechanical properties such as strength, stiffness, creep, and fatigue properties, while the matrix will be generally weaker than the fiber. In many situations it gives protection against crack propagation perpendicular to the fiber direction as shown in Figure 11-16.

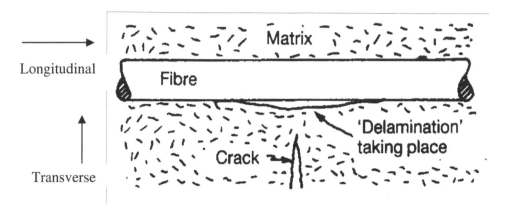

Figure 11-16. General view of fiber in a matrix.

A perfect bonding between fibers and matrix will yield tensile strength of the composite,

$$\sigma_c = \sigma_f V_f + \sigma_r (1 - V_f) \qquad [11\text{-}3]$$

where σ_c is the fracture strength of the fiber material and σ_r the sheer strength of the resin matrix, whilst V_f is the fiber volume fraction. Similarly, the moduli have the relationship

$$E_c = E_f V_f + E_r (1 - V_f) \qquad [11\text{-}4]$$

Also, since $\sigma_r = F_r / A_r$, $\sigma_f = F_f / A_F$, F is loading

$$F_f / F_r = E_f V_f / E_r V_r \qquad [11\text{-}5]$$

All the statements for Equations 11-3 and 11-4 are for the load applied to the longitudinal direction in relation to the direction of the fiber. If the load is applied to the transverse direction to the fiber, then

$$\sigma_c = \sigma_f = \sigma_r = \sigma$$

the strain or deformation of the entire composite is

$$\varepsilon_c = \varepsilon_f V_f + \varepsilon_r V_r \qquad\qquad [11\text{-}6]$$

or since $\varepsilon = \sigma/E$, and divided by σ

$$1/E_c = V_f/E_f + V_r/E_r \qquad\qquad [11\text{-}7]$$

or

$$E_c = \frac{E_r E_f}{V_r E_f + V_f E_r}$$
$$= \frac{E_r E_f}{(1 - V_f)E_f + V_f E_r} \qquad\qquad [11\text{-}8]$$

[Example 11-2]   A continuous and aligned glass-reinforced composite consists of 40 vol% of glass fibers having a modulus of 10×10^6 psi (69×10^3 MPa) and 60 vol% of a polyester resin that, when hardened, displays a modulus of 5.0×10^5 psi (3.4×10^3 MPa).

   (a) Compute the modulus of elasticity of this composite in the longitudinal direction.
   (b) If the cross-sectional area is 0.4 in.^{2} (258 mm^2) and a stress of 7000 psi (48.3 MPa) is applied in this longitudinal direction, compute the magnitude of the load carried by each of the fiber and matrix phases.
   (c) Determine the strain that is sustained by each phase when the stress in part b is applied.

(d) Assuming tensile strengths of 500,000 and 10,000 psi (3.5×10^3 and 69 MPa), respectively, for glass fibers and polyester resin, determine the longitudinal tensile strength of this fiber composite.

(e) Compute the elastic modulus of the composite material described in above Example, but assume that the stress is applied perpendicular to the direction of fiber alignment. (After Calister, Jr.)

(a) The module of elasticity of the composite is calculated using Equation 11-4.

$$E_c = (5.0 \times 10^5 \, \text{psi})(0.6) + (10 \times 10^6 \, \text{psi})(0.4)$$
$$= 4.3 \times 10^6 \, \text{psi} \, (3 \times 10^4 \, \text{MPa})$$

(b) To solve this portion of the problem, first find the ratio of the fiber load to matrix load, using Equation 11-5; thus,

$$\frac{F_f}{F_r} = \frac{(10 \times 10^6 \, \text{psi})(0.4)}{(5 \times 10^5 \, \text{psi})(0.6)} = 13.3$$

or $F_f = 13.3 \, F_r$

In addition, the total force sustained by the composite F_c ma y by computed from the applied stress σ and total composite cross-sectional area A_c according to

$$F_c = A_c \sigma = (0.4 \, \text{in.}^2)(7000 \, \text{psi})$$
$$= 2800 \, \text{lb}_f \, (12,400 \, \text{N})$$

However, this total load is just the sum of the loads carried by fiber and matrix phases, that is

$$F_c = F_f + F_r = 2800 \, \text{lb}_f \, (12,400 \, \text{N})$$

Substitution for F_f from the above yields

$$13.3 \, F_r + F_r = 2800 \, \text{lb}_f$$

or

$$F_r = 195\,\mathrm{lb}_f\,(870\,\mathrm{N})$$

whereas

$$F_f = F_c - F_r = 2800\,\mathrm{lb}_f - 195\,\mathrm{lb}_f = 2605\,\mathrm{lb}_f\,(11{,}530\,\mathrm{N})$$

Thus, the fiber phase supports the vast majority of the applied load.

(c) The stress for both fiber and matrix phases must first by calculated. Then, by using the elastic modulus for each (from part a), the strain values may be determined.

For stress calculations, phase cross-sectional areas are necessary:

$$A_r = V_r A_c = (0.6)(0.4\,\mathrm{in.}^2) = 0.24\,\mathrm{in.}^2\,(155\,\mathrm{mm2})$$

and

$$A_f = V_f A_c = (0.4)(0.4\,\mathrm{in.}^2) = 0.16\,\mathrm{in.}^2\,(103\,\mathrm{mm2})$$

Thus,

$$\sigma_r = \frac{F_r}{A_r} = \frac{195\,\mathrm{lb}_f}{0.24\,\mathrm{in.}^2} = 812.5\,\mathrm{psi}\,(5.6\,\mathrm{MPa})$$

$$\sigma_f = \frac{F_f}{A_f} = \frac{2605\,\mathrm{lb}_f}{0.16\,\mathrm{in.}^2} = 16{,}280\,\mathrm{psi}\,(112.3\,\mathrm{MPa})$$

Finally, strains are computed as

$$\varepsilon_r = \frac{\sigma_r}{E_r} = \frac{812.5\,\mathrm{psi}}{5\times10^5\,\mathrm{psi}} = 1.63\times10^{-3}$$

$$\varepsilon_f = \frac{\sigma_f}{E_f} = \frac{16{,}280\,\mathrm{psi}}{10\times10^6\,\mathrm{psi}} = 1.63\times10^{-3}$$

Therefore, strains for both matrix and fiber phases are identical, which they should be.

(d) For tensile strength TS, we can write

$$(TS)_c = (TS)_r V_r + (TS)_f V_f$$

which, for this particular composite (in the longitudinal direction), is

$$(TS)_c = (10,000\,\text{psi})(0.60) + (500,000\,\text{psi})(0.40)$$
$$= 206,000\,\text{psi}\,(1,420\,\text{MPa})$$

(e) According to Equation 11-8,

$$E_c = \frac{(5\times10^5\,\text{psi})(10\times10^6\,\text{psi})}{(0.6)(10\times10^6\,\text{psi}) + (0.4)(5\times10^5\,\text{psi})}$$
$$= 8.1\times10^5\,\text{psi}\,(5.6\times10^3\,\text{MPa})$$

This value for E_c is slightly greater than that of the matrix phase, but from part (a), only approximately one fifth of the modulus of elasticity along the fiber direction, which indicates the degree of anisotropy of continuous and oriented fiber composites.

11.2 FULLERENE

Recall admentane which was discussed in the previous section. The diamond-like cage molecules, or the diamondoid hydrocarbons form a group of repeating units of diamond lattice (see Figure 11-17).

The last molecule in Figure 11-17, the buckminsterfullerene, or C_{60}, has icosahedron symmetry. It is named after Buckminster Fuller, hence the name fullerene. Inspection of the fullerene molecule structure shows that every pentagon is surrounded by five hexagons. This portion of the structure has the form of corannulene (Fig. 11-18a and b). Another subunit is the structure of the molecule pyraclene which consists of two pentagons and two hexagons (Fig. 11-18c).

Fullerene consists of exactly twelve pentagons and twenty hexagons which appear as a soccer ball (Fig. 11-19). It is made of 60 carbon atoms and has 30 double bonds. The angle between a double bond and its adjacent pentagonal face is

$$\psi = \cos^{-1}[1/2\cos(3/5)\pi \qquad\qquad [11\text{-}9]$$

The angle between two adjacent hexagonal faces is

$$\phi = \cos^{-1}[8/3(\sin(3/10)\pi)^2 - 1] \qquad\qquad [11\text{-}10]$$

and the angle between adjacent hexagonal and pentagonal faces is

$$\theta = 1/2(\pi - \phi) - \psi \qquad\qquad [11\text{-}11]$$

Some physical constituents of the fullerene are listed in the Table 11-6.

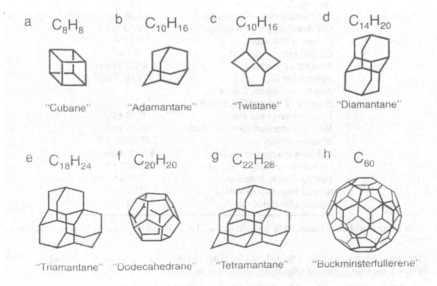

a C_8H_8 b $C_{10}H_{16}$ c $C_{10}H_{16}$ d $C_{14}H_{20}$

"Cubane"   "Adamantane"   "Twistane"   "Diamantane"

e $C_{18}H_{24}$ f $C_{20}H_{20}$ g $C_{22}H_{28}$ h C_{60}

"Triamantane"   "Dodecahedrane"   "Tetramantane"   "Buckminsterfullerene"

Figure 11-17. Some diamond-like and caged molecules.

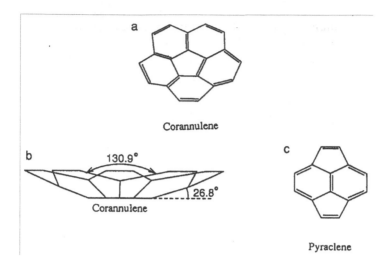

Figure 11-18. (a) Top view and (b) side view of the nonpolar corannulene molecule. (c) A pyraclene or pyracylene molecule. The pyracylene and corannulene motifs are seen on the surface of C_{60} and other fullerenes.

Table 11-6. Physical constants for C_{60} molecules.

Quantity	Value
Average C–C distance	1.44 Å
C–C bond length on a pentagon	1.46 Å
C–C bond length on a hexagon	1.40 Å
C_{60} mean ball diameter[a]	7.10 Å
C_{60} ball outer diameter[b]	10.34 Å
Moment of inertia I	1.0×10^{-43} kg m^2
Volume per C_{60}	1.87×10^{-22}/cm^3
Number of distinct C sites	1
Number of distinct C–C bonds	2
Binding energy per atom[c]	7.40 eV
Heat of formation (per g C atom)	10.16 kcal
Electron affinity	2.65±0.05 eV
Cohesive energy per C atom	1.4 eV/atom
Spin–orbit splitting of C(2p)	0.00022 eV
First ionization potential	7.58 eV
Second ionization potential	11.5 eV
Optical absorption edge[d]	1.65 eV

[a] This value was obtained from NMR measurements. The calculated geometric value for the diameter is 7.09 Å.

[b] This value for the outer diameter is found by assuming the thickness of the C_{60} shell to be 3.35 Å. In the solid, the C_{60}-C_{60} nearest neighboring distance is 10.02 Å.

[c] The binding energy for C_{60} is believed to be ~0.7 eV/C atom less than for graphite, though literature values for both are given as 7.4 eV/C atom. The reason for the apparent inconsistency is attributed to differences in calculational techniques.

[d] Literature values for the optical absorption edge for the free C_{60} molrecules in solution range between 1.55 and 2.3 eV.

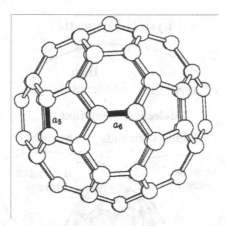

Figure 11-19. The C$_{60}$ molecule showing single bonds (a_5) and double bonds (a_6).

Fullerenes also undergo many chemical reactions. A summary of the reactions in graphical form is shown in Figure 11-20. The unique proportion of the fullerene can lead to the possible application as shown in Fig. 11-21. Some higher homologs of fullerenes such as C$_{70}$, C$_{140}$, etc. are also known (Fig. 11-22).

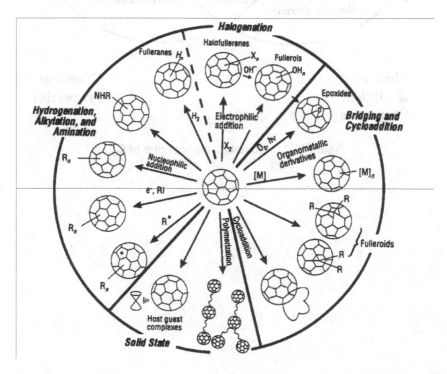

Figure 11-20. Some general categories of reactions known to occur with C$_{60}$ (based on M.S. Dresselhaus et al., 1995). Here R denotes a functional group.

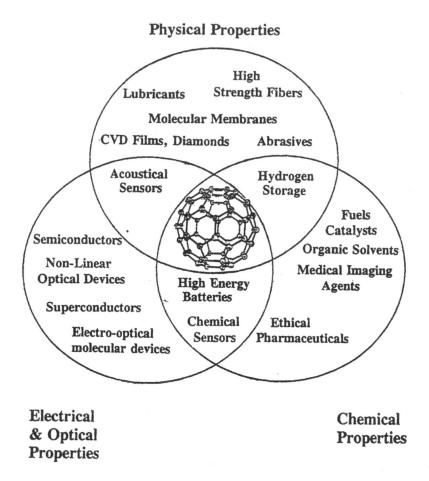

Figure 11-21. Possible applications of C_{60}.

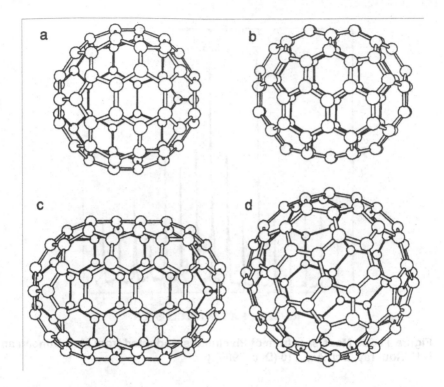

Figure 11-22. (a) The icosahedral C$_{60}$ molecule. (b) the rugby-shaped C$_{70}$ molecule (D_{5h} symmetry). (c) The C$_{80}$ isomer, which is an extended rugby ball (D_{5d} symmetry). (d) The C$_{80}$ isomer, which I an icosahedron (I_h symmetry). (After M.S. Dresselhaus et al., 1995)

11.3 NANOTECHNOLOGY

Nanotechnology deals with various structures of matter having the dimension of the order of nanometer (nm), which is one billionth of a meter. Materials do not behave in their usual manner when they are reduced to this small level of size. **Nanoparticles** are considered to be a number of atoms or molecules bonded together with a radius of < 100 nm. The classification of **atomic clusters** according to size is arbitrary. For example, a cluster of one nanometer radius contains approximately 25 atoms. Usually, the cluster numbers are obtained through mass spectrum data. For example, the lead clusters of 7 and 10 are more stable (see Figure 11-23).

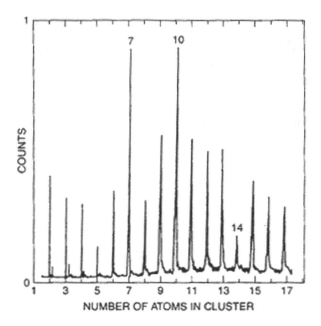

Figure 11-23. Mass spectrum of Pb clusters. [Adapted from M.A. Duncan and D.H. Rouvray, *Sci. Am.* 110 (Dec. 1989).]

These numbers are similar to the structural magic numbers in the sequence of

$$N = 1, 13, 55, 147, 309, 561...$$

These are based on calculations for layers of particles in face-centered cubic particles based on the formula

$$N = 1/3(10n^3 - 15n^2 + 11n - 3) \qquad [11\text{-}12]$$

Some examples of nanoparticles are as follows:

$$Au_{53} (PPh_3)_{12}Cl_6$$

$$Pt_{309} (1, 10\text{-phenanthrolene})_{36} O_{30}$$

$$Pd_{561} (1, 10\text{-phenanthrolene})_{36} O_{200}$$

Actually, fullerene can be treated as a stable carbon cluster of 60 carbons as indicated by the mass spectrum of carbon clusters (Figure 11-24).

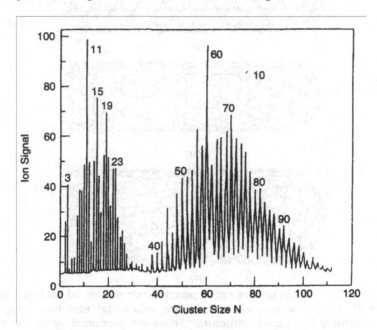

Figure 11-24. Mass spectrum of carbon clusters. The C$_{60}$ and C$_{70}$ fullerene peaks are evident. (Adopted from S. Sugano and H. Koizuni, in *Microluster Physics*, Springer-Verlag, Heidelberg, 1998.)

Another intriguing structure of nano-size is the so-called **nanotube**. One can treat nanotubes as a sheet of graphite rolled into a tube with bonds at the end that close the tube. There is the armchair structure. Other structures are equally important, such as the zigzag structure and the chiral structure. All these belong to the single-walled nanotube type, having a diameter of 2 nm and of length of 200 μm (Figure 11-25). Effectively, they behave as a one-dimensional structure referred to often as **nanowire**.

Multi-walled carbon nanotubes also can be produced as Figure 11-26 shows. Nanotubes not only exhibit semi-conductive properties but also enhance mechanical strength as well, indicated in Figure 11-27.

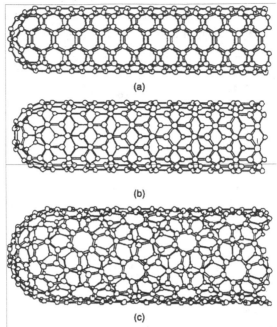

(a)

(b)

(c)

Figure 11-25. Illustration of some possible structures of carbon nanotubes, depending on how graphite sheets are rolled: (a) armchair structure; (b) zigzag structure; (c) chiral structure. These are produced by carbon electrode of 5-20 μm diameter and separated by 1 mm at 500 Torr pressure of flowing He of 20-25 V with Co, Ni, or Fe as catalyst. (After Poole Jr. and Owens, 2003).

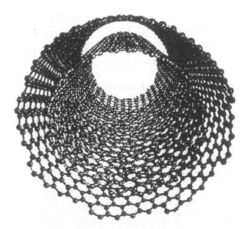

Figure 11-26. Multi-walled carbon nanotube. (Based on Poole Jr. and Owen, 2003)

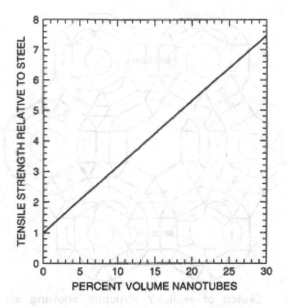

Figure 11-27. Tensile strength of steel versus volume fraction of carbon nanotubes calculated by the Kelly-Tyson formula. The nanotubes were 100 μm long and 10 nm in diameter. (After Poole Jr. and Owens, 2003)

Since most nanoparticles approach the sizes and wavelengths of optical and vibrational spectroscopy, many nanostructures exhibit unique spectroscopic properties. For example, the study of **thermoluminescence** on cadmium sulfide clusters in zeolite-Y cages (Figure 11-28) yields a low peak indicating trapped electron-hole pairs in metastable states (Figure 11-29).

Many nonstructures having particular geometrical patterns find applications in catalysis. One of these is **pillared clays** that can absorb water molecules or positive and negative ions and undergo exchange reactions with the outside species. Usually, Keggin ions in clay can have the pillaring process. Figure 11-30 depicts the positively-charged Al_{13} ion which can form nano-sized pillars. In the center of the Keggin positive ion, there is one tetrahedral group AlO_4. **Keggin ions** (12 polyhedra) and the related **Dawson ions** (18 polyhedra) are essential in catalysis.

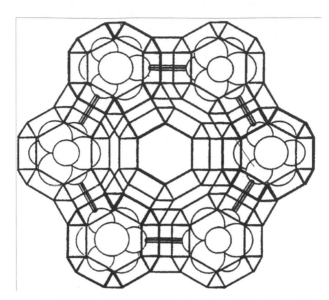

Figure 11-28.   Sketch of zeolite-Y structure showing six sodalite cages (diameter ~0.5 nm) occupied by Cd_4S_4 tetrahedral clusters, and one empty supercage (diameter ~1.3 nm) in the center. (From H. Herron and Y. Wang, in *Nanomaterials, Synthesis, Properties and Application*, A. S. Edelstein, ed., IOP, Bristol, UK, 1996, p. 73.)

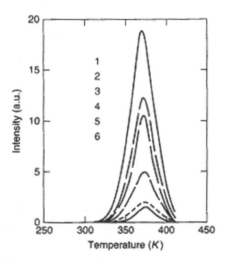

Figure 11-29. Glow curves of CdS clusters in zeolite-Y for CdS loadings of 1, 3, 5, and 20 wt% (curves 1-4, respectively). Curve 5 is for bulk CdS, and curve 6 is for mechanical mixture of CdS with zeolite-Y powder. [From W. Chen, Z. G. Yang, and L. Y. Lin, *J. Lumin.* 71, 151 (1997).]

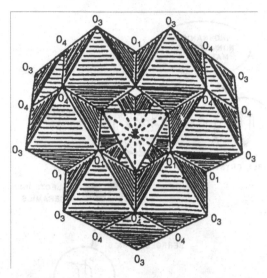

Figure 11-30. Sketch of the structure of the $\{AlO_4[Al(OH)_2H_2O]_{12}\}^{7+}$ Keggin ion with one tetrahedral group AlO_4 in the center position, surrounded by 12 AlO_6 octahedra at the remaining sites where the oxygen of O_1 of the octahedra belong to hydroxyl groups OH, or to water molecules H_2O. [From A. Clearfield, in Moser (1996), Chapter 14, p. 348.]

11.4 CERAMICS — PIEZOELECTRICS AND ELECTROOPTICS

Ceramics have been dated back to the earliest human civilization. Modern high technology devices from material sciences such as positioners, relays, tweeters, loudspeakers, ignitors, shelters, modulators, filters, **positive temperature co-efficient materials** (PTC's), etc. are all derived from ceramics. In general, inorganic non-metallics can be divided into: single crystals, mesomorphic or polycrystalline materials, and amorphorous materials. The ceramics, especially the electronic ceramics, are derived from the polycrystalline materials as indicated by Fig. 11-31. Actually, they are very close among ferroelectronics, piezoelectrics and electrooptics. They all are derived from ceramics.

Piezoelectricity (pressure-electricity) is the phenomenon whereby electric polarization (charge) is generated from a mechanical stress. The converse effect is also exhibited whereas the mechanical movement is generated by the application of an electric field. Both phenomena were observed by crystals such as quartz, tourmaline, zine blende, or Rochelle salt. The mechanism can be

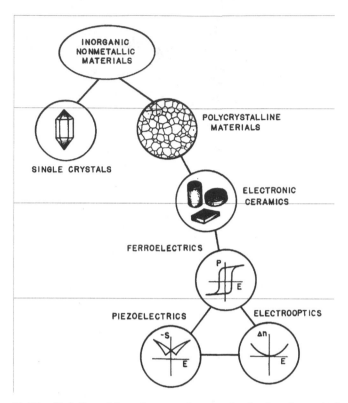

Figure 11-31. Relationship of ceramics and electronic subclasses to all inorganic, nonmetallic solids.

explained by the fact that these crystals of ceramics contain a multitude of tiny elements of dipole (piezoelectric crystallites) at random orientation. These elements will be active (oriented) when certain stress (tensile or compression) is applied directionally during fabrication (the process is called poling). Let **S** be the strain and **T** be the stress, **E** is electric field strength, **P** is polarization, and Δ n is the birefrigencl. The three classes of ceramics, piezoelectrics, ferroelectrics, and electrooptics as illustrated in Fig. 11-31 can be differentiated.

There is an entirely different mathematic convention (notation) used by the ceramic profession. As indicated by Fig. 11-32, the notation of axes for the poled ceramic element not only contains the 1, 2, 3 directions, but 4, 5, and 6 are used to indicate the effects as well. For certain coefficients, a double subscript is used. The first of the two subscripts refers to the electrical direction (electric field or dielectric displacement) and the second refers to the mechanical direction (stress or strain). For example, d_{31}, is the lateral charge or strain coefficient.

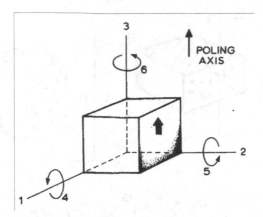

Figure 11-32. Notation of axes for poled ceramic element.

Usually a superscript symbol is used to indicate the quantity of a property that is holding constant when measuring another property. For example, s^E is the compliance at constant electric field (electrode shorted). The following coefficients are often used:

s = compliance

Σ = dielectric constant

d = charge or strain coefficient

g = voltage coefficient

k = electromechanical coupling factor

Some of the relations in electronic ceramics are expressed in matrix form. For example, the direct and converse effects can be generalized as:

$$\mathbf{D} = d\,\mathbf{T} + \Sigma^T\,\mathbf{E} \qquad\qquad [11\text{-}13]$$

$$\mathbf{S} = s^E\,\mathbf{T} + d\,\mathbf{E} \qquad\qquad [11\text{-}14]$$

$$\mathbf{G} = \mathbf{S}\,/\,\mathbf{D} = \mathbf{E}\,/\,\mathbf{T} \qquad\qquad [11\text{-}15]$$

Poling, in relation to piezoelectric ceramics, can be illustrated by Fig. 11-33. Once poled, the ceramic acts like a single entity for the whole body. Fig. 11-33(c) perhaps is a representation of a typical piezoelectric ceramic. Similarly, ferroelectrics contain polarizable elements of dipole. The polarization can be induced by stress spontaneously to form permanent dipoles in structure. The change of polarization with electric field is reversible. Such a hysteresis loop is

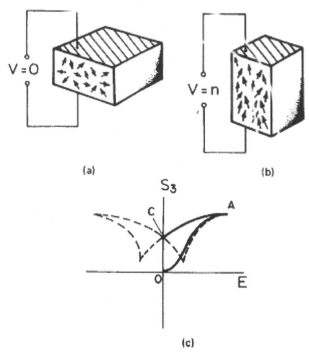

Figure 11-33. Dimensional changes of a ceramic as a result of poling: (a) before poling (virgin state); (b) after poling. The longitudinal strain as a result of poling and switching is illustrated in part (c), where the heavy black line traces the poling strain S_3 from a virgin state 0 to saturation state A and remnant state C.

illustrated in Fig. 11-33(a) and is a unique property of ferroelectric ceramics. The remanent polarization, P_R, is one of the properties of this type of ceramics. The accompanying strain involved in tranversing the hysteresis is also given in Fig. 11-34, referred to as the butterfly loop.

The development of optical ceramics usually depends on the degree of refringence (or light bending) of the optic axis and the other octagonal axis from the crystal.

$$\Delta n = \left| n_e - n_o \right| \qquad\qquad [11\text{-}16]$$

where n_e is the index of refraction of a light wave vibrating in a plane parallel to the optic axis and traveling in a direction perpendicular to it; and n_o is the index of refraction of a wave vibrating parallel to either of the other axes and traveling in the direction of the optic axis. Three most common types of electrooooptic effects are: quadratic bifringence, linear bifrigence, and memory scattering.

These are illustrated respectively in Fig. 11-35(a, b, and c). The transmitted light intensity, I, is as follows:

$$I = I_i \sin^2 \frac{\Delta n t \pi}{\lambda} \qquad [11\text{-}17]$$

where t is thickness of the plate and I_i is incident light intensity and λ is light wavelength. PLZT is a very important class of electronic ceramics and will be explained in more detail.

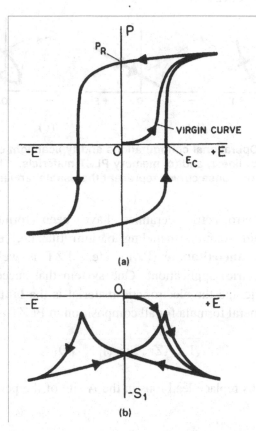

(a)

(b)

Figure 11-34. (a) Typical hysteresis loop for a high-coercive-field ceramic; (b) associated lateral strain (contraction).

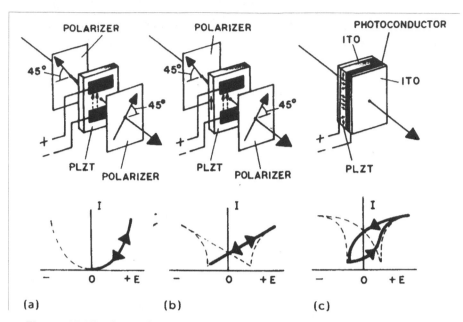

Figure 11-35. Operational configurations and typical light output response of (a) quadratic, (b) linear, and (c) memory PLZT materials. The heavy accented portions of the response curves represent the usable ranges.

Historically, ferroelectric ceramics have been formulated from the composition of solid solution including barium titanate, lead titanate, lead microbate, **lead zirconetetitanate** (PZT), etc. PZT is well known for the transducer (piezoelectric) application. One system that successfully embraces both the piezoelectric and the electrooptic material is the **lanthanum modified PZT** (PLZT). A general formula for all composition in PLZT is

$$Pb_{1-x} La_x (Zr_y Ti_{1-y})_{1-x/4} O_3$$

where lanthanum ions replace lead ions in the A site of the perovskite ABO_3 ion structure in Fig. 11-36.

Traditionally, processing and preparation of PLZT employing ZrO_2, TiO_2, La_2O_3 and PbO as mixed oxide lead to the composition of La/Zr/Ti where Zr/Ti ratio ascends to 100 with percent value of La as doping. For example, in Table 11-7 a number of electromechanical properties are listed for different PLZT compositions.

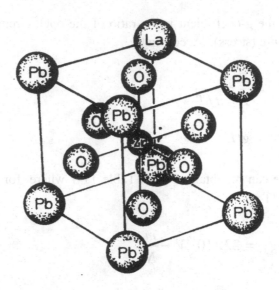

Figure 11-36. Perovskite ABO$_3$ crystal unit cell.

Table 11-7. Electromechanical properties of PLZT compositions.

Composition La / Zr / Ti	k_P	d_{33} ($\times 10^{12}$ m/V)	g_{33} ($\times 10^3$ V-m/N)
2 / 65 / 35	0.450	150	23.0
7 / 65 / 35	0.620	400	21.6
8 / 65 / 35	0.650	682	20.0
4 / 55 / 45	0.670	490	21.6
5 / 55 / 45	0.705	578	18.2
6 / 55 / 45	0.620	417	16.6
7 / 55 / 45	0.565	405	15.1
8 / 40 / 60	0.343	–	–
12 / 40 / 60	0.470	235	12.3
7 / 60 / 40	0.720	710	22.2
Ba Ti O$_3$	0.360	190	12.6

[**Example 11-3**] Calculate the magnitude of the voltage generated from a gas ignitor made of PLZT of a composition of 2 / 65 / 35, assuming the applied stress is 8000psi and the device thickness is 15mm.

Recall that the g-coefficient is the ratio of the open circuit electric field to the applied pressure (stress). Accordingly,

$$g = \frac{E}{T} = \frac{V}{t/T}$$

$$V = gtT$$

The g-value can be obtained from Table 11-7 where, for a composition of 2 / 65 / 35 of PLZT

$$g_{33} = 23 \times 10^{-3} V - \frac{m}{N}$$

Thus

$$V = \left(23 \times 10^{-3} V - \frac{m}{N}\right)\left(10^{-3} m\right)\left(5516 \times 10^4 \frac{N}{m^2}\right) = 1.9 \times 10^4 V$$

This voltage of 19kV is high enough to produce a spark across the conventional $\frac{1}{4}$ in (6mm) gap for combustion of gas.

REFERENCES

11-1. Bernard, A. S., *The Diamond Formula: Diamond Synthesis – a Gemmlogical Perspective*, Butterworth-Heinemann, Woburn, MA, 2000.

11-2. Berthelot, J. M., *Composite Materials: Mechanical Behavior and Structural Analysis*, Springer, New York, 1999.

11-3. Callister, W. D., Jr., *Materials Science and Engineering: An Introduction*, John Wiley & Sons, Inc., New York, 1994.

11-4. Dresselhaus, M. S. G. Dresselhaus, and P.C. Eklund, *Science of Fullerenes and Carbon Nanotubes*, Academic Press, Inc., San Diego.

11-5. Higgins, R. A., *Properties of Engineering Material,* 2nd ed., Industrial Press Inc., New York, 1994.

11-6. D. Koruga, S. Hameroff, J. Withers, R. Loutfy, M. Sundareshan, *Fullerene C_{60}: History, Physics, Nanobiology, Nanotechnology*, Elsevier Science Publishers B.V., Amsterdam, 1993.

11-7. S. Mitura, *Nanotechnology in Materials Science,* Elsevier Science Publishers, Amsterdam, 2000.

11-8. Mosser, R., ed., *Advanced Catalysts and Nanostructural Materials*, Academic Press, San Diego, 1996.

11-9. Poole, C. P., Jr., and F. J. Owens, *Introduction to Nanotechnology*, Wiley-Interscience, Hoboken, NJ, 2003.

11-10. Editors of Scientific American, *Understanding Nanotechnology*, Warner Books, New York, 2002.

11-11. Yen, T. F., "Resonance Topology of Polynuclear Aromatic Hydrocarbons", *Theoret. Chim. Acta (Berlin)*, 20, 399-404 (1971).

11-12. Yen, T. F, "Terrestrial and Extraterrestrial Stable Organic Molecules" in *Chemistry in Space Research* (R. F. Landel and A. Rembaum, ed.), American Elsevier, New York, 1972, pp. 105-153.

11-13. W.D.Callister, Jr. *Material Science and Engineering, An Introduction*, 6th ed. John Wiley, 2003.

11-14. D.R.Askeland and P.P.Phule, *The Scinece and Engineering of Materials*, 4th ed. Thomson Brooks/Cole, Pacific Grove, CA, 2003.

PROBLEM SET

1. A composite composed of long parallel fibers of Kevlar in an epoxy resin matrix is required to have a tensile strength of 1600 MPa. If the strengths of Kevlar and epoxy resin are respectively, 2560 MPa and 80 MPa, what volume fraction of Kevlar must be used in the composite? (After Higgins).

2. A band of polyisoprene is to hold a bundle of steel rods together for up to one year. If the stress on the band is less than 1500 psi, the band will not hold the rods tightly. Design an initial stress that must be applied to a polyisoprene band when it is slipped over the rods. A series of tests have shown that an initial stress of 1000 psi decreased to 980 psi after six weeks. (From Askeland and Phule, 2003).

3. For material relation, the performance index P is important. Usually,

$$P = {\tau_f}^{2/3} \Big/ \rho \; ; \; E = \overline{C} \Big/ P$$

where τ_f is strength, ρ is density, $\overline{C}$ is relative cost (unit cost/iron carbon steel) and E is economical rank. Calculate the performance index and also the ranking of economic scale for the following five engineering materials.

Type	Material	P Mg/m^3	τ_f MPa	$\overline{C}$ $/$
A	Carbon fiber-reinforced composite	1.5	1140	80
B	Glass fiber-reinforced composite	2.0	1060	40
C	Aluminum Alloy	2.8	300	15
D	Titanium Alloy	4.4	525	110
E	4340 Steel	7.8	780	5

CHAPTER **12**

ASPHALT CHEMISTRY

*A*sphalt deals with pavement and construction, which are essential to both civil and transportation engineering. For chemistry, we study microstructure and understand how this relates to the macrostructure of engineering. Asphalt is an organic material and the chemical-physical properties have a great impact on its usage.

This chapter begins with the place of asphalt in a number of naturally occurring bituminous substances and how to define it chemically. The analytical scheme of using solvent parameters for fractionation is presented. The unique properties of mesomorphic substances are thoroughly investigated with the colloidal properties of asphalt. Finally, the average structure of asphalt is explained. Rheological properties are also discussed.

12.1 NATURALLY-OCCURRING BITUMINOUS SUBSTANCES

As early as 3800 B.C., asphalts were used for building stones and blocks and, later on, for making reservoirs, canals, and bathing pools watertight. The major usage of asphalts and bitumens is as binders for aggregates for road pavements, which is extending to modern time. A lesser usage may be water-proof coating or a vitrification medium to encapsulate radioactive materials. The word "asphalt", although derived from Greek, and again was transported from Assyrian and Arcadian, also had an ancient Chinese origin, meaning sticky soil.

In the past, there has been much confusion over the nomenclature of asphaltic or bituminous. Although usage of the terms **asphalt** and **bitumen** are interchangeable in the United States, in Europe asphalt denotes the impure form of the generic material, and bitumen the basic mixture of heavy hydrocarbons free of inorganic impurities.

The American Society for Testing and Materials has promulgated the following standard definitions:

1. *Relating in general to bituminous materials:* Bitumen, a class of black or dark-colored (solid, semisolid or viscous) cementitious substances, natural or manufactured, composed principally of high molecular weight hydrocarbons, of which asphalts, tars, pitches and asphaltites are typical.
2. *Relating specifically to petroleum asphalts:* Asphalt, a dark, brown to black cementitious material, solid or semisolid in consistency, in which the predominating constituents are bitumens that occur in nature as such, or are obtained as residua in refining petroleum.

At the Fifth World Petroleum Congress, the designation naphtha bitumens was proposed for naturally occurring organic materials soluble in carbon disulfide, such as petroleum, coal and related products. Included are the substances commonly called asphalts, which comprise the hard native bituminous ozokerite and asphaltite. The term bitumen was suggested for this family of substances and asphalt for the residual material obtained by the distillation of petroleum. The combination of the two types is sometimes referred to as asphaltic bitumen.

The United Nation's Institute of Training and Research recently formulated a tentative definition of asphalt which considers the origin of this material as a basis for classification: "Natural asphalt is normally composed of organic and mineral matter and contains variable amounts of water. The organic portion is composed of a complex mixture of hydrocarbons, asphaltenes and resins, with significant quantities of sulfur, oxygen and nitrogen compounds, also compounds of vanadium, nickel, iron and other metals. Asphalt is a term commonly applied to a refinery product also called 'petroleum asphalt' and 'refinery asphalt.'"

For convenience, asphalt can be classified into two major groups, natural asphalts and artificial asphalts. Natural asphalts include bituminous materials laid down in natural deposits, such as those in the Trinidad and Bermudez Lakes, and natural bitumens, such as gilsonites and grahamites, which are completely soluble in carbon disulfide. Artificial asphalts mainly include petroleum-derived asphalts and, to a lesser extent, coal tar, water-gas tars, and their pitches. A classification scheme is illustrated in Figure 12-1.

All fossil based oil contains some asphaltics, ranging from 0.1-50%, however, heavy oil contains the highest content. Industrially, asphalt can be manufactured and processed by straight reduction, propane de-asphalting and air-blowing. Specification is based on cement viscosity or residue viscosity from a rolling thin-film oven test. The properties and behavior of asphalts are critically dependent on the nature of the asphalt's constituents. Chemically these constituents consist of hydrocarbons as well as non-hydrocarbons (heterocyclic or nitrogen, sulfur and oxygen-containing compounds). Separation of the various fractions of asphalt is usually based on their different boiling point ranges, molecular weight and solubility in solvents of different polarity. Techniques for obtaining narrow fractions include vacuum distillation, solvent extraction, thermal diffusion, crystallization, etc. individually or in combination, followed by chromatographic separation. In general, the major components in asphalt can be categorized as gas oil, resin, asphaltene, **carbene** and **carboid** as listed in Table 12-1. Usually, carbene and carboid are found in smaller quantities. **Mesophase** can be generated by heat treatment and is seldom found in ordinary asphalt. All of the major components above can be clearly defined by Hildelbrand's solubility parameter range. Both petrolene and maltene are often used in the asphalt literature.

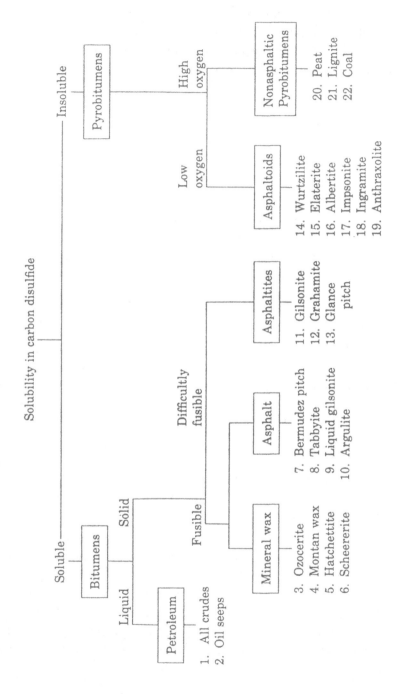

Figure 12-1. Terminology and classification of naturally occurring bituminous substances. (T.F.Yen)

Table 12-1. Definition of different solvent fractions of asphalt and related carbonaceous material[a].

No.	Name	Solvent Characteristics	Range of solubility parameters (δ in hildebrands)[b]	Notes
I	Gas Oil	Propane soluble	Below 6	–
II	Resin	Propane insoluble Pentane soluble	6 – 7	Combined I & II is sometimes called maltene or petrolene
III	Asphaltene	Pentane insoluble Benzene soluble	7 – 9	ASTM use CCl_4 instead of benzene
IV	Carbene	Benzene insoluble CS_2 soluble[c]	9 – 10	ASTM use CCl_4 instead of benzene
V	Carboid	CS_2 insoluble[c]	10 – 11	Combined IV & V is generally referred to as preasphaltene or asphatol
VI	Mesophase	Pyridine insoluble	Above 11	–

[a] Volatile free basis
[b] 1 hildebrand = 1 $cal^{1/2} / cm^{3/2}$ = 2.04 $J^{1/2} / cm^{3/2}$
[c] Because of the flammability of CS_2, pyridine is preferred; fractions 4 and 5 can be combined

Generally, any given asphalt can be separated by solvent fractionation into three major fractions: **asphaltene**, **resin**, and **gas oil**. A short discussion is included here since they are responsible for a number of colloidal properties of the asphalt system.

- **Asphaltenes** are obtained by precipitation in non-polar solvents such as low-boiling naphthas, petroleum ether, n-pentane, isopentane and n-hexane. They are soluble in liquids of high surface tension such as pyridine, carbon disulfide, and carbon tetrachloride but are insoluble in petroleum gases such as methane, ethane, propane, etc., by which they are precipitated (see the Hildebrand's solubility parameter). Commercially, propane is used to separate asphaltenes from asphaltic residues. In terms of chemically fine structure, asphaltene is a multipolymer system containing a great variety of building blocks. The statistical average molecule contains a flat sheet of condensed aromatic systems, which may be interconnected by systems of sulfide, ether, aliphatic chains or naphthenic rings. Gaps and holes in the aromatic system with heterocyclic atoms coordinated to transition metals such as vanadium, nickel and iron are most likely caused by free radicals. The compactness of the aromatic system varies widely as a function of source and temperature. The basic unit sheet of asphaltene usually has a molecular weight of 1000 to 4000, which may form oligomers up to 8.

- **Resins** are soluble in the liquids that precipitate asphaltenes and can be separated from the residua in the maltene mixture by chromatography. They are co-precipitated with asphaltenes in propane de-asphalting processes and are strongly adsorbed onto the asphaltenes. Resins can be released from asphaltene by exhaustive extraction with n-pentane. Resins are usually separated from the residue by desorption with chloroform after chromatography through a silica gel column containing methylcyclohexane as a solvent. Another popular technique is the **SARA (saturated-aromatic-resin-asphaltene)** method, which utilizes two ion exchange resins and a $FeCl_3$-clay-anion exchange resin packing column to retain the resin from the maltene fraction. Resins are considered smaller analogs of asphaltenes, and therefore, their molecular weight is much lower than that of asphaltenes. For this reason, the bulk of the resins can be preparatively separated by gel permeation chromatography as well as by **centrifugal thin layer chromatography**. The resin nuclei contain aromatics substituted with somewhat longer alkyls and a higher number of side chains attached to the rings than asphaltenes have. The combination of the saturated and aromatic characteristics of the resins is what stabilizes the colloidal nature of the asphaltenes in the oil medium.

- **Gas oils** are the lowest molecular weight fraction of the asphalt and serve as the dispersion medium for the peptized asphaltenes. They are soluble in petroleum ether, propane, and most other organic solvents and may be separated into saturates, aromatics, and other hydrocarbon types through column chromatography employing solvents of varying polarity. Molecular weights of the gas oils range below 800, with the majority falling in the region of 360 to 500. Structurally, gas oils consist mostly of **naphthenic**-aromatic nuclei having a greater proportion of side chains than the resins. Alkyl naphthenes predominate and straight chain alkanes are rarely present. The naphthenic content is 15 to 50%, with naphthenics containing 2 to 5 nuclei per molecule. (For naphthenic, see Chapter 3.)

12.1.1 Solubility Parameters

We have discussed that solvent fractionation of asphalt is based on the principle of **solubility parameter**. For example, all the non-volatile portion of fossil fuel derived oil can be separated using the concept. Figure 12-2 is an analytical scheme for obtaining different fractions.

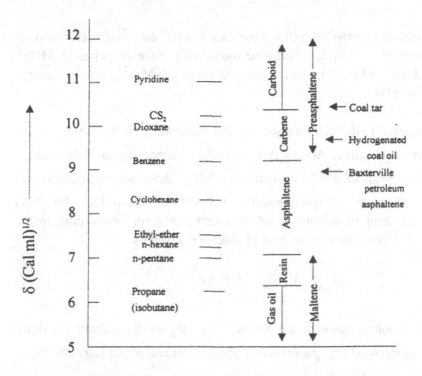

Figure 12-2. Separation of major heavy fractions based on solubility parameters of solvent system. (T.F.Yen, 1998)

From statistical thermodynamics, there is a correlation between the cohesive energy density (potential energy per unit volume) and material miscibility. The cohesive energy density, C is defined as

$$C = \frac{\Delta U_V}{V_l} \qquad [12\text{-}1]$$

where ΔU_V is the energy change for complete vaporization of the saturated liquid to the ideal gas state, and V_l is the molar volume of the liquid. A solubility parameter, δ, is defined as the positive square root of the cohesive energy density

$$\delta = \left(\frac{\Delta U_V}{V_l} \right)^{\frac{1}{2}} \qquad [12\text{-}2]$$

The dimension of the solubility parameter is $cal^{1/2}$ $cm^{-3/2}$ or Hildebrand, or in SI unit, the nearest equivalent being the square root of the mega pascal, $MPa^{1/2}$. ($1MPa = 10^6 Jm^{-3} = 1J\ cm^{-7}$). To convert Hildebrand to $Mpa^{1/2}$, simply multiply by $(4.187)^{1/2}$ or 2.046.

For a material to dissolve in a solvent, the free energy change, ΔG_m, of the process must be negative, $\Delta G_m = \Delta H_m - T\Delta S_m$. Since the entropy change, ΔS_m, is always positive, the heat of mixing, ΔH_m, determines whether or not dissolution will occur. Experimentally, it has been found that, for most compounds, the heat of mixing is either positive or zero. According to the Hildebrand – Scatchard theory, the heat of mixing is given by

$$\Delta H_m = V_m \left(\delta_1 - \delta_2 \right)^2 \phi_1 \phi_2 \qquad [12\text{-}3]$$

where V_m is the total volume of the mixture, ϕ_1, ϕ_2 are the volume fractions, and δ_1, δ_2 are the solubility parameters of the solvent and solute respectively.

Therefore, the heat of mixing will be small or zero and the free energy change negative when $\delta_1 \cong \delta_2$. The assumptions in the above are: (1) forces of attraction are due primarily to dispersion forces, (2) molar volumes of the solute and the solvent are not significantly different, (3) no volume change occurs on mixing, and (4) mixing is random. These assumptions are not generally valid, and many more sophisticated theories have been developed. However, the relation produced is a simple one and is easy to use as a rough guideline.

Evaluation for the values of solubility parameters can be based on the following:

- Heat of vaporization

$$\delta^2 = \frac{(\Delta H_V - RT)}{V} \qquad [12\text{-}4]$$

where ΔH_V is heat of vaporization in cal/g-mole.

- Hildebrand rule

$$\Delta H_V = 0.026T_b^2 + 23.7T_b - 2950 \qquad [12\text{-}5]$$

where T_b is the boiling point in K.

- Critical pressure

$$\delta^2 = 1.25P_c \qquad [12\text{-}6]$$

where P_c is in atm.

- Surface tension

$$\delta = K\left(\frac{\alpha}{V^{\frac{1}{3}}}\right)^{0.43} \qquad [12\text{-}7]$$

where K is 4.1 and α is dyne-cm.

- Refractive index

$$\delta = 30.3\left(\frac{n^2 - 1}{n^2 + 2}\right) \qquad [12\text{-}8]$$

- Group additivity

Where estimation is from the chemical functional groups in structural units and

$$\delta = \left(\frac{\Sigma \Delta E_V}{\Sigma \Delta V} \right)^{\frac{1}{2}}$$ [12-9]

[Example 12.1] Calculate the solubility parameter of ethanol, CH_3CH_2OH.

	Δe	Δv
CH_3	1125	33.5
$-CH_2-$	1180	16.1
$-OH$	7120	10.0
	$\Delta E_V = \Sigma\ 9425$	$\Delta V = \Sigma\ 59.6$

The estimated value is $\left(\dfrac{9425}{59.6} \right)^{\frac{1}{2}} = 12.69$. The actual value is 12.78, both expressed in $cal^{1/2}\ cm^{-3/2}$.

A **mixing rule** has been established that an effective solubility parameter, δ_m of a binary mixture is given by

$$\delta_m = \alpha_1 \delta_1 + \alpha_2 \delta_2$$ [12-10]

where δ_1, δ_2 are the solubility parameters of the solvents and α_1, α_2 are the volume fractions. Usually, a number of effective solubility parameters can be constructed using a number of solvent pairs, such as expressed in Table 12-2. Their ranges can be from 7.0 to 24 hildebrands, or 14.3 to 47.9 $Mpa^{1/2}$.

Table 12-2. List of solvents mixed. $\delta_m = \gamma_1 \delta_1 + \gamma_2 \delta_2$; γ = volume fraction; δ = solubility parameter.

Solvent	Solubility parameter	
	MPa$^{1/2}$	Hildebrands
2-Methyl-pentane	14.3	7.0
Cyclohexane	16.8	8.2
Cyclohexanol	20.3	9.9
HEMA*	25.6	12.5
Water	47.9	23.4

* Hydroxyethyl methacrylate

Different effective solubility parameters can be obtained from the preparation as in Fig. 12-3.

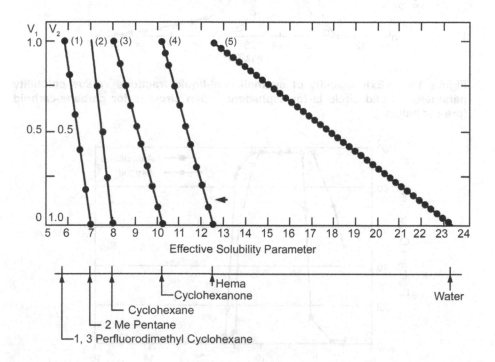

Figure 12-3. Solvent pairs used for obtaining different solubility parameter values between 5.8 and 23.5. Each dot is obtained from experimental mixed liquid pair value.

Samples of asphaltene have been studied. The amount of soluble portions in these solvents pairs can be established by a **solubility parameter spectrum** as shown in Fig. 12-4.

Usually the width of the window can be used to define certain fractions of the asphaltic material. For example, the difference between an asphalt and the asphaltene which is derived from the asphalt can be found in Fig. 12-5.

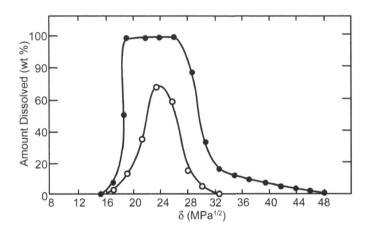

Figure 12-4. Extractability of synthoil coal-liquid fractions versus solubility parameter. Solid circle is for asphaltene, open circle is for carbene-carboid (pre-asphaltene).

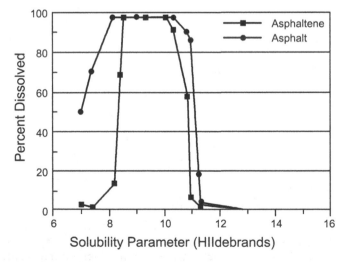

Figure 12-5. The solubility parameter spectrum for West Texas Intermediate/ West Texas Sour asphaltene and asphalt.

Evidently from these studies, the isolation of certain fractions, e.g., asphaltene can proceed from either direction in the procedure for solvent precipitation. From the left direction is known as propane-insolubles for the asphaltene, and from the right is known as the acid precipitation, either by SO_2 or HCl. Another application of the solubility parameter spectrum is the measurement of swelling volume, such as with coal. For example, Fig. 12-6 illustrates the feature of a typical high volatile bituminous coal.

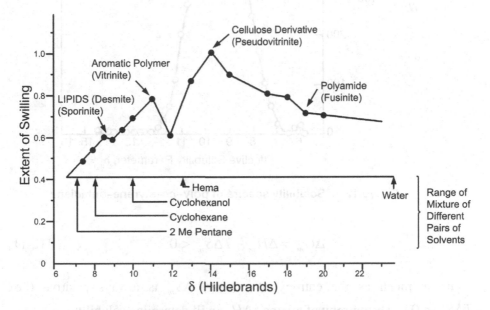

Figure 12-6. Solubility parameter spectrum of a high volatile bituminous coal (PSOC no. 102). The extent of swelling can be obtained from (W_s – W_o / W_o). See Table 12-3 and Fig. 12-2 for the solvent pairs used here. The assignment of the peaks are tentative.

As a note from the swelling data, $\dfrac{W_s - W_o}{W_o}$, where W_s is swollen volume and W_o is original non-swollen volume, related to the binary system, differentiation between a multipolymer (such as copolymer, terpolymer, etc.) and a homopolymer have been made (see Fig. 12-7).

This concept can also be extended to examine whether or not two substances are miscible. In general, the miscibility of two liquids is dependent on the free energy mixing

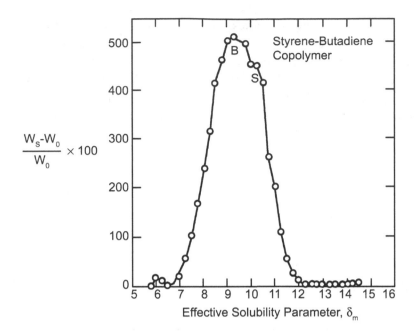

Figure 12-7. Solubility spectra of poly–*co*–styrene–butadiene.

$$\Delta G_m = \Delta H_m - T\Delta S_m < 0 \qquad \text{[12-11]}$$

In as much as the entropy of mixing, ΔS_m is always positive (i.e., $-T\Delta S_m < 0$). The entropy of mixing, ΔH_m will determine solubility

$$\Delta H_m \cong V(\delta_1 - \delta_2)^2 \qquad \text{[12-12]}$$

The closer the solubility parameter values of the two liquids, the smaller the ΔH_m will be, and the decrease in the free energy during mixing will be greater. For immiscibility of two liquids, Hildebrand and Scott have made the condition that

$$\Delta H_m \cong V(\delta_1 - \delta_2)^2 > 2RT \qquad \text{[12-13]}$$

Furthermore, the effective molar volume of the solution can be calculated from the arithmetic mean of the molar volumes of the two liquids, therefore

$$\frac{\frac{1}{2}(V_1 + V_2)(\delta_1 - \delta_2)^2}{2RT} > 1 \quad \text{or} \quad Q > 1 \qquad [12\text{-}14]$$

If $Q > 1$, then the two liquids or two substances are anticipated to be immiscible. These are only a first order estimation or guide.

_____

[**Example 12-2**]   For the design of secondary lithium batteries, the information that is required is whether or not an inorganic electrolyte, lithium hexafluoro arsenate, $LiAsF_6$ is soluble in 2-methylfurane.

Referring to Appendix of Fedor's method for estimation of solubility parameters (hildebrands) and molar volumes (cm^3)

For 2-methylfurane

$$\delta_1 = 8.06 \quad \text{and} \quad V_1 = 100.6$$

For $LiAsF_6$

$$\delta_2 = 10 \quad \text{and} \quad V_2 = 62.7 \quad (\rho = 3.203 \text{ cm}^3/\text{g})$$

For immiscibility

$$\Delta H_m \cong \frac{1}{2}(V_1 + V_2)(\delta_1 - \delta_2)^2 > 2RT$$

at $25°$ C

$$Q = \frac{307}{1183} = 0.26$$

$LiAsF_6$ is soluble in 2-methylfuran at room temperature.

_____

Furthermore, a swollen network can be used for the dissolution of pre-asphaltene (a term used to denote the soluble fraction of fossil materials), such as coal or carbene (carboid) or mesophase molecules. The solubility parameter spectrum approach can predict what portion of material can be dissolved from a multipolymer (sometimes referred to as multimers, in terms with dimer, trimer, etc.).

According to the theory of strain energy considerations, the dilation of a brittle solid can be expressed by

$$S = \frac{2}{9}\left(\frac{1+\upsilon}{1-\upsilon}\right)G\left(\frac{\Delta V}{V}\right)^2 \qquad [12\text{-}15]$$

where S is strain energy per unit volume, υ is Poisson's ratio, G is shear modulus, and V is volume. The fracture of a brittle solid can be expressed as

$$F = \frac{\sigma^2}{2E}$$

where F is fracture energy per unit volume, σ is tensile stress, and E is bulk modulus. Comminution will occur when $S \geq F$. In other words, the swollen network can be broken to cause the bonds and linkages to rupture. Referring to Fig. 12-6, it consists of three maxima. These chosen solvents are based on three maxima at 8.5, 11, and 14 hildebrands and are ethyl benzoate (= 8.2 h), dioxane (= 9.9 h), and phenol (= 14.5 h), respectively. After the coal had swollen for three days, the swollen coals were allowed either to be extracted with tetrahydrofuran (= 9.5 h), or to be pyrolysed. These results are summarized in Table 12-3.

Table 12-3. Solvent extraction or pyrolysis on swollen coal of PSOC 120 (see Fig. 12-3).

Sample	THF extract (% increase due to swelling)	Pyrolysate* (% increase due to swelling)
Coal (control)	12.0 --	26 --
Dioxane-swollen coal	15.5 (29)	31 (16)
Ethyl benzoate-swollen coal	16.9 (41)	37 (42)
Phenol-swollen coal	13.4 (12)	28 (8)

* Carried out at 500° C at 2 mm Hg vacuum pressure

12.2 MESOMORPHIC MATERIAL

Many naturally occurring substances have the unique property of mesomorphism. These materials have limited order in the molecular arrangement. In order to understand these orders, light scattering and diffraction studies have been made especially with x-ray or neutron small angle scattering. Even wide angle x-ray with different repeating patterns of distances is essential. Furthermore, the asphalt is almost like a "bounded mass" or "frozen ice" not in isotropic form. Lastly, this "bounded mass" is formed by random probability by a homologous series of components, particularly the aromatic systems.

12.2.1 Morphological Distances as Structured Parameters with Diffraction and Scattering Studies

Mesomorphic substances are in between amorphous and crystallinic substances. They often possess one or two-dimensional order. Only crystallinic substances have a three-dimensional order.

X-ray high angle diffraction is helpful for quantitative or semi quantitative analysis in the determination of the spacing as well as the intensities of bands in terms of certain structural characteristics. For the homogeneous distributions of structures, the small angle x-ray or neutron scattering at small angles to the primary beam will yield certain valuable information for a given sample for x-ray diffraction; a convenient method is using a monochromatic copper K_α radiation to obtain a reduced intensity. For asphaltics, the intensity and reduced intensities are illustrated by Fig. 12-8.

Some important structural parameters have been obtained from x-ray studies, which are listed below.

Gamma Band ($\sin \theta/\lambda \approx 0.10$ Å^{-1})

The band centered around distances ($2\lambda/\sin\theta$) of 4.0 to 5.5Å is observed for a molten paraffin (m.p. = 60°C) taken at 65°C. This band is attributed to the spacing of disordered aliphatic chains or acyclic rings. When the crystalline paraffin is heated, one can see the peaks that correspond to the intensities of the (110) and (200) packing of the chains disappear as the gamma (γ) band appears. There seems to be no published criterion for distinguishing between saturated rings and saturated chains based on the shape of the band. Apparently, the γ-band is derived from the (110)-band. Figure 12-9 illustrates the geometry of this **interchain distance**.

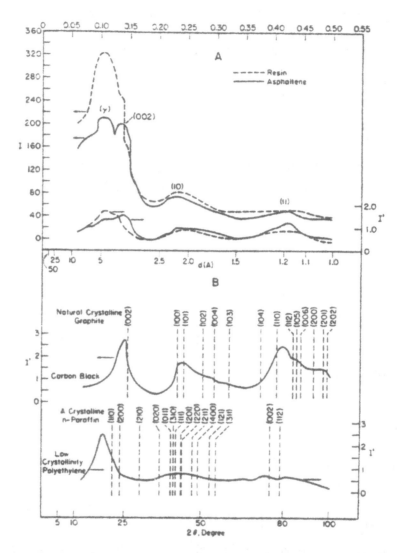

Figure 12-8. X-ray diffraction patterns of mesomorphic substances. (A) Resin and asphaltene fractions of Baxterville crude oil. (B) A carbon black and a low crystallinity polyethylene. The vertical dotted lines indicate the band positions of their fully crystalline equivalents, namely, graphit and a *n*-paraffin. I – original intensity. I' – reduced intensity. According to Yen et al. *Anal. Chem.* **33**, 1587 (1961), Alexander et al. *J. Phy. Chem.* **60**, 646 (1956).

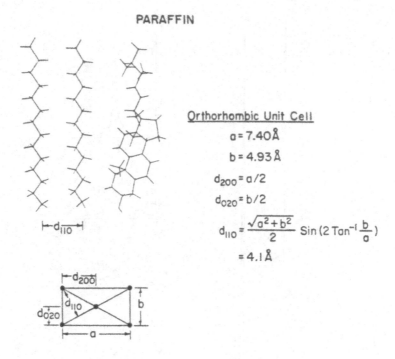

PARAFFIN

Orthorhombic Unit Cell

$a = 7.40 \text{Å}$

$b = 4.93 \text{Å}$

$d_{200} = a/2$

$d_{020} = b/2$

$d_{110} = \dfrac{\sqrt{a^2 + b^2}}{2} \; Sin \left(2 \; Tan^{-1} \dfrac{b}{a}\right)$

$= 4.1 \text{Å}$

Figure 12-9. Geometry of interchain distances from saturated hydrocarbons.

The γ-band has been found to be symmetrical. Although there are no explanations offered as to why it is symmetric, except the scattering from octacosane at elevated temperature, 65-150°C (see Fig. 12-10). This becomes a very important point when the γ-band is separated from the (002) aromatic band by algebraic sum.

The lattice spacing that corresponds to a particular line in a diffraction pattern from a crystalline material is given by the Bragg equation

$$d_\gamma = \lambda/2 \; sin\theta \qquad\qquad [12\text{-}16]$$

There is also an alternative d'_γ value which is found to be 5/4 larger.

$$d'_\gamma = 5\lambda/8 \; sin\theta \qquad\qquad [12\text{-}17]$$

This was based on the diameter for the bundle site of parallel line scatters. Yet a summary of all the interchain distances cannot be found. For d'_γ. See Table 12-4 and Fig. 12-13.

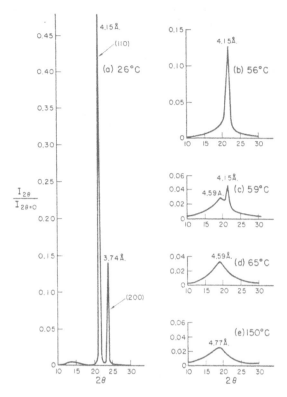

Figure 12-10. Scattering from octacosane at various temperatures. (Modified after Krimm and Tobolsky)

Table 12-4. Calculated interchain distances, d_γ, of crystalline saturated chain and ring compounds based on crystallographic data.

Compound	d_γ, A.
Polyethylene	4.10
Polypropylene oxide	4.26
Poly-3-methyl-1-butene	4.75
Ergosterol + H_2O	4.95
Dihydroergosterol	4.1
Ergosterol-B_3	5.0
Vitamin D_2	5.2
Pyrocalciferol	4.5
Suprasterol I	4.5
Lumisterol	4.5

d_2-Spacing

An examination of the mesomorphic spectra of some poly-α-olefins shows that as the side-chain group becomes larger, e.g., from (Fig. 12-11b) polypropylene to (Fig. 12-11e) poly-α-hexene, a new band emerges. This new band tends to shift from 5.4Å to 6.9, 8.8 and 9.8Å. A plot of some of these d_2-band spacing in relation to numbers of side-chain carbon atoms are shown in Fig 12-13. The occurrences of γ-bands with a d_2-band for many common materials are listed in Fig. 12-12.

(002)-band [(sin θ/λ) ≈ 0.14]

The (002)-band is attributed to the **interplanar spacing** of condensed aromatic rings. Such spacing in natural graphite is 3.35Å, which corresponds to sin θ/λ = 0.15. Geometrically, the interplanar spacing for the fused polyacenes can be depicted as shown in Fig 12-13. The shifts observed in the (002)-band in some compounds are shifts towards larger spacings. They are attributed to actual changes in interatomic spacings or to buckling of planes due to substitution of heteroatoms. Fig 12-13 is for the unit cell of graphite and the interplanar distance is b/2, which is 3.35Å. The interlayer distance d_M for a number of naturally occurring compounds is listed in Table 12-5.

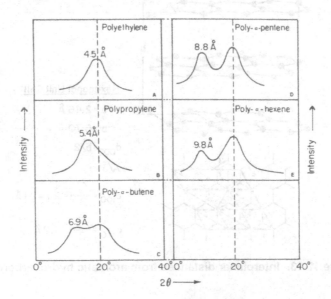

Figure 12-11. X-ray spectra of some amorphous linear poly-α-olefins (After Natta).

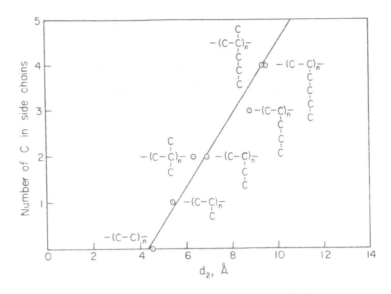

Figure 12-12. Interchain distances of some poly-α-olefins.

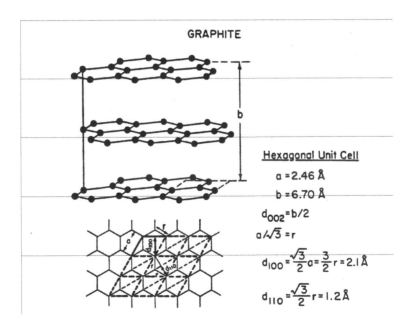

Figure 12-13. Interplanar distances from aromatic hydrocarbons.

Table 12-5. Interlayer distance d_M in certain coals, pitch binder, coke and carbon blacks with some aromatic hydrocarbons.

Substance	d_M, A.
Coal, 80.1%	3.64
Coal, 87.3%	3.54
Coal, 94.1%	3.43
Pitch, electrode binder (U.S. Steel)	3.52 – 3.60
Coke	3.48 – 3.52
Channel black, Spheron 3, Cabot	3.62
Furnace black, Vulcan SC, Cabot	3.54
Graphite, natural	3.34
Ovalene	3.45
Pyrene	3.52

In resins, asphaltenes and related structures, another factor which may increase the distance is buckling of the planes due to imperfections in the hexagonal network resulting from replacement of carbon atoms by such heteroatoms as oxygen, nitrogen and sulfur as shown in Fig. 12-14.

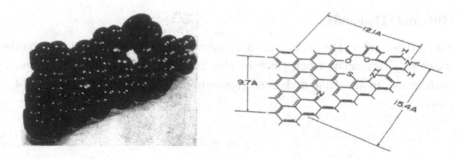

Figure 12-14. A model of a section of a condensed aromatic sheet showing the effect of substitution of carbon atoms by oxygen, nitrogen and sulfur.

In many asphaltenes and related materials, the (004)-band was never found which indicates that the layers are quite limited. Usually from the (002)-band, a cluster diameter is obtained from the width at half maxima

$$L_c = 0.45/B_{12} \qquad\qquad [12\text{-}18]$$

The average number of aromatic layers, M_e can be obtained as

$$M_e = L_c/d_n \qquad \qquad [12\text{-}19]$$

For some naturally occurring substances, the value of M_e can be large due to the impurities present, which can be chelated between layers (Table 12-6). In such a way, the spacing is extended.

Table 12-6. Estimated number of layers per aromatic cluster, M_e, for various carbonized materials

Material	M_e
Saran char	2.0 – 7.7
Sugar char	2.4 – 5.4
Polyvinyl chloride char	4.5 – 33
Petroleum coke, 1000° to 1700° C	4.1 – 33
Graphitized coking coal	12
Pitch	3.2 – 4.0
Coke	4.5 – 5.5
Asphaltenes	5

(10)- and (11)-bands

The (10)- and (11)-bands are the easiest to understand and the hardest to measure. They occur when $\sin \theta/\lambda$ value is around $0.25 – 0.42\text{Å}^{-1}$. These bands correspond to the first and nearest neighbors in ringed compounds (4.2 and 2.4Å).

$$d_{(10)} = 3r/2 \qquad \qquad [12\text{-}20]$$

$$d_{(11)} = 3^{1/2}\, r/2 \qquad \qquad [12\text{-}21]$$

These assignments are consistent with the fact that these bands are always found together with the (002)-band, and possess a characteristic saw-toothed profile, which feature can be clearly seen in the example of a carbon block. This shape has been shown by theoretical calculations to be characteristic of two-dimensional reflections.

From the position, shape and intensities of these four diffuse bands, numerical values for several interesting structural features can be deduced.

Layer Diameter (Size of the Aromatic Sheets)

Approximate values for the average diameter of the sheets of condensed aromatic rings can be calculated by means of Scherrer's crystallite size formula for two-dimensional reflections:

$$L_a = 0.92/B_{1/2} \qquad\qquad [12\text{-}22]$$

where B_{12} is the width of either the (10)- or the (11)-band at half-maximum in units of $(\sin\theta)/\lambda$. A more precise method is to compare the profile of the (11)-band with those calculated by Diamond for near symmetrical condensed aromatic molecules of increasing size (see Fig. 12-15). Tests of this procedure with elongated and bent aromatic molecules show that the values obtained in this case tend to represent the long dimension.

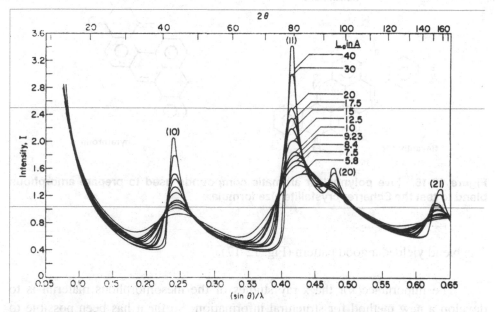

Figure 12-15. Theoretical diffraction patterns for randomly oriented perfect aromatic molecules of various sizes plotted from data published by Diamond.

Yen has verified the Diamond method by using a mixture of polynuclear aromatic compounds to prepare an amorphous blend as seen in Fig. 12-16.

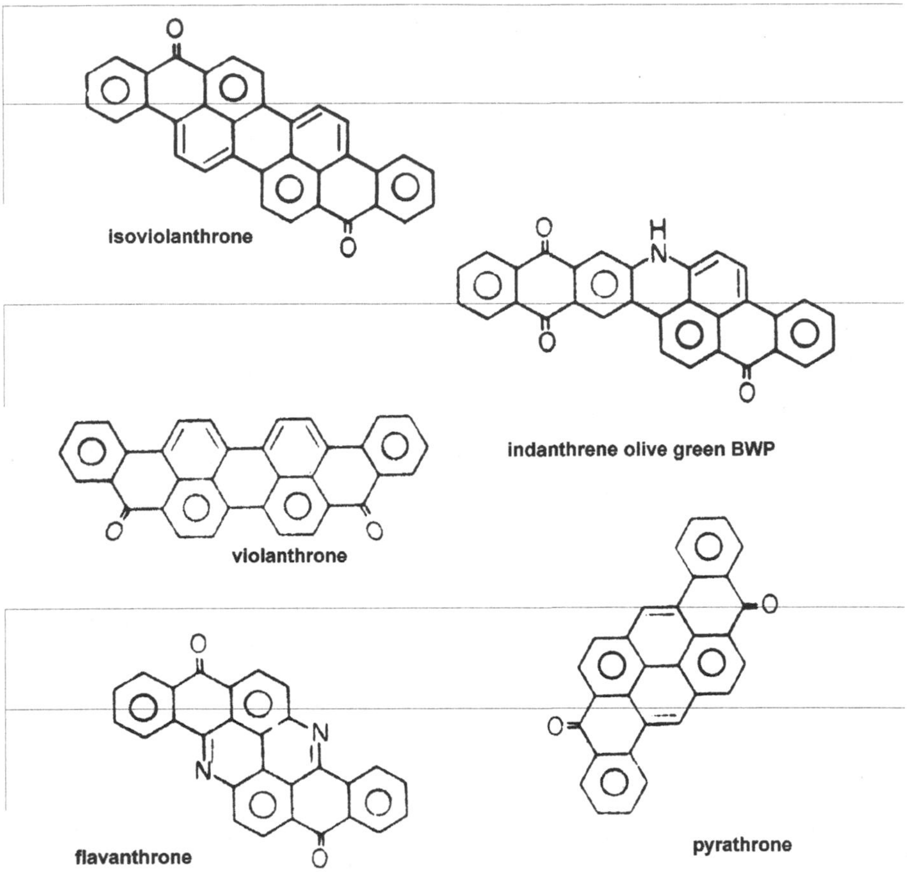

Figure 12-16. Five polynuclear aromatic compounds used to prepare amorphous blend to test the Scherrer crystallite size formula.

The blend yielded a good pattern (Fig. 12-17).

The importance for the x-ray studies of the mesomorphous material is to develop a new method for structural information. So far it has been possible to evaluate what amount of aromatic carbon versus what amount of saturated carbon assuming that the carbon copper

$$C_S = C_P + C_N \qquad [12\text{-}23]$$

in paraffin and naphthalene, the zigzag bonding of carbon is the same and similar.

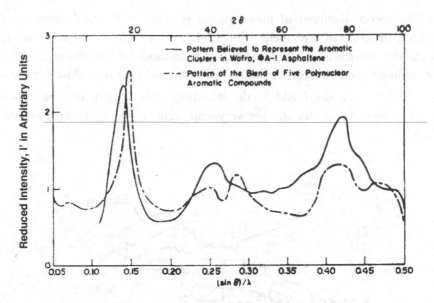

Figure 12-17. Comparison of x-ray diffraction pattern computed for aromatic clusters in petroleum asphaltene with experimental pattern for blend of five polynuclear aromatic compounds of known structure.

If this is the case, then

$$f_a = \frac{C_A}{C_P + C_N + C_A} = \frac{C_A}{C_S + C_A} \qquad [12\text{-}24]$$

$$= \frac{C_A}{C}$$

f_a is called the carbon aromaticity. The C_s can be measured by the intensity of the γ-band and C_A can be measured by the (002)-band.

$$C_S \propto I_r \qquad [12\text{-}25]$$

$$C_A \propto I_{002} \qquad [12\text{-}26]$$

In examination of Fig. 12-8 and adjusting the carbon block to polyethylene ratio to 50:50, the result is rather good. A number of oligomeric polymers have been made to test for this, the result can be found in Fig. 12-19.

The accomplishment of these studies is seen in the establishment of the following structure of asphaltene (Fig. 12-18). For many small angle scattering studies, the emphasis is on the Guinier's method for obtaining a radius of gyration in a solution. Most studies end with the $r = 3 - 5$ nm. Also for the small angle studies, both the $\bar{l}$ and l_c, the inhomogeneity length and the coherence length for the large cluster size are very important for the colloidal properties of asphalt.

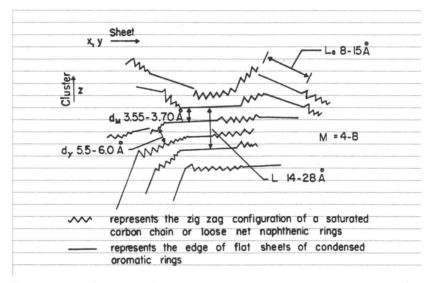

Figure 12-18. Cross-sectional view of an asphaltene model. L_a, diameter of an aromatic sheet [obtained from the width of the (10)- or (11)-peaks]; d_M, spacing between aromatic sheets [obtained from the position of the (002)-peak]; d_γ, spacing between aliphatic chains or sheets [obtained from the position of the γ peak]; L, distance through aromatic sheets in cluster [obtained from the width of the (002)-peak]; M, number of layers per cluster = L/d_M. (After Yen et al.)

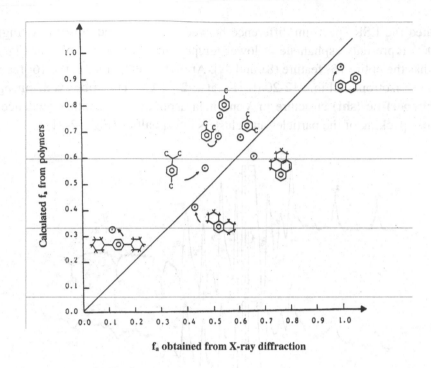

Figure 12-19. Carbon aromaticity values (f_a) from a number of aromatic model polymers.

12.2.2 Anisotropic Solids as Measured by Electron Spin Resonance Probe

Transition metals such as vanadium and nickel are reported to exist with the heavy fractions of petroleum such as asphaltenes. Without exception, almost all the vanadium species exist homogeneous as tetravalent of vanadyl porphyrins within the asphalt matrix. Due to the paramagnetic properties, **electron spin resonance (ESR)** will yield a vanadium spectrum. Most asphalt when examined at room temperature will give a 16 featured spectrum. Under elevated temperature, the 16 features will reduce to an 8 line spectrum which is identical to V^{4+} solution spectrum. In this manner, doubtlessly at elevated temperatures or polar solutions (or under higher Hensen hydrogen bonding solubility parameter effect), the asphalt becomes isotropic. Under usual application conditions, asphalt molecule is with limited freedom for movement (meaning anisotropy). This restricted freedom is sometimes named associated or agglomerated clusters. This associated energy was forced to be within the range of 14-20 kcal/mol. Fig. 12-

indicated the ESR spectrum difference between isotropy and anisotropy. Figure 12-20(a) represents asphaltene at lower temperature in a polar solvent. Fig 12-20(a) has the entire 16- feature ($8\perp$ and $8\parallel$). At higher temperature, the 16- feature becomes isotropic. Fig. 12-20(d) is at 225° C. The observed nitrogen **superhyperfine (shf)** structure in Venezuela asphaltene is another evidence of the close packing of the particles and clusters in asphaltics (Fig. 12-21).

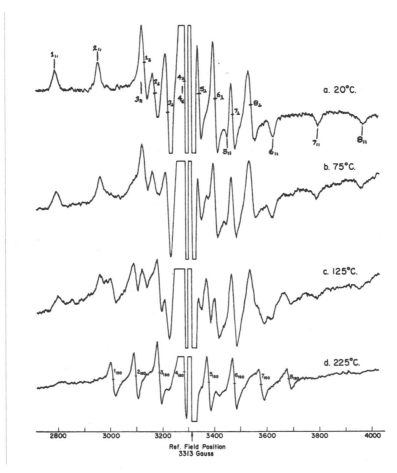

Figure 12-20. Electron spin resonance spectra of asphaltene in benzyl n-butyl ether at different temperatures (all with 3 MW, 6.3 Gauss, 1 sec response of 400 relative gain). (a) 20 °C, (b) 75 °C, (c) 125 °C, (d) 225 °C.

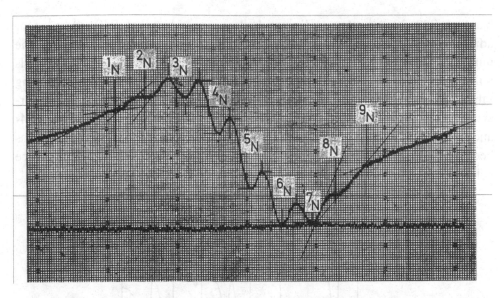

Figure 12-21. Nitrogen superhyperfine structures of asphaltene. The V anisotropic spectrum of 4⊥ is shown and superimposed by the 9-line pattern (small black) =0.2G. (Yen, 1971).

12.2.3 Randomly and Highly Substituted Aromatic Homologs by Infrared Bending Frequencies

Crawford and Miller have estimated the isolated, 2-, 3-, 4- and 5- adjacent hydrogens C-H out-of-plane bending frequencies based on

$$\overline{v} = 1/2(\pi C\mu)^{-1}(k + x_i k')^{1/2} \qquad [12\text{-}27]$$

where,

$$x_i = 2\cos(\pi/i+1) \qquad [12\text{-}28]$$

and,

$$i = \text{Number of adjacent hydrogens}$$

$$k = 0.378 \text{ and k' = -0.057}$$

Typical ranges of these frequencies illustrated by certain compounds as well as asphaltics are exemplified by Fig. 12-22. In general, a band at $880\pm20\text{cm}^{-1}$ represents isolated hydrogen, a band at $830\pm20\text{cm}^{-1}$, two adjacent hydrogens, a band at $780\pm20cm^{-1}$, three adjacent hydrogens, a band at $740\pm20cm^{-1}$ four adjacent hydrogens and a band at $700\pm20cm^{-1}$ five adjacent hydrogens. If the number of adjacent hydrogens on two or more outer rings differs, multiple bands will appear. Uniquely for petroleum-derived asphaltics there are three bending vibrations (1,2,3 type only) which have been further proved by the shift of donor-acceptor complexes (see Fig. 12-23). There is a band at 720 cm^{-1}, between the 4- and 5- adjacent hydrogen positions; by low temperature infrared studies this has been however assigned as $-(CH_2)-_n$ rocking due to the temperature dependence (Fig. 12-24).

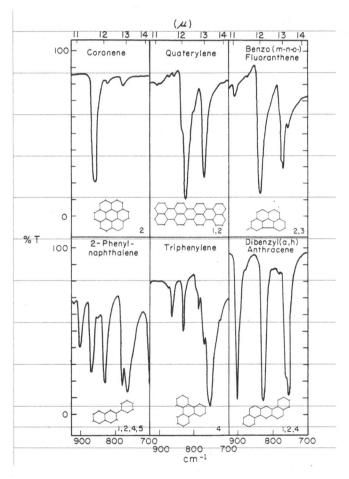

Figure 12-22. C-H out-of-plane bending vibrations in the IR range of some polynuclear aromatic hydrocarbons (numerals at the right of each compound is the type number; (1) isolated; (2) 2-adjacent H; (3) 3-adjacent H; (4) 4-adjacent H; (5) 5-adjacent H; the combined number signifies the existence of multitype number of the compound).

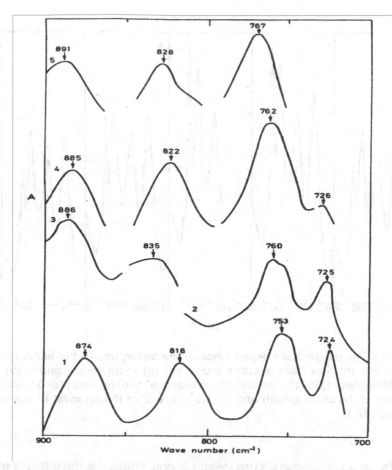

Figure 12-23. Differential infrared spectra of a West Texas asphaltic (1) in nuzol; (2) in methyl iodide; (3) in 5% tetracyanoethylene in benzonitrile; (4) in 2.5% s-trinitrozene-nitrobenzene (1:10); (5) in 2.5% m-dinitrobenzene in nitrobenzene – carbon disulfide (1:10) (discontinued scan = dead region); the extent of the shift can be predicted from Kirkwood-Baner-Magnet rule. (After Yen, 1973)

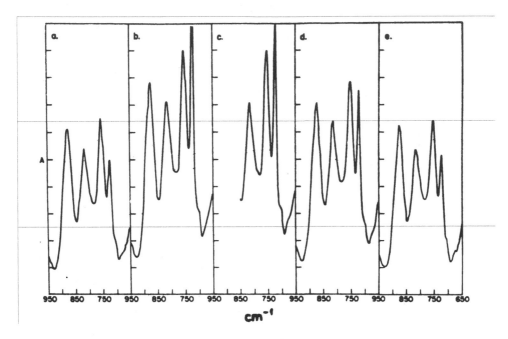

Figure 12-24. The temperature dependence of the absorption in the bending region for asphaltene fraction from a Libya crude oil: (a) room temperature; (b) liquid nitrogen temperature; (c) 15 min after the removal of cooling bath; (d) 0.5 hour after the removal of the cooling bath and (e) 1.5 hours after the removal of the cooling bath. (Yen, 1971)

As far as a polyaromatic core system is concerned, the three bands of C-H out-of-plane stretching will represent more than 50% of the edge hydrogen if the of the outer rings have been substituted. This is from a random probability computation by Yen. Fig. 12-25 is an example that a polyaromatic has been randomly substituted at different positions. P(i) is the number of available (unsubstituted) hydrogens. P_1 is the number of available hydrogens of the isolated, P_2 is for 2-adjacent hydrogens and so on.

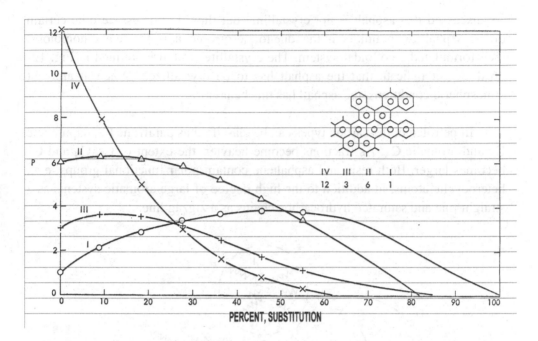

Figure 12-25. Computed probability after certain available hydrogen of a given polynuclear aromatic is substituted by the percentage of substituents. The amount of HI for a given type of C-H bending, I, isolated, II and IV is the adjacent hydrogen types.

Average asphaltic molecules consist of the aromatic portion which is equally substituted (over 50%) oligomers or aromatic polymers. The aromatic is most likely to be π-π association, similar to the porphyrin layer. The size is controlled by layer diameter.

12.3 COLLOIDAL PROPERTIES AND AVERAGE STRUCTURE

As discussed that asphalt is not crystallinic but they do possess definite certain order. Asphaltics are unique in that due to mesotropic, anisotropic and homologs, they formulated a colloidal system. The crystallite structure outlined in Fig. 12-26 does not indicate that the asphalt has to be crystallized. When at ambient atmospheric conditions, the asphalt has this behavior.

In petroleum, the carbon types can be classified as paraffinic C_P, naphthenic C_N and aromatic C_A. As fractions become heavier, the extent of C_P, C_N and C_A becomes larger. Both resin and asphaltene contain polar functional groups, e.g., heterocyclic atoms, in addition to the high degree of large aromatic systems as a straight line and saturated hydrocarbon chains as a zigzag line.

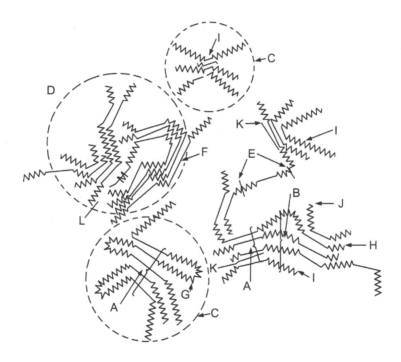

Figure 12-26. Macrostructure of asphaltics. Straight lines denote aromatic constituents; zig-zag line, saturated constituents (aliphatic and naphthenic). (A) crystallite; (B) cahin bundle; (C) particle; (D) micelle; (E) weak link; (F) gap and hole; (G) intracluster; (H) intercluster; (I) resin; (J) single layer, (K) petroporphyrin; (L) metal.

The colloidal structure shown in Fig. 12-26 represents most of the situations involving applications. Each single sheet has its own significance as indicated by Fig. 12-27. Unit sheet is the basis for asphalt. A lengthy vector is about $1.2 - 2.0$ nm which corresponds to a unit molecular weight of $2000 - 2500$.

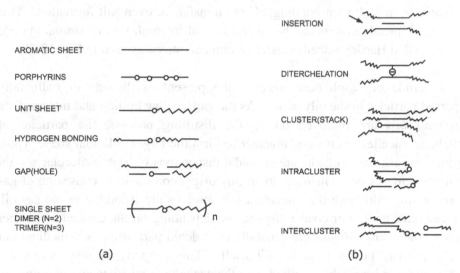

(a)   (b)

Figure 12-27. (a) The basic unit sheet of asphaltene and the related feature of a single sheet. (b) When the unit sheets are associated, fixed stacks are formed within the asphaltene structure. (Polar atoms are indicated by a circle)

Of the three components of asphalt, resin is the most important. It is usually the peptizing agent between asphaltene and gas oil. As indicated in Table 12-7, the peptizing agent is actually a surfactant.

Table 12-7. Colloidal components of asphalt.

Asphalt	1. Asphaltene (micelle)
	2. Resin (peptizing agent*)
	3. Gas oil — Saturates / Aromatics (intermicellar medium)

* peptizing agent can be easily modified by amphiphiles

Asphaltene can exist both in a random oriented particle aggregate form and in an ordered micelle form. In its natural state, asphaltene exists in an oil-external (Winsor's terminology) or reversed micelle. The polar groups are oriented toward the center, which can be water, silica (or clay), or metals (V, Ni, Fe, etc.). The driving force of the polar groups assembled toward the center originates from hydrogen-bonding, charge transfer, or even salt formation. This oil external micelle system can be reversed to oil-internal, water-external micelle (usually called Hartley micelle) under certain conditions as seen in Fig. 12-28.

In crude oil, asphaltene micelles are present as discrete or colloidally dispersed particles in the oily phase. As the various low boiling and intermediate petroleum oils are removed during the distilling process, the particles of asphaltene micelles are massed together to form the larger colloidal sizes. Thus, bitumens or ash-free asphalts are colloidal dispersions of high molecular weight non-hydrocarbons or asphaltenes in an oily dispersion medium consisting of gas oil and resin. Although the asphaltenes themselves are insoluble in the gas oil, they can exist as fine or coarse dispersions, depending on the content of another fraction: the resins. Resins are generally considered part of the oily medium but have a polarity higher than gas oil itself. This property of resin molecules enables them to be easily adsorbed onto the asphaltene micelles and in doing so, they act as the peptizing agent of the colloid stabilizer by change neutralization.

Various physical instrument methods have been used to examine the size and molecular weights of the colloidal ranges for asphalts. This aggregate of asphaltene particles with adsorbed resins is termed supermicelle and is dispersed in the oil medium. Sometimes, oil may be occluded between supermicelles and is called intermicellar media. Micellar structures are predominant in asphalts with high asphaltene content. The growth and ultimate size of the micelles are dependent on temperature and factors such as wax content and the presence of other chemical compounds. The stability of the asphaltene, which depends on the degree of protection offered by the resinous materials, is a fundamental consideration in assessing the properties and behavior of the asphalt.

The homogeneous intermicellar phase and micelles (sol) may be coalesced by peptized native surfactants and gradually enlarging them to a liquid crystal type of gel. This sol-gel interconversion is reversible by temperature, shearing, etc. The other alternative is that the supermicelles may merge continuously to reach a giant floc and thus flocculation and precipitation occur.

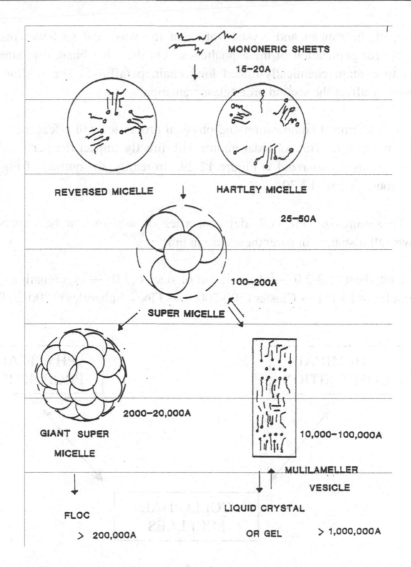

Figure 12-28. Growth of the miceller stages of asphalt to gel or floc and the separation.

The presence of paraffin wax is found to disturb the formation of the colloidal structure of asphalts. Apart from decreasing the aromaticity or solvation power of the maltenes, paraffinic material in asphalts influences the ability of the resinous maltene material to orientate the polar groups onto the asphaltene particle. Sharply defined interfaces are produced between the micelles and the intermicellar oils, and mutual orientation of the micelles is disturbed. In addition, polar resins are adsorbed onto the paraffin wax during cooling of the asphalt and crystallization of the wax, and so fewer resins are available for peptization of the asphaltenes. On the other hand, asphaltenes are found to contain chemically-linked long chain paraffins. The folding of the chains may affect the second order glass transition.

Both chemical composition and physical properties will affect the colloidal nature of asphalt. The colloidal nature will directly impact the performance of the end use as summarized in Figure 12-29. In reality, the outline of Fig. 12-29 can be found in Fig. 12-30.

To summarize, the colloidal properties of asphalt can be expressed as follows (all distances in parentheses are in nm):

Unit sheet (1.2-2.0) $\rightarrow$ Association of stacks (3.0) $\rightarrow$ Aggregation (3.0) $\rightarrow$ Assemblage (10-15) $\rightarrow$ Cluster (200-2000) $\rightarrow$ Flocs, Spherules (1000-20,000).

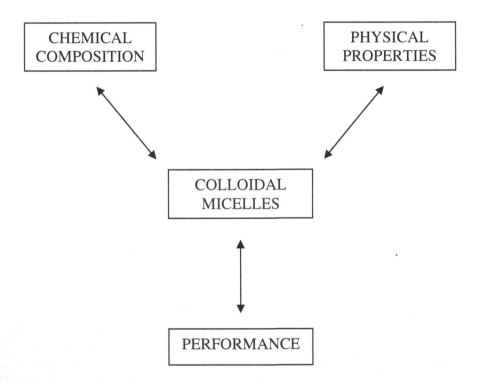

Figure 12-29. Performance of asphalt is controlled by chemical composition and physical properties for the parameters of such studies. See Fig. 12-30.

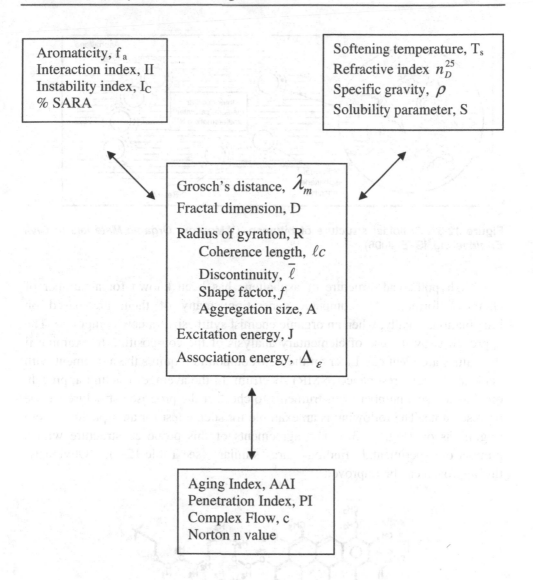

Figure 12-30. Parameters for the block diagram illustrated in Fig. 12-29.

In the US, the Association of Asphalt Paving Technologists (AAPT) and the Asphalt Institute, while in Europe, the European Asphalt Technology Association (EATA) and Central Laboratory for Structures and Roads (LCPC) are for the practical application of asphalt. I introduce Y.Mouton (LCPC); he feels colloidal structure is important (Fig 12-31). A colloidal picture is essential for the application of asphalt.

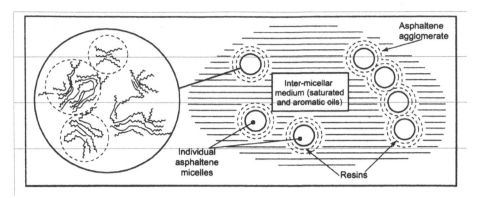

Figure 12-31. Colloidal structure of bitumen. (Y.Mouton, *Organic Materials in Civil Engineering*, ISTE, 2006)

A hypothetical structure of asphaltene has been known for a number of chemical formulas by computer generation; many of them are based on imagination, usually, when an organic chemist synthesizes a new compound. The approach is by the use of elementary analysis of the composition to examine if the values are identical. Later on the new compound requires the agreement with nuclear magnetic resonance (NMR) spectrum. In the average structure approach, one has to use a number of instruments to check if the proposed structure can be representable. The following is an example for such a test for an asphaltene from Lagunillas oil (Fig. 12-32). The agreements of this proposed structure with a number of experimental findings are similar (see Table 12-8). Obviously, this approach can be improved.

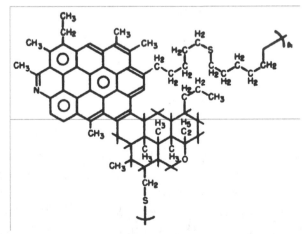

Figure 12-32. Hypothetical molecular formula of an asphaltene from Lagunillas oil. The positions of all substituents are quite arbitrary.

Table 12-8. Structural parameters of asphaltene from Lagunillas crude oil.

				Lagunillas asphaltene ($C_{69}H_{nn}NS_4O$)				
H_L/C_A	=	0.43[a]	(0.44)[b]	L_a	=	9.7 Å	(9.85 Å)
C_A/C_{sat}	=	3.2	(4.3)	C_{sat}/H_L	=	0.71	(0.71)
H_N/C_N	=	1.25	(1.06)	C_{sat}/S_{sat}	=	0.47	(0.50)
f_a	=	0.41	(0.42)	C_N/C	=	0.24	(0.22)
C_P/C	=	0.35	(0.36)	C_{sat}/C	=	0.15	(0.16)
%C	=	84.2	(83.0)	%H	=	7.9	(8.14)
%N	=	2.0	(1.3)	%S	=	4.5	(6.0)
%O	=	1.6	(1.5)	M	=	4750	(4276)
H_a/H	=	0.95	(0.95)	r_M	=	22 Å	(15 Å)
C_{MP}/C_{ea}	=	1.7	(1.1)	H_A/H	=	0.055	(0.046)
H_R/H	=	0.31	(0.31)	A_N	=	576	(524)
F/C_A	=	0.32	(0.60)	H_o/H	=	0.27	(0.25)
H_{sat}/H	=	0.19	(0.21)	C/R	=	6.2	(5.0)
l_a	=	413 Å	(265 Å)	H_N/H	=	0.18	(0.18)

$\rho\ 15° = 5.9 \times 10^{17}$ at 25° (2.3×10^{16}) ohm-cm

[a] Values not in parentheses are obtained from experiments
[b] Values in parentheses are computed based on the structure in the proposed structure

---

[Example 12-3] An asphaltene is allowed to react with an iodine molecule and a stable complex is formed at 3:2. If 11% of iodine to asphaltene is close to the transition of lowest gap energy in resisting, evaluate the unit weight. [see Sill, G.A and T.F.Yen, *Fuel*, **48**, 61 (1969)]

$$UW = 254\ t^{-1}\ R\ (100\text{-}t)$$

t is % of iodine to asphaltene near transition gap and R is stable complex, then,

Unit Weight = 3083.

Assuming t_i is 0.5, then the disk weight of the asphaltene is 1541.

12.4 RHEOLOGY OF ASPHALT

Interestingly, the flow and mechanical properties behave both Newtonian (sol asphalt) and non-Newtonian (gel asphalt). Most asphalts belong to the sol-gel class, which can exhibit either pseudoplastic or shear-thining behavior. For asphalt, the consequence of the dispersions gives rise to a variety of non-Newtonian states for the colloidal particles.

Plasticity→pseudo-plasticity→false body→thixotropy→dilatancy→Newtonian flow

For this rheological section, tensile deformation is used for description. Referring to chapter 9 shear deformation is exchangeable with tensile deformation (See Section 9.4.2).

Based on the colloidal types, apparently the rheological properties of asphalt can have fairly good correlation. Table 12-9 is a summary which can support this viewpoint. In general, three rheological types of asphalts are known: (a) Newtonian (sol) type, (b) viscoelastic (sol-gel) type, and (c) elastic or non-Newtonian (gel) type. For the sol-type, viscosity is independent of shearing stress or time and deformation per unit time is proportional to shearing stress. Micelles of both reversed or Hartley type are free moving. Asphalts made from petroleum are high in aromatics, and those produced from severe thermal cracking belong to this class, especially petroleum from California and Indonesia, both of a geologically young age origin. The quality and quantity of resins available for peptization become important in sol-gel types. Asphalts interconversion of the Hartley micelle to a gel-type structure is in an equilibrium process (Fig. 12-28). This pseudo-plastic flow is exemplified by most asphalts derived from steam and vacuum refined bottoms.

Finally, for the gel-type asphalts, the elastic flow is tied up with a true gel structure of fixed lattice. They exhibit various forms of resilience, thixotropy to some extent or a fixed yield value plastic flow or dilatancy. Air blown or oxidized asphalts belong to this class.

Creep, steric hardening or molecular structuring can be explained based on the sol-gel type of equilibrium process. For example, sol-gel asphalts show various degrees of either glassy brittleness or gel-like elasticity at freezing temperature or below. This factor, along with evaporation of oily components and chemical oxidation is a major cause of hardening of asphalt in pavement.

This hardening occurs rapidly at first but appears to approach a limiting value on prolonged standing. Furthermore, this process is promoted by mineral aggregate surfaces and can be reversed or destroyed by heat, mechanical work, or solvents. Proper control of steric hardening is beneficial, since without it pavement mixtures would not set to produce nontender pavements.

Table 12-9. Some rheological characteristics based on colloidal types of asphalt.

	Sol	Sol-Gel	Gel
Refinery Process	Solvent	Vacuum Distillation	Air-blown
Rheology Property	Newtonian	Viscoelastic	Elastic
Complex Flow, c	1	$0.85 - 1.0$	$0.4 - 0.85$
Asphalt Aging Index	< 0.01	$0.02 - 0.09$	> 0.1
Plastic Flow Index, p	1	$1.0 - 1.4$	> 1.4
Penetration Index, pi	< -1.0	-1	>1.0
Ductility at 25° C (cm)	> 200	160	5.5
ASTM Softening pt., ° C	50	55	65
Elasticity, Relaxation half-time, sec	1	3	13

Some asphalts exhibit Ostwald fluid flow, i.e., they exhibit shear-thinning behavior at moderate shear rate; at low or high shear rate they show Newtonian behavior. These phenomena can be explained by the deformation of coils at intermediate shear rate (as seen in Fig. 12-33). In this case, the dilatant fluids can be explained by particle aggregation under shear as shown by Fig. 12-34.

A number of classically developed empirical flow tests are still in use. Viscosity can be determined by efflux (Saybolt), rotating spindle (Brookfield), inclined-tube (Hoeppler), capillary tube (Cannon-Fenske), coaxial cylinder (falling and rotational), parallel plate, sliding-plate microfilm, etc. Determination of penetration, softening point (ring and ball) ductility (including necking effect), staining (gel asphalts show marked staining), etc., can result in a number of empirical constants useful in characterizing asphalts, e.g., temperature susceptibility index (n), plastic flow index (p), complex flow (c), asphalt aging index (AAI), stain index (SI), penetration index (PI), stiffness modulus (S), etc. Studies in rheology for asphalts, which give meaningful applications to practice, such as pavement, roofing, coating and others, are essential. They can also generate correct testing procedures and measurements for simulating long-term performance.

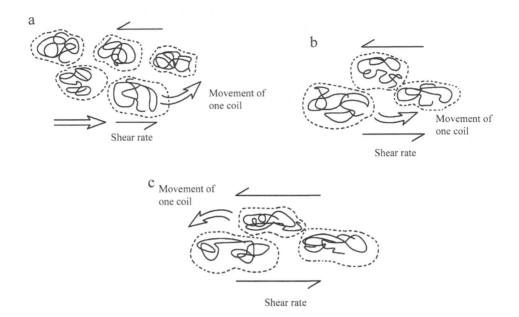

Figure 12-33. Movement of random coils in shear fields: (a) low shear rate, $\mu = \mu_o$, fluid motion does not distort the coils; (b) intermediate shear rate, $\mu 0 > \mu > \mu_s$, velocity gradient elongates the coils in the shear direction; (c) high shear rate $\mu = \mu_s$, coil distortion has reached a dynamic equilibrium between the elongations caused by the shear and the recoil of the molecule. (After E.A. Grulke, *Polymer Processing Engineering*, Prentice Hall, 1994).

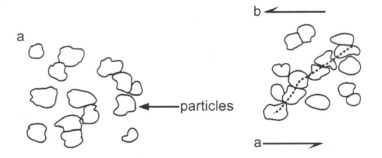

Figure 12-34. Dilatant fluid: (a) fluid at rest, (b) fluid under shear. Velocity gradient causes particles to pack into larger agglomerates that have increased resistance to flow. Dashed line suggests the effective diameter of the particle unit.

REFERENCES

12-1.   T. F. Yen, *Asphaltic Materials*, Reprinted from H.S. Mark, N.M. Bikales, C.G. Overberger and G. Menges, Encyclopedia of Polymer Science and Engineering, Supplement Volume, Second Edition, John Wiley & Sons, New York, 1989.

12-2.   T. F. Yen, Asphaltenes: Types and Sources in *Structure and Dynamics of Asphaltene*, Plenum, New York, 1988.

12-3.   S. S. Pollack and T. F. Yen, Small Angle Scattering Studies of Asphaltics by X-Ray, *American Chemical Society Division of Petroleum Chemistry, Preprints* (1969).

12-4.   Yen et al, Investigation of the Structure of Petroleum Asphaltenes by X-Ray Diffraction, *Analytical Chemistry* (1961).

12-5.   M.A. Sadeghi et al, X-ray Diffraction of Asphaltenes, *Energy Sources*, (1985).

12-6.   Yen et al, A Study of the Structure of Petroleum Asphaltenes and Related Structures by Infrared Spectroscopy, (1983).

12-7.   Y. Mouton, *Organic Materials in Civil Engineering*, ISTE Ltd, Newport Beach, CA, 2006.

12-8.   T. F. Yen and G. V. Chilingar, *Asphaltenes and Asphalts,* Vol 1, Elsevier Science B. V, Amsterdam, 1994.

12-9.   T. F. Yen and G. V. Chilingar, *Asphaltenes and Asphalts*, Vol. 2, Elsevier Science B. V., Amsterdam, 2000.

12-10. J. P. Dickie and T. F. Yen, Macrostructures of the Asphaltic Fractions by Various Instrumental Methods, *Analytical Chemistry,* (1967).

12-11. T. F. Yen and J. G. Erdman, Asphaltenes (Petroleum) and Related Substances: X-Ray Diffraction, Encyclopedia *of X-Rays and Gamma-Rays*, (G. L. Clark, ed.), Reinhold Publishing Corp., New York, 1963.

12-12. T. F. Yen, The Nature of Asphaltenes in Heavy Oil, *1st Pan Pacific Synfuel Conference, Vol. II*, The Japan Petroleum Institute, Tokyo, Japan, 1982.

12-13. S. S. Pollack and T. F. Yen, Structural Studies of Asphaltenes by X-Ray Small Angle Scattering, *Analytical Chemistry*, (1970).

12-14. H. Abraham, *Asphalts and Allied Substances,* Vol. 1, 6th ed., Van Nostrard, Princeton, NJ, 1960.

12-15. E.J. Barth, *Asphalts, Science and Technology*, Gordon and Beach, New York, 1962.

12-16. J.P. Pfeiffer, *The Properties of Asphaltic Bitumen*, Elsevier Science, New York, 1950.

12-17. R.N. Traxler, *Asphalt, Its Composition, Properties and Uses,* Reinhold, New York, 1961.

12-18. J.W. Burger and N.C.Li, *Chemistry of Asphaltenes*, Vol. 195, American Chemical Society, Washington DC, 1981.

PROBLEM SET

1. To make a primary lithium battery using liquid chlorine as the liquid cathode, a soluble salt is required. A suggestion is using tungsten hexachloride with liquid chlorine. Given liquid chlorine gas as $\delta = 0.48$, $V = 46$ and WCl_6 $\delta = 13.27$, $V = 107.43$. Would this work?

2. There are three types of condensed-polynuclear aromatic hydrocarbons. The first type is the linked polynuclear aromatic (very loose). The second is the kata-condensed polynuclear aromatics (having 4 adjacent aromatic hydrogens, H_4). The last one is peri-condensed, the most condensed poly-nuclear aromatic (having only 1- isolated, 2- and 3- adjacent hydrogens, H_4). The three types are shown below.

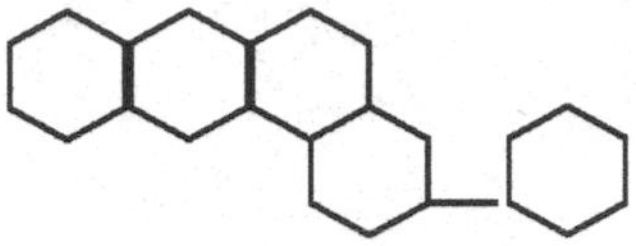

linked- 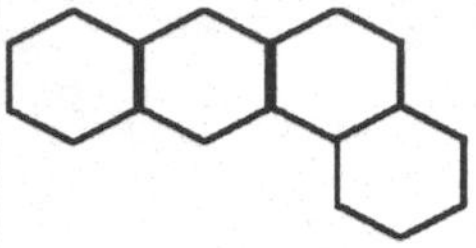   kata-condensed 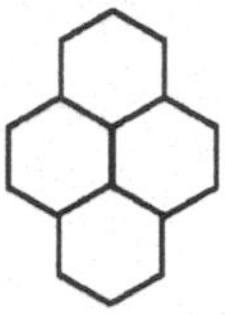   peri-condensed

Can the infrared spectra offer any difference?

3. If the layer diameter of spherm 6 (a carbon black of Cabot Corp.) is 20Å, predict the ultimate carbon analysis for the sample. (Hint: use $C_A = 0.382 L_a^2$)

EPILOGUE

Having been through my idea of transport of chemical science into engineers' usage, we still hope that the perfect linkage can be completed someday. This is my life's wish; an ambitious goal in my life, as each one of us has our own dearest "heavenly treasure"; in ancient China, as poet Yuan Ji termed "splendid person", or modern day Italian poet Ungaretti Giuseppe called "Novia". It can best be described by Fung Yan-Yi that "In dreams the stars departed and reassembled, while awakened it is limited by space-time." I would like to mention the space-time concept. The variation of time and the 3-dimensional time symbolizes the breadth and variation of chemical science knowledge. I have included the following two poems – one by Hungarian Miklós Radnóti, which relates to seasonal changes of the year – *Calendar*; the other is Italian Ungaretti Giuseppe's poem – *The Rivers*. The first poem deals with time while the second poem is related to different geographical contours and the riverine tributaries. If we travel through all the rivers and encounter the golden apple, the silver moon, the red squirrel, the lazy dragonfly and so on, we have already accomplished some goal which we intended to finish.

NAPTÁR (Calendar) by Miklós Radnóti

JANUÁR

Későn kel a nap, teli van még
csordúltig az ég sűrü sötéttel.
Oly feketén teli még,
szinte lecseppen.
Roppan a jégen a hajnal
lépte a szürke hidegben.

JANUARY

The sun rises late. The sky was
filled with a darkness
so black
it almost spilled over.
In the gray cold, dawn
steps on the ice and cracks it.

FEBRUÁR

Újra lebeg, majd letelepszik a földre,
végül elolvad a hó;
csordul, utat váj.
Megvillan a nap. Megvillan az ég.
Megvillan a nap, hunyorint.
S íme fehér hangján
rábéget a nyáj odakint,
tollát rázza felé s cserren már a veréb.

FEBRUARY

The snow flutters, settles to the ground,
then melts; cutting a path,
it trickles away.
The sun flashes. The sky flashes.
The sun flashes and blinks.
Outside, in a white voice,
the flock talks to it.
A sparrow shakes its feathers in that direction and sings

MÁRCIUS

Lúdbőrzik nézd a tócsa, vad,
vidám, kamaszfiús
szellőkkel jár a fák alatt,
s zajong a március.
A fázós rügy nem bujt ki még,
hálót se sző a pók,
de futnak már a kiscsibék,
sárgás aranygolyók.

MARCH

Look, goose pimples on a puddle.
Making noise, March walks around
under the trees with wild,
happy, young winds.
The buds haven't come out, they're afraid
of the cold, spiders aren't weaving their webs,
only little chicken roll around,
like balls of yellow gold.

ÁPRILIS

Egy szellő felsikolt, apró üvegre lép
s féllábon elszalad.
Ó április, ó április,
a nap se süt, nem bomlanak
a folyton nedvesorru kis rügyek se még
a füttyös ég alatt.

APRIL

A breeze cries out. It steps on a tiny piece of glass
and hops away on one leg.
O April, April,
the sun doesn't shine. Little buds
whose noses are always wet don't open
when the sky is filled with whistling.

MÁJUS

Szirom borzong a fán, lehull;
fehérlő illatokkal alkonyul.
A hegyről hűvös éj csorog,
lépkednek benne lombos fasorok.
Megbú a fázós kis meleg,
vadgesztenyék gyertyái fénylenek.

MAY

A petal shivers in the tree and falls.
White smell comes with evening.
A chilly night pours over the mountain.
Rows of leafy trees walk in it.
The little warmth there is hides from the cold.
The candles on wild chestnut trees shine

JÚNIUS

Nézz csak körül, most dél van és csodát látsz,
az ég derüs, nincs homlokán redő,
utak mentén virágzik mind az ákác,
a csermelynek arany taréja nő
s a fényes levegőbe villogó
jeleket ír egy lustán hősködő
gyémántos testü nagy szitakötő.

JUNE

Look around, it is noon now, you'll see a miracle.
The sky is bright, its brow is smooth.
Along the roads the acacias are in bloom.
Suddenly there are gold crests on little streams.
A fat, bragging, lazy dragonfly
with a diamond body writes
flashing signs in the bright air.

JÚLIUS

Düh csikarja fenn a felhőt,
fintorog.
Nedves hajjal futkároznak
meztélábas záporok.
Elfáradnak, földbe búnak,
este lett.
Tisztatestü hőség ül a
fényesarcu fák felett

AUGUSZTUS

A harsány napsütésben
oly csapzott már a rét
és sárgáll már a lomb közt
a szép aranyranét.
Mókus sivít már és a büszke
vadgesztenyén is szúr a tüske.

SZEPTEMBER

Ó hány szeptembert értem eddig ésszel!
a fák alatt sok csilla, barna ékszer:
vadgesztenyék. Mind Afrikát idézik,
a perzselőt! a hűs esők előtt.
Felhőn vet ágyat már az alkonyat
s a fáradt fákra fátylas fény esőz.
Kibomló konttyal jő az édes ősz.

OKTÓBER

Hűvös arany szél lobog,
leülnek a vándorok.
Kamra mélyén egér rág,
aranylik fenn a faág.
Minden aranysárga itt,
csapzott sárga zászlait
eldobni még nem meri,
hát lengeti a tengeri.

NOVEMBER

Megjött a fagy, sikolt a ház falán,
a holtak foga koccan. Hallani.
S zizegnek fönn a száraz, barna fán
vadmirtuszok kis ősz bozontjai.
Egy kuvik jóslatát hullatja rám;
félek? Nem is félek talán.

DECEMBER

Délben ezüst telihold
a nap és csak sejlik az égen.
Köd száll, lomha madár.
Éjjel a hó esik és
angyal suhog át a sötéten.
Nesztelenül közelít,
mély havon át a halál.

JULY

Up there, the clouds are so angry, their bellies ache,
they make faces.
Rainstorms run around barefoot
with their hair all wet.
They get tired, they hid in the ground,
night falls.
The clean body of heat sits above
the bright faces of the trees.

AUGUST

The meadow soaks up in the
blaring light of the sun.
A pretty golden apple
yellows among the leaves.
A red squirrel yelps and up in the proud
wild chestnut tree the thorns are sharp.

SEPTEMBER

How many Septembers have I lived to think about
Lilies and the brown jewel wild chestnuts lie
under the trees. They remind you of Africa's
scorching heat before a cool rain.
Evening makes its bed among the clouds
and a dim light pours over the tired trees.
Sweet Autumn comes with her hair down.

OCTOBER

The wind is golden and cool.
The wanderers sit down.
A mouse is dressing in the dark shed.
Above me a branch shines like gold.
Everything is gold here.
The corn, brought across the Atlantic,
waves its ripped yellow flags.
It is afraid to lose them.

NOVEMBER

Frost has come, it screams on the wall.
The teeth of the dead click. You can hear it.
Up in the dry brown tree, small,
gray, bushy wild myrtles rustle.
A little owl is dropping its prophecies on me.
Am I afraid? Maybe I'm not afraid?

DECEMBER

At 12 the sun is a silver moon,
a dream of itself on the sky.
The fog drifts like a sluggish bird.
Snow falls during the night and
an angel glides through the darkness.
Death comes noiselessly
over deep snow.

"The Rivers" by Giuseppe Ungaretti

Mi tengo a quest'albero	I hang on to this mangled tree
mutilatoabbandonato in questa dolina	abandoned in this sinkhole
che ha il languore	that is listless
di un circo	as a circus
prima o dopo lo spettacolo	before or after the show
e guardo	and watch
il passaggio quieto	the quiet passage
delle nuvole sulla luna	of clouds across the moon
Stamani mi sono disteso	This morning I stretched out
in un'urna d'acqua	in an urn of water
e come una reliquia	and rested
ho riposato	like a relic
L'Isonzo scorrendo	The flowing Isonzo
mi levigava	smoothed me
come un suo sasso	like one of its stones
Ho tirato su	I hoisted up
le mie quattr'ossa	my sack of bones
e me ne sono andato	and got out of there
come un acrobata	like an acrobat
sull'acqua	over the water
Mi sono accoccolato	I crouched
vicino ai mieipanni	beside my grimy
sudici di guerra	battle clothes
e come un beduino	and like a Bedouin
mi sono chinato a ricevere	bent to greet
il sole	the sun
Questo è l'Isonzo	This is the Isonzo
e qui meglio	and here I recognized myself
mi sono riconosciuto	more clearly
una docile fibra	as a pliant fiber
dell'universo	of the universe
Il mio supplizio	My affliction
è quando	is when
non mi credo	I don't believe in myself
in armonia	in harmony

Ma quelle occulte	But those hidden
mani	hands
che m'intridono	that knead me
mi regalano	freely give
la rara	the uncommon
felicità	bliss
Ho ripassato	I went back over
le epoche	the ages
della mia vita	of my life
Questi sono	These are
i miei fiumi	my rivers
Questo è il Serchio	This is the Serchio
al quale hanno attinto	where maybe
duemil'anni forse	two millennia of my farming people
di gente mia campagnola	and my father and mother
e mio padre e mia madre	drew their water
Questo è il Nilo	This is the Nile
che mi ha visto	that saw me
nascere e crescere	born and raised
e ardere d'inconsapevolezza	and burn with unawareness
nelle estese pianure	on the sweeping flatlands
Questa è la Senna	This is the Seine
e in quel suo torbido	within whose roiling waters
mi sono rimescolato	I was mixed again
e mi sono conosciuto	and came to know myself
Questi sono i miei fiumi	These are my rivers
contati nell'Isonzo	reckoned in the Isonzo
Questa è la mia nostalgia	This is my longing for home
che in ognuno	that in each one
mi traspare	shines through me
ora ch'è notte	now that it's night
che la mia vita mi pare	that my life seems
una corolla	a corolla
di tenebre	of darkness

APPENDICES

APPENDIX A

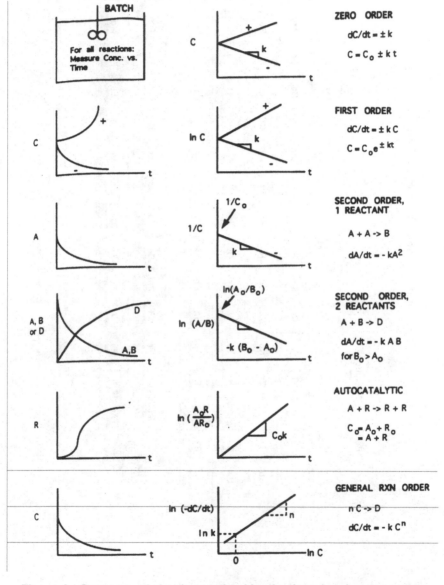

Figure A. Summary of simple reaction kinetics from batch reactor data.

APPENDIX B

IUPAC Periodic Table of the Elements

Notes

- "Aluminum" and "cesium" are commonly used alternative spellings for "aluminium" and "caesium".
- IUPAC 2005 standard atomic weights (mean relative atomic masses) as approved at the 43rd IUPAC General Assembly in Beijing, China in August 2005 are listed with uncertainties in the last figure in parentheses (M. E. Wieser, Pure Appl. Chem., in press).
- These values correspond to current best knowledge of the elements in natural terrestrial sources. For elements that have no stable or long-lived nuclides, the mass number of the nuclide with the longest confirmed half-life is listed between square brackets.
- Elements with atomic numbers 112 and above have been reported but not fully authenticated.

Copyright © 2005 IUPAC, the International Union of Pure and Applied Chemistry. For updates to this table, see http://www.iupac.org/reports/periodic_table/. This version is dated 3 October 2005.

APPENDIX C

Legendre's equation

This differential equation is

$$(1-x^2)\frac{d^2y}{dx^2} - 2x\frac{dy}{dx} + n(n+1)y = 0$$

Legendre's equation may also be written in the form

$$[(1-x^2)y']' + n(n+1)y = 0 \qquad \text{[C-1]}$$

Legendre's equation is a special case of the Sturm-Liouville equation, in which $r(x) = 1 - x^2$, $q(x) = 0$, $p(x) = 1$, and $\lambda = n(n+1)$. Ordinarily, Legendre's equation is considered over the interval $-1 \le x \le 1$. The solutions of Legendre's equation satisfy the orthogonality relationships.

$$\int_{-1}^{1} y_m(x)y_n(x)dx = \begin{cases} 0 & n \ne m \\ k_n & n = m \end{cases}$$

The Legendre series for approximating a function $f(x)$ which is defined over the interval $-1 \le x \le 1$ is

$$f(x) = \sum_{n=0}^{\infty} C_n P_n(x)$$

The coefficient C_n is evaluated by multiplying each term in the preceding expression by $P_m(x)$ and then integrating from -1 to +1. Thus,

$$C_n = \frac{2n+1}{2} \int_{-1}^{1} f(x)P_n(x)dx$$

Laguerre's equation

Another differential equation which arises frequently in engineering practice is the Laguerre equation

$$xy''+(1-x)y'+ny = 0 \qquad n = 0, 1, 2, \ldots$$

or

$$(xe^{-x}y')'+ne^{-x}y = 0 \tag{C-2}$$

This differential equation is satisfied by the Laguerre polynomial

$$L_n(x) = e^x \frac{d^n}{dx^n}(x^n e^{-x})$$

The first few Laguerre polynomials are

$$L_0(x) = 1 \qquad\qquad L_2(x) = x^2 - 4x + 2$$
$$L_1(x) = -x + 1 \qquad L_3(x) = -x^3 + 9x^2 - 18x + 6$$

The Laguerre equation is a special case of the Sturm-Liouville equation, in which $r(x) = xe^{-x}$, $q(x) = 0$, $p(x) = e^{-x}$, and $\lambda = n$. For the interval $0 \le x \le \infty$, then $r(a) = r(0) = 0$, and $r(b) = r(\infty) = 0$, so the solutions of Laguerre's equation are orthogonal with respect to the weight function $p(x) = e^{-x}$. That is,

$$\int_0^\infty e^{-x}y_m(x)y_n(x)dx = \begin{cases} 0 & n \ne m \\ k_n & n = m \end{cases}$$

In particular, it may be shown that

$$\int_0^\infty e^{-x}L_m(x)L_n(x)dx = \begin{cases} 0 & n \ne m \\ (n!)^2 & n = m \end{cases}$$

A function $f(x)$ defined over the interval $0 \le x \le \infty$ may be expanded in terms of Laguerre polynomials as follows:

$$f(x) = \sum_{n=0}^\infty C_n L_n(x) \tag{C-3}$$

The coefficient C_n is obtained by multiplying each term by $e^{-x}L_m(x)$ and then integrating from 0 to ∞. Thus

$$Cn = \frac{1}{(n!)^2} \int_0^\infty e^{-x} f(x) L_n(x) dx \qquad [C-4]$$

There are a great many sets of orthogonal functions. Such functions have numerous and varied applications to engineering problems. An obvious application is that each set may be regarded as forming the basis of a transformation. For example, Eq. [C-4] transforms a function $f(x)$ to a function of n [that is, $C_n = F(n)$]. The corresponding inversion formula for this Laguerre transformation is given by Eq. [C-3].

APPENDIX D

General Data and Fundamental Constants

Quantity	Symbol	Value
Speed of light in vacuum	c_0	299 792 458 m s^{-1} (defined)
Elementary charge	e	1.602 176 53(14) × 10^{-19} C
Boltzmann constant	k, k_B	1.380 650 5(24) × 10^{-23} J K^{-1}
Planck constant	h	6.626 069 3(11) × 10^{-34} J s
	$\hbar = h/2\pi$	1.054 571 68(18) × 10^{-34} J s
Avogadro constant	L, N_A	6.022 141 5(10) × 10^{23} mol^{-1}
Gas constant	R	8.314 472 (15) J K^{-1} mol^{-1}
Faraday constant	F	9.648 533 83(83) × 10^4 C mol^{-1}
Atomic mass constant (dalton, or unified atomic mass unit, $m(^{12}C)/12$)	$m_u = Da = u$	1.660 538 86(28) × 10^{-27} kg
Electron rest mass	m_e	9.109 382 6(16) × 10^{-31} kg
Proton rest mass	m_p	1.672 621 71(29) × 10^{-27} kg
Neutron rest mass	m_n	1.674 927 28(29) × 10^{-27} kg
Permeability of vacuum (or magnetic constant)	μ_0	$4\pi × 10^{-7}$ H m^{-1} (defined) Note: H m^{-1} = N A^{-2} = N s^2 C^{-2}
Permittivity of vacuum (or electric constant)	$\varepsilon_0 = 1/\mu_0 c_0^2$	8.854 187 816... × 10^{-12} F m^{-1} Note: F m^{-1} = C^2 J^{-1} m^{-1}
Bohr magneton	$\mu_B = e\hbar/2m_e$	9.274 009 49(80) × 10^{-24} J T^{-1}
Nuclear magneton	$\mu_N = (m_e/m_p)\mu_B$	5.050 783 43(43) × 10^{-27} J T^{-1}
Landé g factor for free electron	g_e	2.002 319 304 371 8(75)
Fine structure constant	$\alpha = \mu_0 e^2 c_0/2h$	7.297 352 568(24) × 10^{-3}
Second radiation constant	$c_2 = hc_0/k$	1.438 775 2(25) × 10^{-2} m K
Stefan-Boltzmann constant	$\sigma = 2\pi^5 k^4/15h^3 c_0^2$	5.670 400(40) × 10^{-8} W m^{-2} K^{-4}
Bohr radius	$a_0 = 4\pi\varepsilon_0 \hbar^2/m_e e^2$	5.291 772 108(18) × 10^{-11} m
Hartree energy	$E_h = \hbar^2/m_e a_0^2$	4.359 744 17(75) × 10^{-18} J
Rydberg constant	$R_\infty = E_h/2hc_0$	1.097 373 156 852 5(73) × 10^7 m^{-1}

Quantity	Value
Standard acceleration of free fall g_n	9.806 65 m s^{-2} (defined)
Gravitational constant G	6.674 2(10) × 10^{-11} m^3 kg^{-1} s^{-2}
Zero of Celsius scale	273.15 K (defined)
Molar volume of ideal gas. $p = 1$ bar and $T = 273.15$ K	22.710 981 (40) L mol^{-1}
Standard atmosphere	101 325 Pa (defined)
RT at 298.15 K	2.4790 kJ mol^{-1}

PRESSURE CONVERSION FACTORS

	Pa	atm	Torr
1 Pa =	1	9.869 23 × 10^{-6}	7.500 62 × 10^{-3}
1 atm =	101 325	1	760
1 Torr =	133.322	1.315 79 × 10^{-3}	1

Example of the use of this table: 1 atm = 101 325 Pa
Notes: 1 mmHg = 1 Torr : 1 bar = 10^5 Pa

ENERGY CONVERSION FACTORS

	energy E			molar energy E_m	wavenumber $\tilde{\nu}$
	J	eV	E_h	kJ/mol	cm^{-1}
1 aJ	10^{-18}	6.241 509	0.229 3713	602.2142	50 341.17
1 eV	1.602 177 × 10^{-19}	1	3.674 932 × 10^{-2}	96.485 34	8 065.544
1 E_h	4.359 744 × 10^{-18}	27.211 38	1	2625.500	219 474.6
1 kJ/mol	1.660 539 × 10^{-21}	1.036 427 × 10^{-2}	3.808 799 × 10^{-4}	1	83.593 47
1 cm^{-1}	1.986 446 × 10^{-23}	1.239 842 × 10^{-4}	4.556 335 × 10^{-6}	11.962 66 × 10^{-3}	1

Example of the use of this table: 1 eV 'corresponds to' or is equivalent to' 96.485 34 kJ/mol
Note: 1 cal = 4.184 J

Source: The National Institute of Standards and Technology (NIST) reference on Constants, Units, and Uncertainties (2002 values) <http://physics.nist.gov/cuu/constants>

APPENDIX E

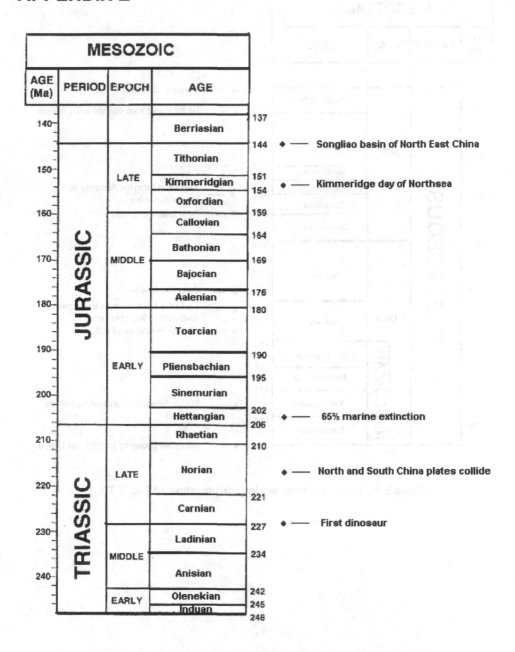

Figure E-a. Mesozoic time scale [continuation of Fig. 7-7(b)].

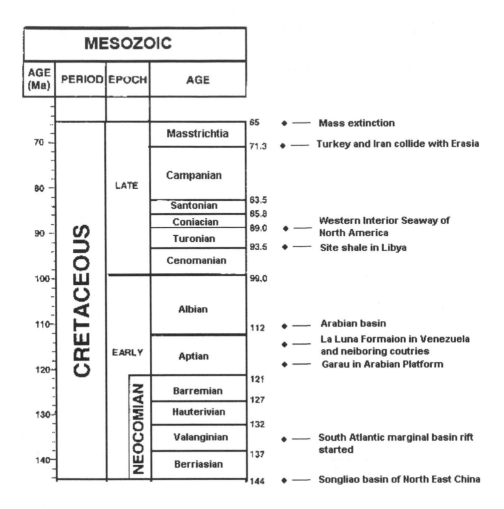

Figure E-b. Mesozoic time scale [continuation of Fig. 7-7(c)].

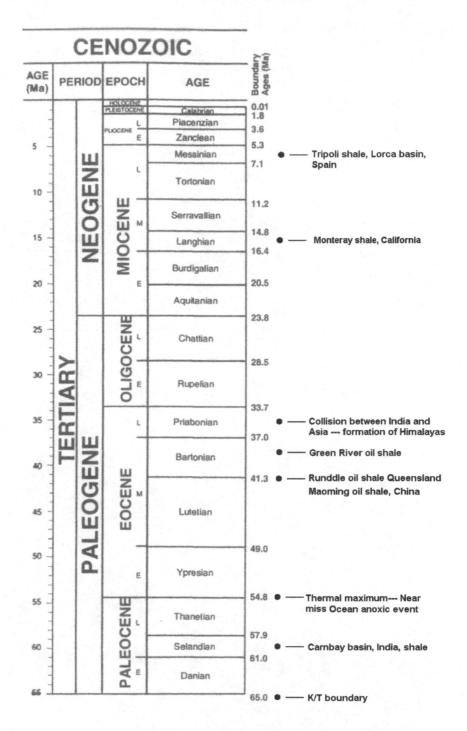

Figure E-c. Cenozoic time scale [continuation of Fig. 7-7(d)].

INDEX